CRITICAL ISSUES
IN MEDICAL TECHNOLOGY

CRITICAL ISSUES IN MEDICAL TECHNOLOGY

edited by

BARBARA J. McNEIL
Harvard Medical School

ERNEST G. CRAVALHO
Harvard-MIT Division of Health Sciences and Technology

 Auburn House Publishing Company
Boston, Massachusetts

The graphic on the cover is derived from a view of the human brain
obtained using nuclear magnetic resonance (NMR) imaging.

Library of Congress Cataloging in Publication Data
Main entry under title:
Critical issues in medical technology.

  Includes index.
    1. Medical innovations—Economic aspects—United
States. 2. Medical innovations—Government policy—
United States. 3. Medical innovations—Social aspects—
United States. 4. Medical innovations—Evaluation.
5. Technology assessment—United States. I. McNeil,
Barbara J., 1941–   . II. Cravalho, Ernest G. [DNLM:
1. Technology, Medical. QY 4 C934]
RA410.53.C75   306'.4   81–7900
ISBN 0–86569–070–7   AACR2

Printed in the United States of America

PREFACE

The Conference on Critical Issues in Medical Technology rose out of a series of discussions at Harvard Medical School and the Massachusetts Institute of Technology on the need to educate health professionals and public policy makers about medical technology—the principles and problems of evaluating it, applying it, and paying for its use. These areas were central to the teaching and research interests of the conference chairpersons, Ernest G. Cravalho and Barbara J. McNeil, who—with the help of a generous gift from the Kieckhefer Foundation, as well as the assistance of Jeffrey Harris, William Knaus, Eugenia Parnassa, and Louise Russell—organized the conference to help meet this need.

Medical technology, a buzz word in contemporary medicine and public policy, has presented us with a series of opportunities, problems, and conflicting objectives. For example, modern medical technologies have an enormous potential for good, but when used improperly can lead to harm. Their safety and efficacy have to be evaluated over both the short term and the long term. As a society we want good technologies made available but, simultaneously, we worry about the costs of providing unlimited access. Concomitant ethical and social problems of a large variety are abundant. These competing considerations make judgments about the role of new and existing technologies extraordinarily difficult.

The goal of the Conference on Critical Issues in Medical Technology was to formulate and try to answer specific questions regarding medical technology in order to extend our current knowledge and define topics and approaches for future research. Among the questions with which it was concerned are:

How do the needs and values of individuals and society as a whole affect medical technologies? This discussion was particularly useful because it uncovered many of the critical questions regarding the social and ethical issues of medical technology. Some of the most important of these are: Are patients passive recipients of medical technologies, or are they active consumers? The entire medicolegal system suggests the latter, but such is often not the case in the clinical setting. Similarly, how should societal preferences influence the development and use of technology? How should conflicts between individual values and societal choice be resolved? The conference concluded that for both individuals and society as a whole, further research is needed on value assessment and risk perception.

What are the main public policy concerns regarding new technologies? Once a technology has emerged and has entered the clinical arena, clarification of competing objectives becomes important. For example, safety, efficacy, ease of access, and costs must be considered in decisions to use and pay for a given technology. They thus enter into the identification of "reasonable and necessary" services for Medicare recipients.

Most participants in the conference agreed that the present regulatory approach to reimbursement has major flaws which markedly impact on medical technology. Among those cited frequently were difficulties in (1) identifying *new* technologies; (2) identifying inappropriate uses of technologies; and (3) modifying behavior so that inappropriate usage diminishes. They also agreed that, from a public policy point of view, the real role of technology assessment lies in the identification of those factors which most strongly influence the effective use of a particular technology; these factors are of primary importance in decisions involving tradeoffs between the benefits and costs of new technologies.

What is it important to know about the origins of medical technology? Information on the development and diffusion of medical technology would aid in tracking emerging and existing technologies and in optimizing their use and availability. Surprisingly, we know very little about these topics. According to some estimates, 80 percent of major technological developments in medicine come from small companies with annual sales of less than $30-$35 million. Such companies may be unable to support detailed studies of the efficacy of the technologies they design or develop. Only a small fraction of new developments arise from industrial giants. With regard to diffusion, Louise Russell's monograph, *Technology in Hospitals* (Brookings Institution, 1979) concisely but completely discusses several big-ticket technologies; but we know virtually nothing about little-ticket items. Several participants in this volume emphasize that little-ticket items may be an even greater culprit than big-ticket items in contributing to rising health care costs.

Edward Roberts discusses empirical research on the development and diffusion of nonbiomedical technologies; this introductory material concisely classifies the influences on innovation (staffing, structure, etc.) and offers insight into possible relationships of developments in other fields to those in medicine.

Once a technology has been developed, how should it be evaluated? This area is among those of greatest misunderstanding by public policymakers and others outside medical practice, yet it is one of the most important. Rational judgments based on such factors as safety, efficacy, and cost must rely on firm data.

Technologies to screen for disease must be evaluated very differently from those that diagnose disease, which must be evaluated still differently from those that are used to treat patients. Although some underlying principles of experimental design transcend the type of technology under discussion, they generally play a smaller role in the evaluation than do problems and biases unique to different types of technologies. For example, a study sponsored by the American Cancer Society regarding the value of screening for various types of cancer illustrates some of the issues that screening programs face. Many examples from studies evaluating diagnostic technologies highlight the need to identify the purpose of that type of technology: extend life or improve its quality. Such research also shows the need to control carefully for differences in instrumentation and biases in interpretation. Finally, although techniques to evaluate therapeutic technologies are generally advanced, we frequently forget that large numbers of pitfalls and biases can still occur, making non-efficacious therapies seem worthwhile or valuable therapies appear inefficacious. In all cases of evaluation, problems associated with incomplete or inadequate data (that is, statistical power) are large.

How do hospitals react to new technologies? Basically, once a new technology becomes available, hospitals must consider its possible adoption from numerous perspectives and must frequently do so on the basis of incomplete data. One important influence for all hospitals is the reimbursement process; positive and negative incentives for providers, patients and third parties make the situation extremely complex and difficult to unravel. Teaching hospitals are also influenced by a unique need to have technologies available so that their teaching abilities do not fall short: for example, a hospital today without computed tomography may find it difficult to attract highly qualified residents in radiology. In all cases, demand for new technology must be balanced against the possibility of skewing the distribution of resources within a particular institution. The Massachusetts General Hospital's recent decision not to introduce cardiac transplantation illustrates the importance that that institution attaches to the diversity of resource consumption—the greatest good for the greatest number. Alexander Leaf, the then-chairman of medicine, summarized the hospital's position by saying, "We are now in an era when the decision to act in one way reduces or forecloses our options to do other things that we may want to do or have to do. In recent years, we have often been admonished that with resources becoming limited, we as a profession must set our priorities or others will do it for us."

What have we learned from assessments of individual technologies? The workshop sessions of the conference included case studies of five technologies. The topics were selected to illustrate major problems in technology assessment and show some of the difficulties in providing a total picture of the costs, risks, and benefits of a new technology.

The case of the intensive care unit illustrates the complexity of identifying patients most or least likely to benefit from a scarce resource. It also indicates the problems associated with attempting to withhold resources from individuals with a *nearly* (but not quite) zero probability of benefiting from them.

The case study on burn care indicates the role of pooled data in evaluating a technology applied to a small subset of the general population and the importance of considering quality as well as quantity of life.

The workshop on end-stage renal disease had a slightly different perspective: enumeration of factors that complicated the conceptualization, design, and implementation of the end-stage renal disease program and its assessment. These matters are particularly relevant today because of the high cost of that program and the possibility that dialysis for nonrenal diseases may become more common.

The case study of antenatal diagnosis emphasizes a frequently ignored aspect of technology assessment—attention to ethical problems—and shows how conflicts between individual values and societal choice can arise. The current controversy regarding screening for neural tube defects through use of the alpha-fetoprotein assay is a case in point.

The artificial heart, the final case study considered at a workshop session, is not discussed much today. However, the case study draws attention to the interaction of medical and nonmedical fields in technology development and acceptance. For example, whether or not to use a nuclear power source for the device was a crucial decision.

Medical technology is a burgeoning field today and poses many important and challenging issues. Over the past decade we have begun to answer some questions regarding the evaluation and optimum use of certain technologies, and we are now just beginning to formulate questions and provide some answers in other problematic areas. This volume is published with the hope of achieving more fully the two original goals of the conference: (1) to raise questions leading to continued discussion about medical technology; and (2) to educate health care professionals and public policy makers about critical issues in medical technology and its assessment.

BARBARA J. MCNEIL

ACKNOWLEDGMENTS

We are particularly indebted to the Kieckhefer Foundation for sponsoring this conference; to our parent universities, Harvard Medical School and the Harvard/MIT Program in Health Sciences and Technology; and to other members of our Planning Committee—Jeffrey Harris, William Knaus, Eugenia Parnassa, and Louise Russell.

We are also indebted to the participants in the conference, whose vigorous interchange made the conference a stimulating experience; to Ms. Rachel Adler and her staff, who helped with its organization; to Ms. Barbara Altman, who provided editorial assistance; and to our secretaries, Catherine Wilheim and Laura Shapiro, who provided assistance in innumerable ways.

B.J.M.
E.G.C.

CONTENTS

CONTRIBUTORS

S. James Adelstein, M.D., Ph.D.
Professor of Radiology and Dean for
  Academic Programs
Harvard Medical School

David Banta, M.D.
Manager, Health Program
Office of Technology Assessment

Clyde Behney, M.B.A.
Senior Analyst
Health Program
Office of Technology Assessment

Betsy Blades, M.S.W.
Medical Social Worker
Baltimore Regional Burn Center
Baltimore City Hospitals

David Blumenthal, M.D., M.P.P.
Executive Director
Center for Health Policy and Manage-
  ment;
Kennedy School of Government
Harvard University

John P. Bunker, M.D.
Professor of Anesthesia and of Family,
  Community, and Preventive Medi-
  cine
Stanford University

John Burckhardt, B.A.
Policy Analyst
Office of Health Research, Statistics,
  and Technology
U.S. Public Health Service

Robert A. Derzon, M.B.A.
Vice-President
Lewin and Associates

David M. Eddy, M.D., Ph.D.
Professor of Engineering-Economic
  Systems and Family, Community
  and Preventive Medicine
Stanford University

Penny Feldman, Ph.D.
Assistant Professor of Political Science
Department of Health Policy and Man-
  agement
Harvard School of Public Health

John C. Fletcher, Ph.D.
Assistant for Bioethics
Clinical Center
Office of the Director
National Institutes of Health

Harvey W. Freishtat, J.D.
Partner
McDermott, Will & Emery

David B. Geselowitz, Ph.D.
Professor of Bioengineering
Pennsylvania State University

Fred Hellinger, Ph.D.
Senior Economist
National Center for Health Care Tech-
  nology
Office of Health Research, Statistics,
  and Technology
U.S. Public Health Service

Anne Kaszuba, B.S.
Health Services Research Center
Johns Hopkins University School of
  Medicine

William A. Knaus, M.D.
Co-Director, Intensive Care Unit
Director, ICU Research
The George Washington University
  Medical Center

Joyce C. Lashof, M.D.
Assistant Director
Office of Technology Assessment

Bernard S. Linn, M.D.
Professor of Surgery and Associate
  Chief of Staff for Education
University of Miami School of Medi-
  cine and VA Medical Center

John A. Locke, M.P.H.
Director of Public Health
Brookline, Mass., Health Department
(formerly Executive Director, New
  England Regional Burn Program)

Edmund G. Lowrie, M.D.
Assistant Clinical Professor of Medi-
  cine
Medical Director, The Kidney Center
Harvard Medical School

Deborah P. Lubeck, Ph.D.
Research Associate
Division of Health Services Research
Stanford University

Bryan Luce, Ph.D.
Senior Analyst
Health Program
Office of Technology Assessment

Joyce Mamon, Ph.D.
Health Services Research Center
Johns Hopkins University School of
  Medicine

Barbara J. McNeil, M.D., Ph.D.
Associate Professor of Radiology
Harvard Medical School
Brigham and Women's Hospital

Andrew M. Munster, M.D.
Associate Professor of Surgery and
  Plastic Surgery
Johns Hopkins University School of
  Medicine
Director
Baltimore Regional Burn Center

Yukihiko Nosé, M.D., Ph.D.
Chairman, Department of Artificial
  Organs
Cleveland Clinic Foundation

Stephen G. Pauker, M.D.
Chief, Division of Clinical Decision
  Making, Department of Medicine
Associate Professor of Medicine
New England Medical Center
Tufts University School of Medicine

Susan P. Pauker, M.D.
Chief of Pediatrics and Medical Genet-
  icist, Cambridge Center,
Harvard Community Health Plan
Director, Genetics Clinic,
Massachusetts General Hospital
Instructor in Pediatrics,
Harvard Medical School

William S. Pierce, M.D.
Associate Professor of Surgery
Pennsylvania State University

John B. Reiss, J.D., Ph.D.
Attorney
Baker & Hostetler

Richard A. Rettig, Ph.D.
Senior Social Scientist
The Rand Corporation

Edward B. Roberts, Ph.D.
David Sarnoff Professor of the Management of Technology
Chairman, Technology and Health Management
Alfred P. Sloan School of Management
Massachusetts Institute of Technology

Gerson Rosenberg, Ph.D.
Research Associate in Surgery
Assistant Professor of Bioengineering
Pennsylvania State University

Louise B. Russell, Ph.D.
Senior Fellow
The Brookings Institution

David L. Sackett, M.D., M.Sc. Epid.
Professor of Clinical Epidemiology and Biostatistics and of Medicine
McMaster University

Charles A. Sanders, M.D.
Executive Vice President
E. R. Squibb & Sons
(formerly General Director, Massachusetts General Hospital)

Leonard D. Schaeffer, B.A.
Executive Vice President
Student Loan Marketing Association
(formerly Administrator, Health Care Financing Administration)

Alan J. Snyder, B.S.
NSF Fellow
Bioengineering Program
Pennsylvania State University

W. Vickery Stoughton, M.B.A.
President
Toronto General Hospital
(formerly Director, Peter Bent Brigham Hospital)

J. Michael Swint, Ph.D.
Associate Professor of Economics
University of Texas School of Public Health

Laurence J. Tancredi, M.D., J.D.
Associate Professor of Law and Psychiatry
New York University

George E. Thibault, M.D.
Assistant Professor of Medicine
Harvard Medical School
Massachusetts General Hospital

Sankey V. Williams, M.D.
Henry J. Kaiser Family Foundation Faculty Scholar in General Medicine
Department of Medicine
University of Pennsylvania

Robert J. Wineman, Ph.D.
Program Director, Chronic Renal Disease Program
National Institute of Arthritis, Diabetes, and Digestive and Kidney Diseases
National Institutes of Health

Richard Zeckhauser, Ph.D.
Professor of Political Economy
Kennedy School of Government and Center for Health Policy and Management
Harvard University

Part One

THE DEVELOPMENT
OF TECHNOLOGIES

THE DEVELOPMENT OF BIOMEDICAL TECHNOLOGIES

by Edward B. Roberts

My focus in this paper is on describing influences that affect the development of successful biomedical technology-based innovations, including both invention and implementation or exploitation of new medical technologies, taking into account some consideration of the diffusion of those technologies into practice. My emphasis is empirical, rather than anecdotal or speculative, and I shall refer to a number of studies that shed light on the influences on development. This empirical emphasis, however, lies largely outside the biomedical arena, on studies of technologies developed in other fields. The reason for this is not bias on my part; rather, it reflects the lack of such investigation in the biomedical field.

Very little empirical research has in fact been carried out on the processes that affect the development of biomedical technologies. Only the Comroe/Dripps study[1] in the area of cardiology and pulmonary advances represents a major and substantive work focusing on development processes. (The TRACES study[2] contains some relevant cases as well but consists of a biased self-serving sample and therefore lacks objective outcomes.) Regarding diffusion of medical technology, the classic study on the diffusion patterns of a single drug in a limited number of communities, carried out some twenty years ago by Coleman, Katz, and Menzel[3] still represents a primary substantive work in the field. Later works by Gordon,[4] Kimberly,[5] and others have added to this base. Current focus on the development of biomedical technology consequently requires examination of empirical efforts on technological development largely from nonbiomedical fields, with speculation on the transferability of ideas to the biomedical area.

Recently the National Institutes of Health sponsored a conference entitled "The Development and Dissemination of Biomedical Innovations," with the subtitle "Foundations for Program Development."[6] Recognizing the lack of systematic empirical research in the field, the NIH sponsors were concerned that this lack restricted the basis on which either biomedical research programs or policy formation relating to biomedical research and technology could be advanced. The need for the federal government, as well as foundations and other research sponsors, to lay the basis for a more extensive effort in this field seems clear.

The Meaning of Innovation

Generally the process of innovation, as distinct from merely invention, takes into account all steps up to the first utilization of a new or improved product, process, or practice. In the biomedical area differentiation among these several possible areas of technological innovation is necessary because innovation in clinical practice (for example, surgical technique, diagnostic approach, or therapeutic regimen), may well be entirely different from innovation in clinical devices or drugs. The former area has not yet received even the most cursory attention from empirical researchers. Innovation requires invention plus exploitation. Within the term "exploitation" I include a wide variety of activities: the appraisal or evaluation of the technology; the focusing of technological development efforts toward particular objectives; the movement or transfer of research results; and the eventual broad-based utilization, dissemination, and diffusion of those research outcomes. When we look at technological innovation, these perspectives are potential areas of managerial and/or policy concern.

Empirical research literature on innovation can be subdivided into an overlapping set of typologies. A classification of innovations that have been examined might be as follows:

(1) Product vs. process
(2) Radical vs. incremental
(3) New item vs. modification
(4) Industrial vs. consumer good
(5) Services

Concerning (1), most studies on differences between product and process technology focus on the former area. (2) In exploring differences between radical technological developments and the incremental changes that occur more frequently, other than the study by Comroe/Dripps,[1] the few innovation studies focused specifically on the biomedical area have, for the most part, taken anecdotal evidence from radical change areas and attempted to draw broad-based policy conclusions with regard to the handling of overall technological development.[7] Such a practice contributes to an erroneous impression that a productive biomedical innovation is the same as a major breakthrough or Nobel prize-winning discovery, whereas studies of technological developments in nonbiomedical areas indicate that incremental rather than radical innovations dominate research and developmental outcomes. (3) We may also examine research carried out on efforts resulting in new items or new practices versus modifications and improvements of old practices. Here modification and upgrading activities, in contrast to the creation of new entities, dominate most fields of endeavor. (4) Addressing differences in innovation patterns found between industrial and consumer goods, medical devices and prescription drugs fall into the general category of "industrial goods"—that is, they are products developed and turned over to professionals for further use, as opposed to over-the-counter remedies sold directly to the consumer. (5) Finally, we might try to examine innovations with regard to services. Although medical services surely comprise a large segment in this category, a review of the literature reveals few meaningful empirical studies of innovation or developmental activities in medical service delivery. Thus when we try to document empirically the influences on the development of biomedical technologies, we find that data are either fragmented, sparse, or lacking altogether.

Major Influences in Innovation

Influences on innovation can be identified, described, and grouped as follows: (1) staffing, (2) structure, (3) strategy, and (4) supporting systems. I shall discuss the first three of these influences on development: (1) the kinds of people utilized and their various contributions to technological development; (2) the structural issues affecting development linkages and development effectiveness; and (3) some strategic questions. (A more complete coverage might discuss the kinds of supporting systems that are helpful to organizations in the search for effective development of technology.)

What I shall describe as tentative conclusions from this discussion, drawn largely from outside the biomedical field, will actually pose some researchable questions. Indeed I should like the reader to ask throughout, as shall I, whether conclusions on innovation drawn from nonbiomedical fields apply appropriately to questions of technological development within the biomedical sphere. These unanswered questions may then become part of an explicit agenda for further concentrated research.

Staffing

Critical Roles.   In an initial review of staffing, a variety of studies in other fields suggests that a number of key people have critical roles in achieving successful innovation. Five different role clusters[8] emerge:

1. Idea generators (idea "havers" vs. idea "exploiters")
2. Entrepreneurs/product champions
3. Program managers/business innovators
4. Gatekeepers/communicators (technical, market, manufacturing)
5. Sponsors/coaches

Innovators themselves, as well as the literature, mention creative people and creative contributors.[9] If we focus on idea generators, empirical research quickly points out a significant difference between idea-havers and idea-exploiters—the differences between people who create ideas and those who apply the ideas they have generated. These differences have been documented in studies of university laboratories, academic departments, and industry.

Entrepreneurs, known in some of the empirical literature as "product champions," also play key roles in achieving successful innovation. They champion change and innovation, take ideas and attempt to move them forward in organizations to gain their adoption.[10] A significant number of entrepreneurs can be found in the biomedical field, although in academia the labeling of a colleague as an entrepreneur is often considered pejorative. It is hardly pejorative, however, to be called such in industry. Indeed, from the perspective of economic history, Schumpeter[11] has referred to the entrepreneur as "the engine of economic growth and development." (If only academic administrators understood how these "engines" are to be fueled, oiled, maintained, and targeted, rather than stalled and throttled!)

The third necessary contributor to development is referred to in the literature as the program manager, sometimes regarded as the "business innovator," the person who handles the supportive functions of planning, scheduling, business, and finance relating to development activities of his or her technical colleagues.[12]

Gatekeepers or special communicators also play important roles in contributing to development by providing bridging messages that link information sources from the outside world to the inside world of developmental activities. These human bridges tie together technical, marketing, and manufacturing sources of information to the potential users of the information, and their efforts have been subject to extensive study; the work by Thomas Allen[13] is especially

Table 1 **Peters/Roberts Findings on Having and Exploiting Ideas Whose Scope Is Outside the Laboratory**

Laboratory	N	Claimed Such Ideas		Acted on Ideas	
		#	%	#	%*
Lincoln	161	72	45	25	35
Instrumentation	138	75	54	24	32
Total	299	147	49	49	33

* Percentage of those who had claimed such ideas.

noteworthy. In the medical area the gatekeeper's role is essentially the only one that has received empirical attention in the literature, beginning with the study of Coleman, Katz, and Menzel,[3] which identified this role as critical to the processes of diffusion of new drug entities throughout the medical community.

Finally empirical studies in the area of technological development identify the role of the sponsor or the coach. This more senior person does not carry out the research, does not directly champion change, but does provide the encouragement, support, facilitation, and help in bootlegging activities that are necessary to aid junior people in their attempts to move technological advances forward in an organization.[10]

Idea Generators vs. Entrepreneurs. To illustrate the research on staffing influences on development, let me review some empirical indications of these role differences from a study[14] we conducted in two major MIT laboratories, in which we attempted to demonstrate (see Table 1) the distinction between having ideas and exploiting ideas. (The sources cited in references 8–14 document additional evidence on the other three key roles, which we shall not further discuss here.)

Note that 49 percent of the laboratory scientists and engineers claimed to have ideas that lay outside the major area of interest of the laboratory and that had commercial implications. Only 33 percent of those who claimed such ideas attempted any action, and the study offered the widest range of choices for making such a claim. Fully 67 percent of these academically employed scientists and engineers wholly ignored what they said was a significant development.

We then investigated three major academic departments at MIT, which included 66 faculty members, statistically distributed to include appropriately all major ranks, and assessed essentially the same kind of behavior.[15] As indicated in Table 2, 47 percent of the faculty who claimed to have ideas that they felt had commercial merit did nothing about those ideas, not even to the point of attempt-

Table 2   Roberts/Peters Findings on Degree of Academic Exploitation of
          Commercially Oriented Ideas

Degree of Commercial Exploitation	Academic Ideas	
	N	%
None	32	47
Weak	10	15
Strong	26	38
Total	68	100

ing to publish. They certainly did not undertake more substantial action, such as looking into prototype development, commercial applicability, or transfer possibilities. Furthermore only 38 percent of the academicians studied had made strong efforts to move their ideas forward to the point of use. In a current study at MIT, Finkelstein[16] is attempting to replicate this research among several samples of academic clinicians to gain a better understanding of the patterns of idea generation versus idea exploitation in the clinical academic community.

Data concerning those within the MIT faculty group who undertook to exploit their ideas reveal a number of characteristics well documented over the years in prior studies of the characteristics of entrepreneurs.[10] Differences between the faculty who attempted to implement their ideas and their larger number of colleagues who did little or nothing with their ideas emerge with regard to birth order, publishing efforts, and patenting experience. Characteristics of those in the former group include: (1) being first-born children (in this case, sons); (2) earlier activities indicating certain modes and patterns of persistence, including book authorship; (3) prior patenting experience; and (4) an understanding of the financial community and an awareness of sources of financial support. Most faculty members did not have these characteristics: most were not first-born sons and had not authored books; few had earlier patents. Our findings indicate that the faculty member who takes an idea and tries to move it forward is quite different behaviorally and sociologically from his or her colleagues who have as many ideas but who do essentially nothing with those ideas. An understanding of these differences is important in seeking insights into the development of new technology. Are similar critical differences to be found affecting biomedical innovation?

Structure

Structural considerations can determine how successful developments are initiated, how effective technical solutions are found for the problems at hand, and how research results are transferred or exploited. I refer to these as structural issues because the answers to questions posed here frequently lie in relationships between an organization carrying out a technological development and linked or supporting organizations with which they are working. I shall discuss the following:

1. Relationships to sources that motivate initiation of innovative activity.
2. Relationships to sources of effective technical solutions.
3. Relationships to channels for successful exploitation.

Sources of Initiation of Innovation. Multiple sources are seen as motivating successful technological development: technology-push vs. market-demand/need-pull; customer demand vs. manufacturing demand; user role; and regulatory role (whether as stimulus or inhibition). The literature reveals extensive controversy among empiricists who favor the technology-push theory versus those who support the market-demand or need-pull theory of innovation.[17] The latter group believe, and I shall indicate some supporting evidence, that factors of mission, need, or demand dominate in motivating those activities which eventually produce the most successful technological development. The alternative technology-push theory, employed by the former group, asserts that focusing upon underlying research, and attempting to push technology where technological opportunism seems to exist, eventually will result in significant technological development. The Comroe/Dripps study[1] was intended, in part, to provide some greater support for the technology-push theory in opposition to the mission-based or need-pull view. Indeed, evidence assembled by Comroe/Dripps provides some reasonable bases for questioning whether the data from other studies actually apply to biomedical technology.

Table 3 lists data from several different studies that draw the same conclusion that market-pull is dominant.[18] The studies were carried out in the United States and in Great Britain, with different sampling approaches from different industries over different time periods. Yet the data show the same result—that 60 to 80 percent of successful innovations seem to have been initiated by activities which represent market-pull or need-oriented forces. The Comroe/Dripps study did not find these studies to be incorrect, but rather that no clean or neat conclusion could be drawn, and that other sources of initiation, like technology-push, were also important. The industrial innovation literature clearly claims that the perception of need that generates response seems to be the principal motivating factor behind successful innovation. More studies relevant to biomedical endeavors are required to clarify the possible conflict here.

Other researchable questions also exist concerning the motivation

Table 3   Innovations Stimulated by Perceptions of Market Needs

Investigator	Subject of Study	N	% from Market, Mission, or Production Needs
Baker et al.	Corporate Research Laboratory	303*	77
Carter/Williams	British Board of Trade	137	73
Goldhar	"Industrial Research"	108	69
Langrish et al.	Queen's Awards	84	66
Myers/Marquis	5 Industries	439	78
Sherwin/Isenson	"Hindsight" Weapon Systems	710†	61
Tannenbaum et al.	Materials	10	90
Utterback	Instruments	32	75

SOURCE: Redrawn from J.M. Utterback (1974). Innovation in industry and the diffusion of technology. *Science* 183:622.

* Ideas for new products/processes.
† Research events used in 20 developments.

leading to initiation of successful technological development efforts. For example, I think that the special role of the users of eventual technology, as opposed to the developers, has great but largely undocumented importance in the area of biomedical innovation. Finally of course the regulatory role needs to be taken into account. Some have suggested that regulation acts as a stimulant of innovation, but I would imagine that those who are involved with biomedical technology would find that a difficult argument to make.

What is the government's role as a stimulus to biomedical innovation? Evidence found in other markets has suggested that government regulation can act as a stimulus for initiating successful innovation, but a careful look at that evidence shows that the regulatory role stimulates technological innovation primarily in areas of environmental and safety regulations.[19] When the government pronounces, "You cannot, unless it meets the following specifications," then innovation often takes place suddenly to assure meeting those specifications. In the area of biomedical innovation, the government's role is not to set performance standards but to regulate interference. The biomedical regulatory process—certainly in the area of drug innovation, where most data exist—adds enormous costs to the development of new technologies, while adding significant time delays. The evidence indicates increasingly that abusive regulatory behavior, particularly in the United States, even denies efficacious entities to clinical practice. Regulation seems primarily to have negative influences on development processes, although I admit that regulation may help to separate good from bad outcomes. From the perspective of developmental quantity and cost, time delay and time requirement, however, regulation is an inhibitor of biomedical innovation.

Today many researchers seek to determine and enhance the measurement of efficacious outcome; presumably regulation in the biomedical area is directed primarily toward contributing to the efficacy purpose being sought. We still lack evidence demonstrating that the net effect of regulation is supportive, even with respect to the determination of efficacy. The most recent research results that I have seen on drug development and the regulatory process were presented by Wardell.[20] He traced the drugs introduced in the United States and Great Britain from 1972 to 1976 and suggested that some of the regulatory excesses of the United States were weakening, bringing the United States closer to Great Britain in a number of areas of market-available biomedical technological development. The Wardell data, however, also reveal significant medical areas in which U.S. practices of drug regulation continue to have questionable influences upon innovation.

Sources of Solutions. Once we identify a need and initiate a program attempting to meet the need, the next set of researchable questions in the processing of a successful technological development concerns possible sources of technical solutions to those needs. Where are the answers found? The following are a number of different sources of technical solutions to need-generated problems: ideas from both inside and outside the organization; personal experience/contact; scientific literature; original vs. adopted/adapted solutions; and users vs. manufacturers.

Distinctly different sources of ideas in any organization can be found both internally and externally. In fact, much research on industrial innovation demonstrates that technical answers to major problems come from outside the organization in which the work is underway. Personal experience and contacts are used more often as information sources for problem solving than is the scientific literature. In most studies systematic use of the literature as a source of technical information for solving problems has been demonstrated to be relatively ineffective, despite the good intentions of publications and computer-based information retrieval services (such as those available through the National Library of Medicine) to provide for better utilization of organized research information. Consideration should be given to the contrast between original solutions and previously developed solutions that are adopted or adapted, especially in the later stages of an innovation cycle. When modifications and improvements upon an earlier development occur, adoptions and adaptations of the work of others seem to dominate as the sources of technical solutions. Finally, the contributions of the user,

Table 4   Key Information Inputs

Innovator Received Key Input from:	No. of Cases	Percent
Inside the Firm		
Printed Materials	9	2
Personal Contacts	25	4
Own Training and Experience	230	41
Formal Courses	1	0
Experiment or Calculation	40	7
	305	54
Outside the Firm		
Printed Materials	33	6
Personal Contacts	120	21
Own Training and Experience	39	7
Formal Courses	8	2
	200	36
Multiple Sources	62	11
Total	567	101

relative to the producer, to effective technological idea generation might prove to be important in the area of biomedical technology.

A study by Myers and Marquis,[21] which supports the evidence described above on the sources of information for successful technical solutions, was carried out in five industrial fields with 120 firms and involved the study of 567 innovations. Table 4 lists the various sources of technical solutions to the problems dealt with in these successful commercial innovations. Within the organization doing the work, personal contact accounted for 4 percent of sources and personal training and experience 41 percent. Personal contacts outside the organization gleaned another 21 percent, and personal training and experience that took place outside the firm, usually prior to joining the organization, was the source of another 7 percent. Thus personal contacts generated 25 percent of total solutions; personal training, experience, and background elicited an additional 48 percent, clearly dominating the sources of technical insights and information that were applied to solving the problems involved in successful technological development. Other activities listed accounted for relatively little and, while I hesitate to emphasize this, formal courses accounted for only 2 percent of the useful information applied to the technological developments.

Furthermore, the evidence from study results presented in Table 5 indicates that the user who has the need frequently solves his or her own problem and immediately incorporates the solution in his or

Table 5   User vs. Manufacturer Role as Source of Industrial Innovation

Investigator	Industry	N	Percent by Source		
			User	Mfr.	Other
Berger	Engineering Polymers	6	0	100	
Boyden	Plastics Additives	16	0	100	
Enos	Petroleum Processing, Major	7	43	14	43
Freeman	Chemical Processes/Equipment	810	70	30	
Knight	Computers, 1944–1962				
	Improved Performers	143	25	75	
	Radical Structures	18	33	67	
Lionetta and von Hippel	Pultrusion Machinery	13	85	15	
Peck	Aluminum Industry				
	Joining	52	17	50	33
	Finishing	27	33	48	19
	Fabricating	76	30	49	21
	Alloys	39	3	79	18
von Hippel	Scientific Instruments				
	First of Type	4	100	0	
	Major Improvements	44	82	18	
	Minor Improvements	63	70	30	
von Hippel	Semiconductor and Electronic Assembly Manufacturing Equipment				
	First of Type	7	100	0	
	Major Improvements	22	63	21	16
	Minor Improvements	20	59	29	12

SOURCE: Redrawn from E. von Hippel (1978). "A Review of Data Bearing on the User's Role in Industrial Innovation." Working Paper 987-78, MIT Sloan School of Management, Cambridge, Mass.

her own use. In the large series of studies listed, many but not all demonstrate the significant contributions of the user of innovation.[22] The first two studies listed, both in the plastics area, reveal no contributions of users; manufacturers of the final product dominate that area of innovation. But in other areas—petroleum and chemical equipment, the computer industry, specialized machinery, the aluminum industry, scientific instruments, and semiconductor and electronic manufacturing process—a heavy percentage of innovations were created by the users of the products or processes, rather than by those in the business of manufacturing the resulting innovations. In each of these cases, a user arrived at the successful solution, implemented it first in his or her own organization for personal use, and often made copies available to others on request. Later a manufacturer discovered the successful development and use, fully adopted the solution, made the engineering modifications as needed, and then produced the innovation in large volume, distributing and diffusing it to industrial customers or to the consuming public.

In the only study of this type that I am aware of in the field of biomedical technology, von Hippel and Finkelstein[23] sought the sources of innovation in test methods embodied in medical laboratory clinical analyzers. The study focused on the design of a physical piece of equipment and how it encourages or discourages scientific manipulation, experimentation, and eventual further contribution to the process of innovation. The DuPont clinical analyzer was compared to the Technicon machine, and each of the diagnostic test methods employed was studied as to origin. For the Technicon SMAC, of the 20 most significant test methods employed, 14 were successfully developed and implemented initially by users of the equipment. Another method was developed by a reagent manufacturer, a firm supplying chemicals for use with the Technicon equipment. Only 4 of the 20 test method innovations in the Technicon SMAC were generated by Technicon itself. One more test method was questionable, due to multiple sources of contribution. In contrast, all of the 18 test methods constituting principal utilization of the DuPont clinical analyzer were developed by DuPont itself; no contributions were made by the users.

Without further evidence available, I can only speculate that additional systematic studies in areas of biomedical tools and devices development would show that users, not manufacturers, dominate the sources of technology innovation. I would guess that the role of the manufacturer of biomedical devices is primarily to adopt and distribute the product and not to generate ideas or develop successful solutions to the problems involved in those ideas. However, no empirical research is available to support this speculation.

Channels for Exploitation.   Other research questions relate to the linkages for transferring results of a successful technical development that is still internal to an organization. The first movement toward the utilization and implementation of technological development generally takes place on a trial basis, especially in medicine, before evidence begins to accumulate that might induce widespread diffusion and dissemination. Effective linkages are needed between research laboratories and product-line departments, in such organizations as pharmaceutical companies, and between universities and industry if the university is to be a significant source of original technological development in the biomedical field. Yet no meaningful empirical research exists to indicate which patterns of linkage and transfer currently dominate or which channels might be more effective than others. One of the most significant problems of achieving successful biomedical development may be that, despite exceptions,[24] effective channels do not exist between the universities and medical schools, in particular, and the industrial establishment for transferring research results into products. The bridges used for achieving such transfers are many and varied—procedural, human, and organizational—but again, evidence for what occurs is not very broad based.[25] For example, Young[26] has indicated the ineffectiveness of clinical trials as a mechanism for effecting the transfer of research results into broad-based clinical practice, while Levy and Sondik[27] have assembled evidence in support of the utility of clinical trials as a transfer mechanism. Mechanisms for transferring research results into utilization need further empirical investigation,[28] especially with respect to biomedical innovation.

Strategy

Strategy issues are the last cluster of variables that influence the development of technology, and raise the following questions: How does the stage of a technology affect the pace and the nature of technological innovation? Who does the technological innovation? Is successful innovation carried out more by large companies, by small companies, or by individual inventors? Is it done more in universities, in industry, or in governmental laboratories? Is it done more by outsiders to a given industry? For example, one theory describes innovation by the invasion of outsiders who focus on the problems of a field that is not theirs and who impose innovative change upon that field.[29] There are also strategy-relevant questions pertaining to the amount of technological change actually embodied in significant innovations. How costly is the process of technological development itself? What is the role of patents and trade secrets in these areas? Appropriate answers to these strategy questions in the area of biomedical development should lay important foundations for policy development. All attempts to formulate policy, to regulate, or to influence emerging or existing technologies, whether through mechanisms of financing or acceptance of technology, implicitly assume answers to these and other strategy questions. I assert yet again that very little empirical research exists in any of these areas.

One controversial finding suggests that individual inventors and small firms are principal contributors to product and even process innovation, particularly of a radical nature, and especially in early stages of a technology. Table 6 contains an array of data from large-scale industries which indicate that small companies and individual inventors tend to play a rather dominant role in the generation of the key innovations in these fields.[30] Medical innovations were not included in the table, due to lack of available data from systematic research.

Table 6   Sources of Key Innovations

			Percentage from	
		N	Major	Small Firms
Investigator	Study		Firms	and Inventors
Enos	Petroleum Refining, Basic Major			
	Innovations	7	0	100
Hamberg	Steel	11	36	64
Hamberg	Major Innovations, 1946–1955	27	<33	>67
Jewkes	Major Innovations, 1900–1945	61	<50	>50
Peck	Aluminum, Major Innovations	7	14	86
Tilton	Semiconductors, Major Product			
	Innovations	—	54	46

In a study of 77 companies generating hundreds of innovations, Utterback and Abernathy[31] divided the phases of development of a technology into three stages. Stage I was the initial stage of a new field of technology; Stage II was its later development; and Stage III was the maturation phase of a technological area (see Table 7). Examination of the size in sales volume of companies which made leading contributions to the technological innovations in these various stages of development demonstrates that small companies with less than $10 million in sales clearly dominate innovation in the Stage I aspect of a new technology. These companies are the principal contributors to major product and process change in the early stages of development of a new field. By the time a technology reaches Stage III, the role of small companies, although still important, is no longer dominant. Indeed, large companies, indicated here by a sales volume of over $100,000,000, tend to dominate the patterns of technological innovation in mature fields. I believe the same process is occurring

Table 7   Utterback-Abernathy Findings on Firm Size and Successful Innovations (Myers-Marquis Data on 77 Firms)

Sales	Stage I		Stages II and III	
($000,000)	N	%	N	%
Unclassified*	12	23	8	32
<10	18	34	0	0
10–100	6	12	2	8
>100	16	31	15	60

$X^2 = 11.2$, $p < 0.01$

Source: Assembled from data in J.M. Utterback, and W.J. Abernathy (1979). A dynamic model of process and product innovation. *Omega* 3:654–656.

* Most of those companies listed as unclassified according to sales are privately-held small companies who would not release their sales data. For the most part, their sales are under $10,000,000.

Table 8   Utterback-Abernathy Findings on Technological Change in Successful Innovations (Myers-Marquis Data on 77 Firms)

Degree of Invention Required	Percentage Distribution		
	Stage I	*Stage II*	*Stage III*
Little	14	19	33
Considerable	41	50	48
"Invention" needed	45	31	19

$X^2 = 19.1$, $p < 0.001$

SOURCE: J.M. Utterback and W.J. Abernathy (1979). A dynamic model of process and product innovation. *Omega* 3:651.

in the area of biogenetic technology where small companies are the dominant contributors to innovation. Again systematic studies are needed to evaluate the role of the small firm and the individual inventor in bringing new technologies to fruition in the biomedical area.

The degree of technological change embodied in a successful innovation is also an important question. In Table 8 data on the 77 companies presented in Table 7 indicate that significant technological change was embodied in 45 percent of the innovations that occurred during the Stage I aspect of a technology.[31] In Stage III, after a technology has been well established and well accepted, only 19 percent of those technological successes embody invention to a meaningful extent; most of the successes in late stages of a technology involve merely incremental technological change.

The final empirical evidence—data on the role of patents—is used to raise further questions about the issues of biomedical technology and to begin to suggest ways in which it may differ from other areas. Studies of the role of patents, with occasional exceptions such as for Polaroid and Xerox, demonstrate that patents have very little influence on the successful development or commercialization of technologies. Data presented in Table 9, however, demonstrate that the area of pharmaceuticals is significantly different from all others, in that patent royalties in pharmaceuticals represent a larger relative contribution to profits than royalties in any other industrial area.[32] Studies on the percentage of research and development budgets influenced by the potential availability of patents produce a similar result: The pharmaceutical industry is distinguished from almost all other fields by being very concerned about patent availability in motivating the direction, character, and extent of research support.[33] This evidence indicates one way in which biomedical innovation differs from other areas, although I speculate that many similarities also exist.

Table 9   Importance of Patents

	Wilson (1971 Royalty Data, U.S.)	Taylor-Silberston (1968 Royalty Data, U.K.)	
Industry	Royalties Paid as % of Sales	Royalties Paid as % of Sales	Industrial Activity
Chemicals			Chemicals
Industrial	0.244	0.042	Basic
Drug	0.745	0.635	Pharmaceutical
Other	0.034	0.044	Other Finished and Specialty
Machinery	0.051	0.255	Mechanical Engineering
Electrical	0.13	0.182	Electrical Engineering

SOURCE: Redrawn from E. von Hippel (1979). "Appropriability of Innovation Benefit as a Predictor of Functional Locus of Innovation." Working Paper 1084–79, MIT Sloan School of Management, Cambridge, Mass.

Summary and Questions

I would like to summarize the influences on the development of technologies, drawn primarily from outside the biomedical arena.

1. Five staffing roles influence the development of successful technological innovations. The strongest evidence exists on the need for idea exploiters or entrepreneurs, as distinct from idea generators or idea havers. An important need exists for the facilitators of communication, the information gatekeepers, and the senior management helpers whom I identify as sponsors or coaches.

2. Structure exercises important influences on innovation. Ties to the market-motivating forces have been found to be primary in effecting eventual successful technological development. Linkages to outside information sources for initial ideas, technical solution ideas, and whole-solution adoption of ideas are frequently key to the technological development process, with personal contacts and experience as the major mechanisms by which these linkages are effected successfully. Adopted innovation in general, as well as the special role of the innovative user as a source of eventually adopted innovation, require more attention. Little evidence exists on the differential effectiveness of various channels of research results transfer which, in my judgment, is the most critical stage of technological development. At the present time we have only anecdotes describing how this transfer process operates effectively. As I see it, the role of government regulation inhibits innovation in the biomedical

area, and I hope my statement will stimulate some controversy and response.

3. With respect to strategic questions, pilot studies show that different forms of innovation dominate at different stages of the technology cycle. Individual inventors and small firms seem critical, especially early in a field. Patents, seen as insignificant elsewhere, are important to the pharmaceutical industry. This does not, however, necessarily mean that patents are important to the medical device industry, given the many distinctions among innovation in the areas of drugs, devices, and clinical practices. Again the availability of more careful research would enable possible distinctions to be highlighted.

The findings cited here are based heavily on empirical studies taken mainly from outside the medical field. Some differences between research and technology in the biomedical field and endeavors in other fields deserve note, however.

1. The uncertainties involved in natural science research and development are far greater than in R&D dependent upon the physical sciences, and include problems of biological variability as well as efficacy determination. These problems are far more significant for biomedical research and technology than for similar efforts in other fields.

2. The federal government is highly involved in the sponsorship of research in the biomedical technology area, but, in contrast with the situation involving defense, the government is not a direct customer for the implementation of the R&D results. This shift in the research sponsorship role has very important consequences for successful development and use of biomedical technology.

3. Relative to any other scientific-technical field, the highest degree of academic involvement in and domination of research execution (relative to industry, for example) exists in the biomedical technology area. This fact clearly needs to be taken into account in considering possible differences between the biomedical area of technological development and other areas.

4. The highest extent of government regulation of product acceptability and product diffusion is encountered in the biomedical area.

5. Strong emotional factors affect involvement with innovation in and utilization of products, processes, and practices that influence health and life. This should be remembered in the attempt to understand the influences on successful biomedical technology development.

A major program of federal and foundation-supported research is needed on the influences involved in all stages of development and dissemination of biomedical innovation. This program should be initiated promptly in order to begin building an empirical basis for managerial direction, policy formation, and regulatory action in regard to medical technology.

Endnotes

1. Comroe, J.H., Jr. and Dripps, R.D. (1977). *The Top Ten Clinical Advances in Cardiovascular-Pulmonary Medicine and Surgery, 1945–1975*, DHEW Publication No. (NIH) 78–1521, January 31. Washington, D.C.: U.S. Department of Health, Education and Welfare.
2. *Technology in Retrospect and Critical Events in Science* (TRACES). Prepared for National Science Foundation by the Illinois Institute of Technology, December 15, 1968.
3. Coleman, J.S., Katz, E., and Menzel, H. (1966). *Medical Innovation: A Diffusion Study*. Indianapolis, Ind.: Bobbs Merrill Co.
4. Gordon, G., and Fisher, G.L. (1975). *The Diffusion of Medical Technology*. Cambridge, Mass.: Ballinger Publishing Co.
5. Kimberly, J. (1978). *Hospital Adoption of Innovations in Medical and Managerial Technology: Individual/Organization and Contextual/Environmental Effects*. Final report to National Science Foundation, Yale University.
6. Roberts, E.B., Levy, R.I., Finkelstein, S.N., Moskowitz, J.A., and Sondik, E.J., eds. (1981). *Biomedical Innovation*. Cambridge, Mass.: MIT Press.
7. Battelle Memorial Institute. *Interactions of Science and Technology in the Innovative Process: Some Case Studies*. NSF-C667, March 1973.
8. Roberts, E.B. and Fusfeld, A.R. (1980). "Critical Functions: Needed Roles in the Innovation Process." Working Paper 1129–80. Cambridge, Mass.: MIT Sloan School of Management.
9. Pelz, D.C. and Andrews, F.M. (1966). *Scientists in Organizations*. New York: John Wiley & Sons. For more recent findings, see Andrews, F.M., Innovation in R&D organizations: Some relevant concepts and empirical results. In *Biomedical Innovation*, E.B. Roberts et al., eds.
10. Roberts, E.B. (1968). Entrepreneurship and technology, *Research Management* 11:249–266.
11. Schumpeter, J. (1934). *Theory of Economic Development*. Cambridge, Mass.: Harvard University Press.
12. Marquis, D.G. and Rubin, I.M. (1966). "Management Factors in Project Performance." Working Paper, MIT Sloan School of Management, Cambridge, Mass.
13. Allen, T.J. (1977). *Managing the Flow of Technology*. Cambridge, Mass.: MIT Press. Also Rhoades, R.G., Roberts, E.B., and Fusfeld, A.R. (1978). A correlation of R&D laboratory performance with critical functions analysis, *R&D Management* 9:13–17.

14. Peters, D.H., and Roberts, E.B. (1969). "Unutilized ideas in university laboratories," *Acad. Man. J.* 12:179–191.
15. Roberts, E.B. and Peters, D.H. (1981). Commercial innovations from university faculty. *Research Policy.* 10:108–126.
16. Finkelstein, S.N., Scott, J.R., and Franke, A. (1981). Diversity as a contributor to innovative performance by physicians. In *Biomedical Innovation*, eds. E.B. Roberts et al.
17. Mowery, D. and Rosenberg, N. (1979). The influence of market demand upon innovation: A critical review of some recent empirical studies. *Research Policy*, 8:102–153.
18. Utterback, J.M. (1974). Innovation in industry and the diffusion of technology. *Science* 183:622.
19. Allen, T.J., Utterback, J.M., Sirbu, M.A., Ashford, N.A., and Hollomon, J.H. (1978). Government influence on the process of innovation in Europe and Japan. *Research Policy* 7:124–149. Gerstenfeld, A. (1977). Government regulation effects on the direction of innovation. *IEEE Transactions on Engineering Management, EM-24*, No. 3:82–86.
20. Wardell, W.M. (1978). The drug lag revisited. *Clin. Pharm. Ther.* 24:499–524.
21. Myers, S. and Marquis, D.G. (1969). *Successful Industrial Innovations*, National Science Foundation NSF 69–17. Washington, D.C.: U.S. Government Printing Office.
22. von Hippel, E. (1978). "A Review of Data Bearing on the User's Role in Industrial Innovation." Working Paper 987–78, MIT Sloan School of Management, Cambridge, Mass.
23. von Hippel, E. and Finkelstein, S.N. (1979). Analysis of innovation in automated clinical chemistry analyzers. *Science and Public Policy*, February, p. 31.
24. For an example of one potentially important exception, see the discussion of the Harvard-Monsanto long-term medical and biological research program in Industry R&D renews the old campus ties, *Chemical Week*, February 21, 1979, pp. 38–39.
25. Roberts, E.B. (1979). Stimulating technological innovation: Organizational approaches. *Research Management* 22:26–30.
26. Young, D.A. Communications linking clinical research and clinical practice. In *Biomedical Innovation*, eds. E.B. Roberts et al.
27. Levy, R.I., and Sondik, E.J. (1981). The management of biomedical research: A case in point. In *Biomedical Innovation*, eds. E.B. Roberts et al.
28. Roberts, E.B., and Frohman, A.L. (1978). Strategies for improving research utilization. *Technology Review* 80:32–39.
29. Schon, D.A. (1963). Innovation by invasion. *International Science and Technology*.
30. Enos, J.L. (1962). *Petroleum Progress and Profits: A History of Process Innovation*. Cambridge, Mass.: MIT Press. Hamberg, D. (1963). Invention in the industrial research laboratory. *J. Political Econ.* 71:95–115. Hamberg, D. (1966). *R&D: Essays on the Economics of Research and Development*. New York: Random House. Jewkes, J., Sawers, D., and Stillerman, R. (1958). *The Sources of Invention*. London: Macmillan Publishers Ltd.

Peck, M. (1962). Inventions in the postwar American aluminum industry. In *The Rate and Direction of Inventive Activity*, ed. R.R. Nelson. Princeton, N.J.: Princeton University Press, pp. 299–322. Tilton, J.E. (1971). *International Diffusion of Technology: The Case of Semiconductors*. Washington, D.C.: The Brookings Institution.

31. Utterback, J.M., and Abernathy, W.J. (1975). A dynamic model of process and product innovation. *Omega* 3:639–656.

32. von Hippel, E. (1979). "Appropriability of Innovation Benefit as a Predictor of Functional Locus of Innovation." Working Paper 1084–79. Cambridge, Mass.: MIT Sloan School of Management.

33. Taylor, C.T., and Silberston, Z.A. (1973). *The Economic Impact of the Patent System*. Cambridge, England: Cambridge University Press.

Part Two

HOSPITALS AND NEW TECHNOLOGIES

ADOPTION OF NEW TECHNOLOGIES IN HOSPITALS

by Charles A. Sanders

The role of technology in the provision of health care often evokes strong and conflicting viewpoints. Some think technology is cost provocative, overzealously utilized, and often nondefinitive or of marginal usefulness in its application. Others have a more sanguine approach to technology's role in medicine, believing that technological development is a necessary and beneficial accompaniment to the evolution of a more effective system of health care. Truth can be found in both viewpoints; the issue is more of degree than of substantive differences regarding the value of technology and its application. The basic question concerns the way society governs or should govern its abundant technology rather than how it should be governed by it.

Six Basic Premises About the Role of Technology

First, technology is basically good. It has improved the quality of patient care, extended or saved lives, and in most instances enhanced the quality of life. The continued development of technology is an ineluctable creative force in human nature. It is unlikely that society could shut down development even in the face of the most zealous regulations or restraints from any source. Continuing commitment to technological innovation—halfway technologies in many instances —is vital to the ultimate goal of curing *and* preventing disease.

Second, controversies surrounding technology, at least in the field of health care, relate to three issues, one of which is the application of marginal or poor technologies. Harvey Fineberg[1] did a study of gastric freezing, showing its rapid acceptance and diffusion, followed by its demise, all in a period of some four years. Although many people were treated by gastric freezing, it was an ineffective technology.

The second issue concerns abuse in utilization of technologies,

25

often good technologies. The best example of this, of course, is CT scanning, the most revolutionary medical technology of the past twenty-five years. This procedure affords quick, accurate, and safe diagnosis of many types of diseases, some of which had previously eluded diagnosis by all other techniques short of exploratory surgery. However, there is no question that CT scanning has been inappropriately applied and overzealously used, and the charges for it have been excessive in some instances.

The third issue surrounding technology and creating confusion is that almost any new technology is cost provocative in the absence of specific tradeoffs. All too frequently none of the older technologies can be dropped while a new technology is brought on line. Thus new technology is usually an additive phenomenon. The positive side of this equation may be better diagnosis or therapy; the negative side is added cost as the technological basket grows fuller.

Fourth, it is unlikely that one means will be found to identify, evaluate, or determine the dissemination of the technologies. The originating sources of new technologies and the paths they follow into use are too disparate to be dealt with by any single methodology. The questions are different for *emerging* technologies, for *new* technologies that are on the market but have not been widely disseminated, and for *existing* technologies which need to be applied as new technologies are introduced. Evaluation of technologies is therefore multifaceted and requires a range of approaches, some of which include governmental regulation; consensus-seeking conferences, such as those sponsored by the National Center for Health Care Technology; clinical trials; and consumer involvement. Problems exist in regard to involvement, timing, utilization, and decision making. How can we deal with an outcome which displays efficacy and safety of a technology, and a role in health care of unpredictable magnitude? For example, the importance and potential utility of penicillin were easy to project. CT scanners were being widely used before their impact was really fully appreciated. Is there a role for the principle of *caveat emptor* in this particular setting, providing the system for utilization and reimbursement is modified to promote it?

Fifth, the ultimate adoption and utilization of the approach selected to evaluate a technology will be determined by economics. Do sufficient funds exist to pay for it? It is doubtful that the National Health Planning Act and its implementation will play more than a complementary role in determining adoption, dissemination, and utilization of technologies. A large part of the economic equation will be the role of positive or negative incentives for providers, patients, and third parties. The system must have risk sharing which includes the possibilities that some may make money, some may lose money, and some sanctions may be available for inappropriate capital expenditures or technology utilization. Each of these possibilities has an impact.

Furthermore, the cost effectiveness of a *therapeutic* technology, when equated with enhancement of the quality of life, is a satisfactory equation. Alternatively, cost effectiveness of a diagnostic technology should not require a favorable outcome to be acceptable. Such judgments serve only the purposes of the strong advocates of cost containment and ignore questions relating to ease of application, safety, precision, and shortening or eliminating hospitalization in making an accurate diagnosis.

Sixth, appropriate utilization of technology is compatible with both good patient care and cost containment.

Finally, technology is blamed, and often inappropriately so, for the runaway costs in health care. Such "blame" ignores the fact that there is more in our technological basket than ever before. Government action has consequently been sought more and more frequently as a means to control costs and in particular to control technology. Ironically, as will be discussed, the government should and must bear a significant share of responsibility for the current situation. To prevent further compounding of this error, it should be emphasized that a *sine qua non* for the future is an effective cost containment system prior to the institution of any form of National Health Insurance.

The Current Situation

The foregoing premises underlie the environment in which technology is developed, disseminated, and judged, but additional pivotal factors in the current policy and institutional arenas should be considered. First, the system of reimbursement for health care in this country is obsolete. Cost reimbursement per se is a disincentive to the economy. Second, the practice of medicine is not compatible with the free market development that operates in other sectors. For example, the physician controls the referrals and the utilization of 80 percent of the dollars that are spent; medicine could not be practiced in any other way. Third, some competition is possible, as with Health Maintenance Organizations (HMOs) or Independent Practice Associations (IPAs), but the potential is limited. Probably no more than 25 to 30 percent of the medical market can be penetrated by such organizations. Both labor and management are working toward a goal which will foster competition but, despite this effort, patients really do want their own physicians. Fourth, cost containment measures adopted by the federal government have not worked in the past. The economic stabilization program of 1972 to 1974 was very far from satisfactory, and it is possible that the seeds of the 1980 inflation were sown in that period by that program. Regardless of individual views, cost containment efforts at the federal level have been ineffective. There is no reason to assume that pursuit of similar efforts in the future will differ. In contrast, the voluntary effort of the American Hospital Association, which has been in effect for more than three years, has shown considerable promise during this time. Nonetheless, the voluntary effort creates questions as to the number of providers who are deferring costs or expenditures at this time to keep this effort in line with the Consumer Price Index (CPI). Inevitably such restraint will result in requirements for substantial expenditures for new capital equipment or the imposition of additional operating costs which will result in an abrupt increase in hospital costs very much like that seen in 1975 after the Economic Stabilization Program (ESP) was discontinued.

The closing or merger of hospitals actually represents the ultimate means of cost containment. Such actions result from adverse economic circumstances affecting particular institutions and not from any specific health planning actions. In some instances closings have been desirable. However, it could be argued that the public has not been well served, particularly where deinstitutionalization of patients in the mental health area has been presented as the most innovative form of care when, in fact, a lack of money determined the policy.

Government regulators can be "manipulated." Examples include the laetrile issue, the inability to affect the adoption of technology in physicians' offices, and the unfettered use of the "little" technologies, for example, automated blood tests and chest x-rays. Other technologies such as renal dialysis are supported by special interest groups, which suggests that once such a group starts to work (often pursuing a very laudatory goal for a particular constituency), it can have a tremendous influence on Congress. In the face of such political heat it is unlikely that the federal government will refuse to comply. On the other hand, rate regulation of the type found in Massachusetts would appear to hold a lot of promise. This system has exercised considerable influence in holding down incremental costs in hospitals by restraining the percentage increase allowed annually for hospital revenues. However, the Massachusetts Rate Setting Commission has direct control only over the Medicaid budget. That budget has increased from $500 million in 1975 to over $1 billion in 1980. In light of this experience, the effectiveness of the regulatory process could be questioned when it has specific responsibility for administering tax dollars. Also Medicaid funding does not fully cover the cost of care despite the enormous increase in the budget over the past five years. As costs have risen, the reimbursement formula appears to be derived from the numbers of patients covered divided into the available dollars. In this instance the state government pays out only as much money as it has and leaves the short-fall to the private sector. Similar behavior is seen at the federal level as well, manifested by proposed cutbacks in rates of Medicare reimbursement with no reduction in benefits. This performance is not a compelling argument for more governmental control.

Specific Technologies

Three examples clearly illustrate the problems encountered in predicting the rate of adoption, dissemination, and utilization of a technology.

Hyperalimentation

As recently as twelve years ago effective nutrition could not be sustained through the administration of intravenous fluids; solutions of glucose and water simply could not deliver enough calories to the patient to meet the normal daily requirement. As a result, patients maintained on intravenous fluids fell into negative nitrogen balance; their healing processes and their ability to resist infection were seriously affected. In 1967 research demonstrated that high concentrations of nutrients administered into a central vein of the body, the superior vena cava, would not only maintain metabolic balance but would allow the patient actually to gain weight during the period of administration. The research problem to be solved involved determining the appropriate composition of the fluids, the rapidity and route of administration, and the evolution of necessary measures to prevent infection both locally and systemically. Close monitoring of various electrolytes and metabolites was essential.

After the initial reports of success from other research studies,[2] hyperalimentation was introduced in 1970 to the Massachusetts General Hospital by an individual physician entrepreneur without prospective review. At that time the staff consisted of one physician. Approximately two patients per week were treated at a cost of $30 per day with an average duration of treatment of 20 days. In 1979 the staff numbered nine physicians, the number of patients had increased from 2 to 58, the cost from $30 to $96 per day, and the duration of treatment remained the same. There was, and still is, a heavy reliance on the pharmacy back-up.

If we examine the effectiveness of this technology in terms of patient care, the totality is very difficult to measure. In many ways the technology is all pervasive, affecting almost all of the components of the hospital either directly or indirectly. For example, hyperalimentation allows physicians to bring into the hospital a patient who may be grossly undernourished from a chronic ulcer which has precluded intake or retention of food. Administration of hyperalimentation restores that patient to positive nitrogen balance and surgery can be performed under the most favorable circumstances. Not only is mortality reduced, but morbidity is shortened. The patient is returned to work within a reasonably short period of time. Other examples include patients who have had significant amounts of their gastrointestinal tracts removed surgically. Ordinarily such a significant loss of intestine leads to death. Now, because sick patients can be maintained by hyperalimentation for a prolonged period while the remaining intestine assumes new functions, such patients can be weaned from hyperalimentation and returned to a useful life.

In 1970 the direction of hyperalimentation was impossible to determine. We did not know how much space, personnel, and equipment the hospital would have to allocate to it, or how many people would be benefited. This technology has grown as its techniques and applications have been steadily refined. Nonetheless, hyperalimentation is but an ancillary technology, to be put into the equation of patient care. While it is not definitive, this technology buys time until more definitive therapies can be employed.

Respiratory Therapy

Another ancillary or nondefinitive type of technology is respiratory therapy, which has followed an unusual course of development. The purpose of this technology is to make people breathe better—breathe better if they have emphysema, or breathe better postoperatively. Respiratory therapy began in the late 1940s at the Massachusetts General Hospital (MGH). Initially two physicians treated 10 to 20 patients per day and carried out about 6,000 procedures per year at a cost of $15 to $20 per procedure. The cost per patient course of therapy was unknown because collecting statistics in those days was not an important priority. By 1979 the staff had grown to a total of 73 people: 3 physicians, 55 respiratory therapists, 12 chest physiotherapists concerned primarily with postoperative patients, and 3 support personnel. The number of patients per day had risen to 250 to 300 and the number of procedures to 184,000 per year. Cost per course of treatment has remained reasonably constant and is now about $150. This ancillary therapy actually has diminished in volume over the last six to seven years as indications for chest physiotherapy and respiratory therapy have been refined. The number of procedures per year has dropped from 234,000 in 1973 to the 184,000 mentioned previously. This technology currently appears to be in a reasonable state of equilibrium.[3]

Cardiac Surgery

Cardiac surgery is a definitive therapy and offers some interesting contrasts to respiratory therapy. This technology first started at the MGH in 1957, when cardiopulmonary bypass became available on a regular basis. Thoracic surgeons began to shift their operating focus from the lung to the heart. Initially no more than six to eight open heart procedures were performed per month. With the advent of another new technology, the introduction of prosthetic valves, volume increased from three patients per week in 1963 to over one thousand per year in 1976, a figure which remains constant to this day. Costs have risen to be sure, but the interesting change is the diminution in intraoperative monitoring of the body functions. Now, only arterial and venous pressures and periodic arterial blood gases are monitored, in contrast to a much more comprehensive protocol, which included constant display of the electroencephalogram, employed in the early days.

The techniques and skills of carrying out the entire procedure—cardiac surgery itself in addition to intensive care unit therapy, including respiratory care—have resulted in a shorter hospital stay for the patient. The increase in total costs has been fueled by a number of factors, including inflation, more sophisticated technology in the operating room and ICU, and the need for more personnel to service technology. Still, increments in costs have been limited by the trade-offs mentioned above. In addition, a conscious choice has been made at the MGH not to expand the cardiac surgery program in terms of the total number of procedures done each year. This choice is dictated simply by competition for hospital beds from physicians in other specialties. A hospital that runs 95 to 97 percent full during the course of the week does not have much extra room to take on additional procedures. This has had a salutary effect in that a queue of sorts has developed, producing patient selection for application of all forms of technology. In the case of cardiac surgery, operations are performed on those patients who have a reasonable chance of being helped. Alternatively, it is not difficult in such a milieu to deny surgery to the hopelessly ill. The former attitude was that "since the patients will die anyway, we should go ahead and perform surgery in the hopes of making the patient better, remote though that possibility might be." In most instances, experience has shown (as the queue has lengthened) that physicians are able to distinguish between those patients who will survive and be improved and those patients for whom surgery will be a useless exercise.

In discussing cardiac surgery, the cardiac transplantation decision made at MGH in early 1980 provides some insights into the future.[4] The hospital's technological capacity and expertise to perform cardiac transplantation was never in question, nor was financial reimbursement for the procedure addressed. The question was what impact the program would have on the allocation of resources going to other, larger numbers of patients already being treated with existing technologies. A number of technologies being applied were only halfway measures at best and perhaps no better than cardiac transplantation. Nonetheless, they were being applied effectively to the extent they could be. The hospital felt that to reallocate a significant amount of resources to a few patients, no more than six per year, raised serious ethical questions. The Board of Trustees ultimately made a decision not to proceed with the transplant program. It was, in all likelihood, a precedent-setting decision. More decisions of this kind will have to be made, whether by trustees, physicians, or others, as our technological capacity inevitably increases.

Decisions involving programs that have a direct impact on patient care can be made more effective than those involving the ancillary areas. Addressing questions relating to the impact of cardiac surgery was easier conceptually than making projections about hyperalimentation or respiratory therapy. In other settings such examinations may be virtually nonexistent due to the dominating personality of a single entrepreneur or the unfettered entrepreneurial spirit in a hospital determined to compete more effectively in its community. Judgments about individual situations in these circumstances are difficult to make; the lack of a process to assess the consequences of technological adoption and utilization evokes concern.

Discussion

What are we trying to accomplish, given the heterogeneity among hospitals in dealing with technology? In my opinion we are trying to provide high-quality, comprehensive medical care at a reasonable cost to all of our citizens. This may be provided by a single institution, or, more likely, by a series of health providers whose capacities cover the spectrum of primary, secondary, and tertiary care, alongside programs for health promotion and early detection of disease. Each of these areas will employ some form of technology in fulfilling its mission. As our society continues to deal with the complex, multifaceted problems of health maintenance and health care, we must work toward developing a process to examine the impact of new and existing technologies on the providers and their communities.

When a hospital adopts a new technology, knowledge of its effectiveness and safety are absolutely essential to the hospital's decision makers. Such input comes from both external and internal sources.

External Sources

The Food and Drug Administration has played an important role in questions relating to the safety and efficacy of any technology. In addition, the Office of Technology Assessment of the Congress has conducted a number of useful studies. Another resource, the National Center for Health Care Technology, has been established recently but is still in the germinal stages. Potentially the Center has a bright future from its large legislative mandate, but like many government programs, does not have sufficient funding. The Center has shown great promise in terms of providing information about past, present, and future technologies, and has been in the unique position of being able to coordinate all governmental agencies involved in technology assessment. The Institute of Medicine has made a contribution by studying different types of technology. Lastly scientific literature and meetings have been available to evaluate technology. Such information has affected introduction and dissemination by addressing issues relating to effectiveness, safety, coverage, and potential impact. Still the impact issue is poorly developed at this time and susceptible to local, state, and federal political activities.

Dissemination throughout the system is determined by medical utility and the question of financial coverage. Health planning may play a role if the technology is large enough. Again, however, the threshold of the health planning law is $150,000. Because physicians' offices are not involved at the present time, its impact upon the little technologies and upon those activities carried on outside the hospital is limited.

Utilization is also affected externally by the activities of the Professional Standards Review Organization (PSRO) and internally by the hospital Utilization Review Committee (URC). The PSROs have not reached their full stage of maturity and there is real concern about whether they ever will. Of the two, the Utilization Review Committee is the most productive. This approach has done more than any other to ensure that the hospital is properly utilized. However, the issue of technology utilization, large or small, is not addressed. A recent requirement issued by the Bureau of Health Insurance related to standing orders will have a great impact on the utilization of smaller technologies. Medicare will no longer pay for standing orders without question. Simply stated, if a patient is being admitted, the orders to be carried out must be written *and* signed, indicating that the physician considered the special needs of the patient. Greater savings may emerge from this program than from many others that have been employed to date.

Internal Sources

In response to requirements of the Joint Commission on the Accreditation of Hospitals (JCAH), hospitals have developed an internal process for medical audits to ensure that the quality of care is appropriate, an approach which, however, does not address the issue of assessing the appropriate use of technology. Recently the JCAH has dropped this requirement in favor of a more comprehensive audit program which is well motivated but of unproven practicality. In looking at ancillary utilization of existing technology, the hospital industry has been virtually silent. To address issues surrounding appropriate utilization, a Medical Care Practices Evaluation Unit has been established at the MGH. Some of its findings suggest that such studies are practical and useful in establishing standards and criteria for diagnosis and therapy.[5] In general, however, ignorance abounds in this area, although, potentially, expenses could be curtailed by more judicious use of existing technology. In the past two years the cost of ancillary care has surpassed the *per diem* cost. While the *per diem* cost is emphasized ($180 to $200 per day in the state of Massachusetts), it now costs over $200 a day just to serve the average *ancillary* cost.

A Resource Allocation Board (RAB) was established several years ago at the MGH to deal with the problem of technologic introduction. Five physicians and four administrators make up the Board, which is essentially an internal certificate-of-need program. The threshold for consideration of new technologies is $25,000, which is probably too low at this time. The RAB requires review of any new program or significant extension of existing programs. The director of the hospital refers the technology programs under consideration to the RAB, and while the RAB findings are not binding, information and an assessment upon which to make a judgment are provided. Some applications are rejected, although most are adopted. The great advantage, even with ultimate approval, is that the final proposals are more sophisticated and more carefully thought out, and a process for future evaluation is put into place to determine whether the program or technology is effective in its application. Such a process is useful, but the major factor in determining adoption on a routine basis is obviously whether or not sufficient money is available to pay for it, either from third parties or other sources.

In dealing with the medical care system as a whole, the behavior of hospitals, individual providers, and consumers will clearly have to be modified if technology is to be utilized most effectively. This is the fundamental problem in a time of limited resources, and its solution will require development of a mechanism which draws on the natural instincts and forces of the human race to make it work. An incentive must be created to make a group of people in a hospital setting, who have acted as individuals, work together to utilize their resources most wisely. In prior years, when resources were unlimited, those difficult decisions were not required. Now, working together in the true institutional interest must take precedence over all else, requiring a change in physician attitudes and practices that may only be achieved by some form of risk sharing. The HMO, especially the IPA model, is probably the best example of the principles involved: in that setting providers share both gains and losses, a broader participation in the planning and utilization of technology occurs, and the involvement of the physician is vital to the process. However, HMOs are not acceptable to all patients and physicians, nor do they address the significant problem of our ever-increasing elderly population.

How do we develop a system, other than the HMO, that achieves the necessary sharing of responsibility by physicians with the hospital? There are at least two solutions. First, the necessity for hospitals to stay within a fixed annual budget should be recognized. While the physicians are not involved directly in such a process, indirectly they are forced to utilize technologies as wisely as possible to maintain the availability of varied resources in the hospital for the benefit of their patients. However, the ultimate cooperation of the physician can only be attained by linking the physician payment system to this process. That approach is difficult and emotionally charged, and will evolve through time and perceived need, not through revolution.

The second major control mechanism is through more effective utilization review, a process involving not only use of beds but ancillary services and technology as well. The twofold process has the advantage of maintaining decision making at the local or hospital level. Also, a hospital is required to define its mission. The process provides the basis for self-sustaining systems which can be built or changed to meet the individual needs of particular institutions and mechanisms to ensure appropriate utilization of the resources. There is no escaping the necessity to limit expenditures for providing health care in hospitals, which now constitute 40 percent of our annual expenditure for health care. Progress is slow but the system is changing in response to the forces at work. Inflation does show that we can provide high quality health care without the availability of unlimited resources.

In summary, most of the problems we face in dealing with technology result from its success, not its failure. Any failure is to be found in our inability to govern technology wisely, with the consequence that it can then govern us. The key terms for efforts toward maximizing effective use are limited resources, shared responsibility, incentives, and innovation. In these challenges lie opportunities to build a better society and a better system of caring and curing. Alternatively we face the unacceptable, mindless application of technology, the loss of individuality, and ultimately the simple, sterile question of how much and what kind of health care we can buy with a limited amount of dollars.

Endnotes

1. Fineberg, H.V. (1979). Gastric freezing—a study of diffusion of a medical innovation. In *Medical Technology and the Health Care System: A Study of the Diffusion of Equipment-Embodied Technology*, pp. 173–200. Washington, D.C.: National Academy of Sciences.
2. Dudrick, S.J., Wilmore, D.W., Vars, H.M., and Rhoads, J.E. (1969). Can intravenous feeding as a sole means of nutrition support growth in the child and restore weight loss in an adult? An affirmative answer. *Ann. Surgery* 169:974–984.
3. Personal Communication. Dr. Henry Pontoppidan, Massachusetts General Hospital.
4. Knox, R.A. (1980). Heart transplants: To pay or not to pay. *Science* 209: 507.
5. Thibault, G.E., Mulley, A.G., Barnett, C.O., Goldstein, R.L., Reder, I.A., Sherman, E.L., and Skinner, E.R. (1980). Medical intensive care: Indications, interventions, and outcomes. *N. Engl. J. Med.* 302:938.

MEDICAL COSTS AND TECHNOLOGY REGULATION: THE PIVOTAL ROLE OF HOSPITALS

by W. Vickery Stoughton

Health care expenditures nationally are close to $200 billion or 10 percent of the gross national product. The expenditure of health care dollars is increasing rapidly and government participation has grown from less than 2.5 percent before 1966 to over 42 percent currently.[1] The health system is under continuous and mounting pressure from third-party payers, industry, and individual consumers to control costs. On the cutting edge of this pressure is the federal government, which currently finances health care of over 50 million Americans, as well as public programs that provide an average 55 percent of hospital revenue and 24 percent of physician services.[2]

To reverse the escalating levels of spending, numerous regulatory programs have been developed which have changed health care delivery methods and are likely to effect even more significant changes in the future. Legislators and regulators have directed the controls toward the more highly structured segments of the health care system—hospitals. Institutions are expected gradually to self-regulate health care delivery through utilization review, Professional Standards Review Organizations (PSROs), and audit committees. Along with demand regulation, prices have been regulated through ceilings on reimbursement for routine patient care, and supply through determination of need programs and the National Health Planning and Resources Development Act. Because of these regulatory controls, hospitals can no longer operate in a free market environment.

In many areas in the United States, the emphasis on capitalism and free market interaction between supply and demand have been tempered by movements toward socialistic programs and more regulatory control. This control has been prevalent for long periods of

37

time in the public utility business and public transportation. Hard lessons learned through these regulatory efforts have suggested that no pat answer exists as to whether intervention in a free market system will produce the desired outcome. Intervention can be a product of the political system. Adam Smith wrote about the interventionist attempting to correct social problems:[3]

> *He imagines that he can arrange members of society with as much ease as the hand arranges the different pieces upon a chessboard— but that in the great chessboard of human society every single piece has a principle of motion of its own altogether different from that which the legislature might choose to impress upon it. If the two principles coincide and act in the same direction, the game of human society will be successful. If they are opposite or are different, the game will go on miserably and society—in the highest degree of disorder.*

Public utilities and public transportation may be examples of the "degree of disorder" seemingly imposed by regulatory efforts and the negative effect that these controls have had on technology development.[4,5] Regulators argue, however, that failures in other areas are not necessarily applicable to health care because the health industry does not operate under the same economic principles as do profit-oriented industries. Furthermore, relationships between hospitals, physicians, third-party payers, and patients prevent free market interaction of supply and demand choice.

The growth of federally financed health programs has created a structure whereby certain providers (primarily hospitals) have had to work with the major third parties (primarily the federal government) to develop contractual arrangements for the provision of health care to subscribers. In this role the federal government functions as any other lobby group attempting to influence legislation. Sometimes there is consensus in these efforts; more often, however, the "highest degree of disorder" prevails. Within this framework the outcome over time is weighted in the direction of the federal position for three reasons: it is the major buyer of health care services; it exerts significant influence on the legislative process; and it translates law into written regulation. This last factor gives the federal government enormous power which the health industry can counter only by working with the regulators in the developmental state or by challenging the regulations through the courts.

Given the enormous increase in expenditures on health care and the government's major role in paying for these services, the federal commitment to better control of health care costs should be clear. This commitment has evolved concurrently with federally financed health care programs. The Medicare and Medicaid acts caused an increase in the use of health care services by segments of the population (the elderly and indigent) which, because of financial considerations, previously had only limited access to these services. An environment of increasing demand was further complicated by the constant changes of improved technology, resulting in more specialization in the delivery of health care services. As technology developed and knowledge advanced, costs of health care began to rise at a much faster rate. For example, the National Center for Health Statistics reported in 1979 that $12.7 billion was spent on health in 1950; $26.9 billion in 1960; $74.7 billion in 1970; and, by 1978, $192.4 billion.[1] Diagnostic capabilities and treatment modalities associated with improved technologies all required better educated, more specialized clinicians and support staff who relied on sophisticated technology to provide both routine and complex treatment procedures that were only imagined 20 years ago. None of these advances was achieved without cost impacts which, when combined with the increased demand for services resulting from federally financed health insurance programs, led to an industry with an alarming growth rate in terms of financial obligations and responsibility for delivery of care.[6]

In a political system shifts in policy are preceded by shifts in public beliefs, attitudes, and opinions. Forecasting techniques based on economic modeling, opinion polls, and utilization data suggest that a 30-year surge in expanded spending for human welfare services has come to an end.[7] Forecasters are predicting that federal health spending will remain static or decrease; that fixed costs associated with ongoing programs (Medicaid) will dry up discretionary monies; that Medicaid assistance will erode because eligibility levels have not been adjusted upward for inflation by the states; and that increasing numbers of patients without insurance will be seeking health care services.

Patients without coverage have typically turned to hospitals for care. Given the current political environment in which inflation, energy, defense, and unemployment seem to be the most prominent issues, health care will be under pressure from two sources. The federal government will be concerned with the appropriate expenditure of tax dollars. Consumers who are ineligible for federal support and unable to pay for health care services will worry about sources of their care.

If technological advancements continue and demands for services by an aging population increase, hospitals will play a pivotal role in developing and implementing a process of regulating medical costs without severely prohibiting technology advancement in health care services delivery. All hospitals work toward maintaining availability of new technology. Historically the efforts of hospitals in this endeavor have been competitive, independent, and costly. One study shows that technology change is responsible for up to 50 percent of the rise in hospital costs.[8] Regulation of hospital costs therefore means control over technology development, dispersion, and utilization. Attempts to achieve this control through the certificate-of-need process have failed to produce the desired savings.[9] Some experts have suggested that efforts to reduce hospital costs should not be focused on the big technologies; rather, they argue that better control over all technologies should be established through changes in reimbursement, educational efforts, and the development of appropriate incentives designed to develop more selective use of technology. The view that little technologies result in high costs is supported by a report from Johns Hopkins.[10]

A knowledge of the component costs of hospitals, methods of reimbursement, and effects of technology on the costs and revenues of hospitals is necessary to understand the impact of big technology control versus better control over all technologies.

Hospital Costs and Revenues

The cost breakdown that follows is based on composite costs of a teaching institution and the allocation of these costs to groupings specified by third parties. These groupings are based on routine, ancillary, and ambulatory costs categories. Within these major categories, cost centers are identified by departmental activities, which are further broken down by direct and indirect costs associated with departmental functions. This cost system is, in turn, based on traditional accounting methods allocating costs to activity centers.

Direct costs assigned to an activity center include salaries and

Table 1 Sample of Indirect Costs and Their Corresponding Allocation Basis

Indirect Costs	Basis of Allocation
Depreciation	Number of square feet
Plant maintenance	Number of square feet
Malpractice insurance	Physician compensation
Laundry	Number of pounds
Housekeeping	Hours of service
Cafeteria	Percentage of employees in each department
Dietary	Number of meals served to patient units
Nursing administration	Nursing hours
Medical records	Percentage of time spent
Social service	Number of cases
Admitting	Patient days
Finance and data processing	Accumulated costs

wages, fringe benefits, supplies, and purchased services. Indirect costs and their corresponding allocation basis, as determined against activity centers, is shown in Table 1. These costs are not necessarily assigned to each activity center. For example, social service cost allocation is based on the number of cases in inpatient and outpatient areas that are designated cost centers; social service costs would therefore not be assigned to radiology, surgery, or other areas.

Based on a theoretical hospital with an annual operating budget of $60 million, using allocations similar to hospitals providing medical/surgical services, the following breakdown is representative:

A. *Routine Costs*
   Inpatient Medical, Surgical,   34.4%
   Intensive Care Services
B. *Ancillary Services*   52.4%
C. *Ambulatory Care*   13.2%
   Emergency Services
   All Other Ambulatory Services

A similar breakdown of revenue is representative:

A. Routine Revenue   32.5%
B. Ancillary Revenue   60.4%
C. Ambulatory Care   7.1%

Normally revenue generation will be budgeted at a level well in excess of costs, depending on a hospital's experience with contractual allowances, bad debts, and free care. As an example, a $60 million expense budget may require an $80 million revenue budget if the financial objective of the hospital is a net income of $500,000. Hospital revenues are typically tied up in receivables from 60 to 90 days and longer. To keep on top of operating expenses and current in-house payment for services purchased, the projected net income should be sufficient to provide the working capital requirement.

A more complete breakdown of routine, ancillary, and ambulatory care costs and revenues is detailed in Tables 2, 3, and 4, respectively, and must be calculated against the $60 million in expenses and the $80 million in revenue to gain an appreciation for the amount of actual revenue over cost as generated by different cost centers.

Table 2 Costs and Revenues for Inpatient Services

Inpatient Services	*% of Costs*	*% of Revenue*
Medical/Surgical	25.9	27.1
ICU	8.5	5.4
Total	34.4	32.5

Table 3   Costs and Revenues for Various Ancillary Services

Ancillary Services	Costs	Revenues
Surgical	10.4%	8.6%
Laboratory (including Blood Bank)	19.3	27.9
IV & Medications	5.6	7.6
Medical Supplies	1.0	1.5
Respiratory Therapy	1.3	3.3
EKG/EMG/EEG	1.2	1.3
Diagnostic Radiology	7.2	6.0
Radiation Therapy	1.5	1.2
Nuclear Medicine	1.1	1.0
CT Scanner	1.0	0.3
Dialysis	2.1	1.0
PT/OT	0.7	0.7
Total	52.4%	60.4%

Table 4   Costs and Revenues for Ambulatory Services

Ambulatory Services	Costs	Revenues
Emergency	4.0%	2.1%
Other Ambulatory	9.2%	5.0%
Total	13.2%	7.1%

Reimbursement Impact

In a rational economic system a straightforward cost/revenue analysis should be possible to show the impact of the cost of new technology in a hospital and the corresponding cost to the public. Based on my personal experience with a tertiary care hospital, new technology and its installation can have an impact on length of stay, related ancillary services, and direct inpatient costs. Cost/benefit analysis is therefore complex for technologies used on hospital patients.

What makes each analysis even more complex is the current reimbursement procedure requiring an analysis of technology that takes into account the mix of third-party payers. Because the major third parties have different reimbursement formulas for ancillary, routine, and ambulatory costs, the cost of technology to the institution and to the public is affected by this mix of patients and their insurance coverage. Estimates of use and cost/benefit of new technology must therefore consider the intricacies of reimbursement formulas, the breakdown of patients using the technology by insurance coverage, and the effect this usage pattern has on an institution's capabilities for generating revenues.

Before considering the possible impact of technology on the hospitals and its reimbursement constraints, it might be well to review briefly the reimbursement procedures of the major third-party payers.

Medicare requires that a hospital calculate total charges for each ambulatory and ancillary service and the percent of charges applicable to Medicare patients treated at the hospital. The hospital is reimbursed for this percent of fully loaded costs (direct and indirect). For the inpatient, Medicare reimburses the fully loaded costs divided by total patient days of the institution. This cost per patient day is multiplied by the number of patient days related to subscribers.

Blue Cross pools all ancillary charges together with all ambulatory charges and determines the Blue Cross percent of total charges for these services. This percent is then applied to fully loaded ancillary and ambulatory costs. Blue Cross reimburses for inpatient services using the same formula as Medicare.

Medicaid requires that inpatient and ancillary costs associated with inpatient services be calculated and audited. The audited costs for the two preceding years are averaged and then divided by the number of Medicaid inpatient days experienced at the institution two years ago. This base number is then increased by an inflation factor established by regulating authorities and used to add two years inflation increases to the base number.

Medicaid normally requires that total ambulatory service and associated ancillary costs of one year ago be divided by the number of Medicaid visits. The number is increased by a one-year inflation factor, the amount of which is established by regulatory authorities. However, the Secretary of HEW can waive reimbursement regulations in favor of an experimental state program. For example, in Massachusetts the formula for ambulatory and associated services is based on charges.

Commercial insurance pays charges for all services covered.

Additional factors associated with cost ceilings imposed on institutions take priority over the formulas used: Blue Cross reimburses for a portion of the bad debt and free care hospital experiences; certain costs can be approved as a "cost beyond control" of an institution and therefore are not affected by the cost ceilings or by inflation factors.

Impact of New Technology

Given the complexity of the reimbursement methodology, an institution must evaluate the number of patients treated, categorized according to the type of insurance coverage, in order to make an educated analysis of the impact of new technology on the costs to the hospital and the public. These attempts result in a very complex matrix in which payer mix and third-party reimbursement are compared with projected utilization. As a result, efforts to determine the cost impact of technology are not particularly straightforward; in fact, they might be termed adventuresome.

Impact of Big Versus Small Technology

If appropriately enforced, the federal government can have an impact on the distribution of big technology. The Certificate of Need (CON) process can impact technology development in private industry. A good example of this is the CT scanner. The CON has curtailed the proliferation of scanners in states where it is used more rigidly. In at least one case, the CON has required providers to evaluate the efficacy of the technology in terms of the benefit derived for the diagnostic test, that is, from the viewpoint of information as well as appropriateness of use when compared with other available technologies. The fact that one technology may be more efficacious for some uses but less effective for others is important knowledge and could be used as the basis for supporting or limiting proliferation of the technology. The impact of this type of control on private industry and further development of a new technology remains to be determined.

An approved CON licenses the procedure and establishes an institution's right to charge. Big technology draws attention, and even an approved CON does not necessarily end the regulatory process. For example, approval was given to the Brigham and Women's Hospital in Boston for the operation of a full body scanner. This approval was tied into an efficacy study to be conducted over three years. Immediately certain third parties claimed the scanner would be reimbursed according to the usual and customary ancillary formula but that reimbursement would be further limited to patients in specified diagnostic categories. As a result the institution had to conduct negotiations concerning appropriate diagnoses, which would have been fine if the efficacy study had been completed; in fact, it had just been started. The hospital, realizing that new technology changes, leased the equipment for five years. The third parties assigned a seven-year depreciable life, thereby reimbursing at less than cost. A difference of opinion developed over regulations which specified that volume used for determining allowed costs was 3,000 scans. This regulation was developed following a study of head scans. Since the CON included an efficacy study of body scans, it became apparent that volume would not be realized unless the institution focused on head scans rather than body scans. The third parties arbitrarily determined that a reasonable aggregate expense of operating the scanner was X. In fact, industry standards for staff, supplies, and maintenance resulted in an accurate reflection of cost to be X *plus*. The plus factor was disallowed. In this case, the reimbursement of this new technology resulted in legal expenses and court actions. The day before the court hearing, a settlement was negotiated which recognized the lease and volume considerations. However, over two years later, the hospital still had not been reimbursed retroactively. Nevertheless, the third parties reimbursed the hospital for the legal expenses, which exceeded the original costs being debated.

This case is mentioned in detail because, irrespective of the benefit that results from certain regulations, many examples of poor coordination and cooperation are evident among the regulatory agencies. Each agency seems to feel that achieving the maximum leverage possible to control health costs is the appropriate goal. Obviously the regulatory system should not function in this manner, and I will discuss this point later.

The CT scanner was incorporated into hospital expenses as a separate cost center. The total cost of the scanner and its operation was 1.0 percent of the cost to the hospital and 0.3 percent of the revenue because of the reimbursement problems (see Table 3). If revenue is generated in excess of costs, real revenue in this case was about $120,000 less than costs. In relation to the reimbursement formula, the cost impact on the institution and the health system is dependent on third-party mix and on the application of regulatory process on an institution's charge and cost structure.

The distribution of big technologies can be controlled by the CON process and, in fact, the cost of operating CT scanners nationally is less than half of one percent of the expenditure for health care prior to 1977.[11] Controlling the use of new technology by requiring evidence of efficacy before proliferation has been suggested.[12,13] I support this concept even though regulatory inconsistencies are apparent. Beyond the example I have discussed, the application of CON controls from one state to the next is not uniform and is often influenced by the political process. The emphasis on control of big technologies should not be on more regulations but on a more defined structure that will ensure uniform application of criteria based on efficacy studies and objective measurements of need.

The reimbursement process also needs immediate attention. A

brief analysis of clinical technology is appropriate for a better understanding of reimbursement reform needs.

Small Technology Control

Control over a clinical technology, as opposed to a high-cost piece of equipment, only recently has come under the scrutiny of federal regulations. Emphasis on controlling major clinical technologies, such as those involving cardiac surgery, dialysis, transplantation, and vascular surgery, is growing.[12,13,14]

No effective controls exist over small technologies. As pointed out in an article by Moloney and Rogers,[15] the small technology component of clinical practice is a significant part of the cost of the large clinical technology. As an example, the cost to the health care system for clinical laboratory tests represents about 8 percent of the total expenditures.[16]

Predicting the result of introducing a new clinical technology without again focusing on third-party mix is difficult because of the reimbursement formulas that exist. However, the formulas notwithstanding, no incentives exist to use less patient days, less laboratory tests, or less diagnostic x-rays. Focusing on laboratory procedures as a cost center, I showed one experience in which the laboratories represented 19.3 percent of the hospital costs and 27.9 percent of the hospital revenue. Utilization of laboratory services is not regulated; yet, some studies suggest that it should be. In fact, one effort at a Massachusetts hospital showed that a 40+ percent reduction in laboratory tests was achievable through educational sessions between house staff and attending physicians who discussed appropriateness of clinical laboratory orders.[17] The relevance of this finding to other hospitals, teaching or nonteaching, is not known, but the opportunity for a 40+ percent reduction in laboratory charges certainly makes future studies worthwhile. Also significant was that the finding for this study was provided by a major third party working with hospital staff. Speculating that a 15 percent reduction in laboratory services might be achievable throughout acute care hospitals, the savings to the system could be about 3 percent of hospital costs and 4 percent of hospital charges. In 1978 national health expenditures for hospital care represented approximately $76 billion.[1] A 3 to 4 percent savings of these expenditures directly related to small technology use is not insignificant.

Studies of the inappropriate use of drugs are well documented.[18–20] In FY 1978 the wholesale cost of antimicrobials used on inpatient services of one Massachusetts hospital totalled 41 percent of the pharmacy's drug costs. Total charges for the year were $1.4 million. Approximately 55 percent of the 10,500 total average yearly admissions are surgical, and a conservative estimate would be that 20 percent (1,155) of these patients receive a course of antimicrobials. There is no evidence to support the use of antimicrobials for prophylactic purposes beyond 48 hours. However, 70 to 80 percent of the antimicrobials prescribed in most hospitals are administered after the initial 48-hour period.[21] These figures, extrapolated to the same Massachusetts hospital, indicate that prolonged prophylaxis resulting in toxicity rate at 5 percent[22] and an increased length of stay may add $450,000 a year to patient charges.

Numerous other examples suggest that the small technologies have as much, if not more, impact on health care expenditures as have the large technologies. These examples suggest that control over large technology distribution, whether clinical or diagnostic, is only part of the answer to the issue of health care costs. Another part of the equation is control over small technology utilization. Appropriate utilization, whether related to large or small technology, is a factor in efforts to achieve cost-effective, quality health care.

Prescription for the Future

If appropriate utilization is to become an issue, incentives must exist for the hospital and its staff to manage the utilization of resources. Three steps should be followed by the hospital industry and regulators to assist physicians in developing utilization controls.

First, uniform data must be collected in a systematic fashion so that providers can become aware of treatment outcomes and related costs. This information must be used to develop cost-effective care models which are clinically sound.

Second, providers must develop a data base of information that will incorporate appropriate clinical care models into daily hospital operations so that uniformity of treatment of similar diagnostic problems is achieved.

Third, reimbursement systems must be changed so that providers are rewarded for achieving the first and second steps above. Health professionals have pointed to the lack of incentives in the system for reducing services; however, less service is not always appropriate and can be disastrous in the practice of medicine. A reimbursement system that recognizes the application of cost-effective care procedures is needed. The incentives within the system should be tied to physician reimbursement, as the decisions that ultimately result in resource allocation and utilization are made by the physicians. The current system rewards for time, effort, and outcome but not necessarily for efficient use of resources.

A system could be developed whereby a physician's minimum income level would be established, based on a specified volume of activity, with a bonus available on a case-by-case basis if certain parameters of utilization were maintained. Hospitals would have to monitor this activity and judge appropriate utilization. Since utilization would affect hospital income, the reimbursement system for hospital services would have to be changed as well. A global budget to each hospital, initially based on volume considerations but then paid irrespective of volume, would create an environment in which hospitals would not push for more patients but would be more inclined to work with physicians on the use of resources. If the hospital budget included bonuses given only in accordance with specified utilization parameters, then a system rewarding for less service rather than more, while promoting institution and physician cooperation, might be realized. This is only one example of the potential affect of positive incentives on utilization. Other systems could be developed to create a complimentary approach from hospitals and physicians, based on positive rather than negative incentives.

Summary

Over the years the legitimate role of government in health care delivery has been recognized and the importance of provider/regulator dialogue and cooperation has been emphasized.[23,24] More recently the need for collaboration on meaningful and objective evaluations of technology development and use has been promoted.[25,26] Ultimately technology cost is directly related to utilization which involves the physician decision-making authority of the physician. Any attempt to control physician decisions through regulations developed by interventionists will be disastrous. Strong opposition to the hospital's role as regulator will occur if interventionist controls are imposed on the health system. However, part of a hospital's social responsibility should be to regulate the use of accepted clinical practice models reflecting cost-effective care techniques. Moloney and Rogers[15] have suggested the need for incentives for the wise use of technologies and importance of appropriate educational efforts which emphasize cost effective care.

The example of the CT scan reimbursement negotiation discussed above reflects an interventionist attitude which will be fought by providers. Clearer provider incentives are needed to develop clinically accepted utilization models, establish new educational approaches, and change reimbursement systems. Federal involvement in health care delivery will continue to evolve and become more sophisticated. Providers can influence the degree of this involvement. Both groups have a social responsibility that includes maintaining balance and equity in the system through cooperation and self-regulation.

Endnotes

1. *Health United States*. 1979. U.S. Department of Health, Education, and Welfare, National Center for Health Statistics.
2. Gibson, R.M. and Fisher, C.R. (1978). National health expenditures fiscal year 1977. *Social Security Bulletin*, 41, No. 7.
3. Smith, A. 1976. *Wealth of Nations*. Vols. I and II. Westport, Conn.: Arlington House Reprint.
4. Noll, R.G. (1975). *Consequences of Public Utility Regulations of Hospitals' Controls on Health Care*. Washington, D.C.: National Academy of Sciences.
5. Frielander, A.F. (1969). *The Dilemma of Freight Transport Regulation*. Washington, D.C.: Brookings Institution.
6. Gibson, R.M. (1979). National health expenditures, 1978. *Health Care Finance Review* 1:1–36.
7. Blendan, R. and Schramm, C. (1980). Presentation to Robert Wood Johnston staff by senior program consultants, February 26.
8. Feldstein, M. and Taylor, A.K. (1977). "The Rapid Rise of Hospital Costs." Executive Office of the President, Council on Wage and Price Stability. Washington, D.C.
9. Sulkever, D.S. and Bice, T.W. (1979). *Hospital Certificate of Need Controls: Impact on Investment, Costs and Use*. Washington, D.C.: American Enterprise Institute for Public Policy Research.
10. Zuidema, G.D. (1980). The problem of cost containment in teaching hospitals: The Johns Hopkins experience. *Surgery* 87:41–45.
11. Institute of Medicine (1977). "Computed Tomographic Scanning: A Policy Statement," pp. 1–54. Washington, D.C.: National Academy of Sciences.
12. Heysset, R.M. (1977). *Controlling Health Technology: A Public Dilemma*. Sun Valley Forum. DHEW Publication (PHS) 79–3216, pp. 262–272.
13. Preston, T.A. (1977). *Coronary Artery Surgery: A Critical Review*. New York: Raven Press.
14. Luft, H.S., Bunker, J.P., and Enthoven, A.C. (1979). Should operations be regionalized? *N. Engl. J. Med.* 301:1364–1369.
15. Moloney, T.W., and Rogers, D.E. (1979). Medical technology—A different view of the contentious debate over costs. *N. Engl. J. Med.* 301:1413–1419.
16. Fineberg, H.V. (1977). *Clinical Chemistries: The High Cost of Low-Cost Diagnostic Tests, Medical Technology; The Culprit behind Health Care Costs?* Sun Valley Forum. DHEW Publication (PHS) 79–3216, pp. 144–165.
17. Martin, A., Wolf, M.A., Thibodeau, L.A., Dzau, V., and Braunwald, E. (1980). A trial of two strategies to modify the test-ordering behavior of medical residents. *N. Engl. J. Med.* 303:1330–1336.
18. Kunin, C., Tuposi, T., and Cragi, W.A. (1973). Use of antibiotics. *Ann. Intern. Med.* 79:55.

19. Meiring, P. and Briscoe, R.J.N. (1968). A survey of systemic antibiotics prescribed in a general hospital. *South African Medical Journal* 42:836.
20. Roberts, H.W. and Visconti, J.A. (1972). The rational and irrational use of systemic antimicrobial drugs. *Amer. J. Hos. Pharm.* 29:828.
21. Seidl, L.G., Thorn, G.F., Smith, J.W., and Cluff, L.E. (1966). Studies on epidemiology of adverse drug reactions. *Bul. Johns Hopkins Hosp.* 119:299.
22. Simmons, H.E. and Stalley, P.D. (1974). Trends and consequences of antibiotic use in the United States. *JAMA* 227:1023.
23. Derzon, R.A. (1976). "Legitimate Role of Government in the Private Health Services Delivery System." Michael A. Davis Lecture, Graduate School of Business, University of Chicago.
24. Hiatt, H. (1978). "The Hospital as a Change Agent." Third Annual Parker B. Francis Foundation Distinguished Lecture.
25. Fineberg, H.V. and Hiatt, H. (1979). Evaluation of medical practices. *N. Engl. J. Med.* 301:1086–1091.
26. Schwartz, W.B. and Joskow, P.L. (1978). Medical efficacy versus economic efficiency: A conflict in values. *N. Engl. J. Med.* 299:1462–1464.

Part Three

PRINCIPLES AND PROBLEMS IN EVALUATING TECHNOLOGIES AND ASSOCIATED COSTS

PITFALLS AND BIASES IN EVALUATING SCREENING TECHNOLOGIES

by David M. Eddy

Physicians, program directors, and health planners must answer many questions when they design and evaluate screening programs. Consider the case of screening for breast cancer. Should women have a physical examination and a mammogram every year? If so, at what age? Perhaps the mammogram could be recommended biennially instead of annually, saving some money that might be used to advertise and bring more people in for screening. How serious is the radiation hazard of mammography? If the x-rays cause new cases, will the earlier detection resulting from mammography outweigh the number of new cases generated and still leave a woman with a net benefit?

In answering these and similar questions, many different factors must be considered. These include the pathophysiology of the disease (for example, how rapidly it develops and the duration of time it is asymptomatic); age- and sex-specific incidence rates; mortality rates; risk factors; the effectiveness of the tests; the order and frequency with which the tests are given; the person's age; effectiveness of therapy; complications of the various screening tests; workups of patients with false-positive test results; and the costs, both financial and nonfinancial.

Errors in Evaluating Screening Programs for Breast Cancer

Given this large number of factors and the complicated ways they interrelate, many possible errors and biases occur in the evaluation of screening programs. Perhaps the most common class of errors is oversimplification; it is tempting to ignore most of the factors and focus on one or two. A second type of error is faulty probabilistic reasoning. A physician often has information about the probability that a patient with a particular disease will have a particular test result. For example, the true- and false-positive rates of tests present information of this type.* Clinical decisions, however, require assessing the chance that a person with a particular test outcome has a particular disease, which is an entirely different probability. A third class of errors may be the result of drawing conclusions from a biased sample; for example, a practitioner may draw conclusions about the value of screening many asymptomatic people for a disease from observations of a few patients who come into his office with the disease. A fourth type of error is the misinterpretation of the outcomes of uncontrolled clinical studies. Additional errors can be made by economists and analysts when they perform formal quantitative analyses of screening programs. Finally, some important biases can affect the interpretation of randomized controlled trials.

Some of these biases and errors will be discussed and examples will be drawn from the literature on screening for breast cancer. Specific examples cited are not intended to criticize individuals, but are selected to illustrate the complexity of the screening problem and to show how difficult it is to avoid errors.

Oversimplification

As an example of oversimplification, consider the following statement that appeared in a book on mammography published in the 1960s. The author wrote: "The work involved to effectively screen women in the proper age group (over 35) is staggering to contemplate. This should not be a deterrent." He concluded that "the application of mammography as a screening examination is very logical and, to me, worthwhile."[1] Aside from an unamplified statement that paraprofessionals could be used to read the x-rays, this was the only discussion of any of the factors that should be considered in making such a recommendation. No mention was made of the effectiveness of the screening test, the number of cases to be detected, the impact on life expectancy or case-survival rates, the risks, the radiation hazard, false positives, or any other factors.

Unfortunately this is not an unusual example. Another comes from an article that described the use of thermography to detect a breast cancer in a particular woman whose mammogram and physical ex-

---

* The true-positive rate is the probability that the test will be positive, given that a cancer is present. The true-negative rate is the probability that it will be negative, given that a cancer is not present.

amination had been negative. There is little doubt that this particular woman was helped by the use of this test; however, the author concluded the article with a very general statement that "the thermogram fulfills the basic requirements for a mass screening test. It is convenient and safe. It is quickly performed . . . averaging twelve minutes in experienced hands."[2] These appear to be the only criteria considered in determining the appropriateness of a test for mass screening.

Another type of oversimplification is to choose a single outcome and try to optimize it. One of the most commonly chosen outcomes is the accuracy or true-positive rate of the screening test. For example, breast cancer in asymptomatic women can be detected in several ways, and it is well known that each test sometimes can detect cancers that are undetectable by the others. There is, then, a strong temptation to perform all the tests in all women and do a biopsy if any one of them is positive. Since this amount of testing increases the accuracy of the package of tests, it might seem reasonable. As stated by one group of authors, "For maximum diagnostic accuracy, both modalities (physical examination and mammogram) are necessary in every case."[3] However, using only physical examination for low-risk women could reduce the risks and save resources that could be used with greater effect elsewhere—an option that is missed in the attempt to maximize accuracy. Other investigators wrote: "For the best results in breast cancer detection, both clinical examination and mammography should be combined. Although the accuracy of either method [alone] is only 88 percent, each potentiates the other so that cancers overlooked by one examination are likely to be recognized by the other."[4]

Another factor frequently used to determine whether a screening test should be recommended is its ability to detect cancers in asymptomatic people. If all of the better-known tests that have been shown to detect cancers in asymptomatic people are included in a screening program at the usually recommended ages and frequencies,* the cost of screening would be about $20 billion, which is approximately twice the direct cost of detecting and treating all cancers.† There are several other types of oversimplification, but these examples suffice to demonstrate this class of problems.

---

* Examples are: chest x-ray, sputum cytology, Pap test, pelvic examination, general physical examination, sigmoidoscopy, mammogram, breast physical examination, stool occult blood test, endometrial tissue sample, urine cytology, urinalysis, and a complete blood count and differential.
† The direct costs of cancer in 1975 were estimated to be $5.3 billion.[5] Today the figure is closer to $10 billion.

Probabilistic Reasoning

A second general type of error occurs in probabilistic reasoning. Consider the following passage from a letter that discussed the use of mammography to screen for breast cancer.[6]

> *The accuracy of the examination of mammography is reported to be between 80 percent and 90 percent . . . Even if we conclude that accuracy is 85 percent generally . . . , then that means that 15 percent of the women x-rayed will wind up with incorrect interpretations of the findings, or, more likely, their mammograms will simply fail to demonstrate the disease. This means that 15 percent of the women will be given a false sense of security if they are told their x-rays are normal, if indeed they already have cancer.*

The first sentence raises several important issues: the accuracy of a screening test cannot be encoded meaningfully in a single number. At the very least we need to know the true-positive rate and the true-negative rate. When mammography is used for the differential diagnosis of a woman with a breast mass, these two numbers are both about 85 percent, which probably explains why this author referred to a single number. For many other tests, however, the two rates are quite different and must be considered separately. A second problem is that a true-positive rate of 85 percent is observed when the test is used to diagnose women with advanced disease. In the screening setting we are concerned with the detection of tiny cancers, and the true-positive rate for these cases is much lower, about 60 percent.

Even using 85 percent as the true-positive and true-negative rates, does this mean that 15 percent of women given x-rays will receive incorrect interpretations of the findings? An incorrect interpretation can occur in two ways: A woman can have cancer even though her test is negative, or she might not have cancer even though her test is positive. Using a reasonable number for the probability that an asymptomatic woman in a screening program will have cancer, we can calculate the probability of the first case to be about 0.00075. The probability of the second case (that she does not have cancer but has a positive test result) is about 0.14925. Since these two numbers total 0.15, the author's statement is correct—so far. Unfortunately he then stated that "more likely, their mammograms will simply fail to demonstrate the disease."[6] The probability that a cancer is missed (0.00075) is not more likely than the probability of a false-positive result (0.14925). It is approximately 200 times *less* likely.

The author's estimate of the probability that a woman will have a false sense of security is also incorrect. A woman will have a false sense of security if she has a negative mammogram and cancer. This probability is about 0.00089, and the probability of a false sense of security is therefore approximately 140 times less than the author thought (0.15). Finally, the author's last statement is even more confusing. The probability that women have cancer, "if indeed they already have cancer" is 1, not 0.15.

Drawing Conclusions from a Biased Sample

In evaluating the efficacy of a screening program, it is natural to look to clinical practice for insights. For example, to determine the need for annual Pap tests, we might want to examine the clinical records of women who have invasive cervical cancer (ICC), review their past history of Pap tests, and determine how many had a negative test within the past year. If 50 percent had negative tests within a year of developing ICC, it is tempting to conclude that the false-negative rate of the Pap test is 50 percent, and that even with annual testing, 50 percent of the cases will be missed. This conclusion would be faulty, and would result from drawing conclusions about one population from observations of a different population—in this case drawing conclusions about asymptomatic people from a highly bi-ased selection of patients with invasive cancer. These women who are observed in a clinician's office represent women who "slipped through the net." They developed invasive cancer because, for one reason or another, their premalignant lesions were not detected, and in each case we would expect to find that: (1) they did not have a Pap test; (2) they had the test but it was misread; or (3) they had a Pap test but it was negative. By design, this source of information (examining the records of women with invasive cancer) will never show the thousands of women who had early cases of dysplasia or carcinoma *in situ* which were detected prior to becoming invasive.

To learn the false-negative rate of the Pap test and to assess the value of screening at various frequencies, we must examine prospec-tively a large population of asymptomatic people—not only the few patients who walk into an office with invasive cancer but also the thousands of individuals who were screened successfully. Conclu-sions based primarily on observations of clinical practice can easily cause a misjudgment of the value of a screening program.

Interpreting Uncontrolled Studies

In the analysis of cancer screening programs the results of uncon-trolled studies are usually reported in terms of the case-survival rates of the patients detected through screening, that is, how long these patients live after the time of diagnosis. While there is very valuable information in this type of data, interpretation of case-survival rates can be confused by several factors, including the lead-time bias, length bias, patient self-selection, and overdiagnosis.

An example of the use of case-survival rates to determine the value of screening is provided by a statement published in the Federal Registry by the Occupational Safety and Health Administration (OSHA). OSHA advocated the use of sputum cytology and chest x-rays to screen people exposed to industrial carcinogens because "the use of cytologic examination of sputum in addition to chest x-rays is advisable since individuals with negative x-rays and positive cytology may have a higher survival rate following surgery."[7] The fact that a higher proportion of screened people survived five or ten years after surgery, compared to unscreened patients, does not necessarily indicate that screening led to an overall decrease in mortality.

The Lead-Time Bias. The facts that a test *can* detect a condition before it is detectable by other means, that it tends to detect cancers in earlier stages, and even that it delivers higher case-survival rates do not necessarily mean that the test will increase the chance for a "cure" or prolong a patient's life. It is possible that the time of detection and the stages of detection have no effect on the course of disease, and earlier detection may only move forward the time of diagnosis, without moving back the time of death.

The interval between the moment a condition is actually detected and the moment that this condition would have been detected otherwise (by the patient through the observance of signs and symptoms) is known as the "lead time." Because of the lead-time effect, a comparison of five- or ten-year survival rates in screened or unscreened populations can be misleading. By itself, the earlier diagnosis will increase the lead time of detection, and thus the interval between a patient's diagnosis and death, which in turn will increase the probability that the patient is alive five or ten years after the time of detection (that is, the five- and ten-year survival rates). Thus, observations that screening improves survival rates from the time of detection do not necessarily mean that screening actually prolongs life, and estimates of the effectiveness of screening must take the lead time into account.

Patient Self-Selection Bias. Persons who elect to receive early detection tests may be different from those who do not, in ways that could affect their survival from a disease like cancer. For example, they could be more health conscious; more likely to control risk factors such as smoking, diet, and sexual habits; more alert to the presence of signs and symptoms of disease; more adherent to treatment; or simply healthier and better able to combat their disease. Any of these factors could produce longer survival from cancer in ways independent of early detection. The observance of higher survival in screened patients as compared with the general population could in fact be due more to the selection of patients than to the effect of early detection.

This bias can be adjusted for by tracking the survival rate of all patients offered early detection tests, including those who reject the offer, and comparing it to the survival rate of a comparable group not offered early detection. The bias is much more difficult to correct for in retrospective studies. A chart review may show that patients diagnosed through early detection have earlier stage cancers than those who sought care on their own, but it is extremely difficult to identify the patients who were offered but refused an early detection protocol, or the "interval cases" who were dutifully examined but their cancer was missed.

Length Bias. A third problem encountered in evaluating screening programs is that cancers detected by a test in a periodic early detection program tend to have longer preclinical intervals than average. (Conversely, the interval cases tend to be cancers with shorter preclinical intervals.) The preclinical interval is the period between the time a screening test *could* detect a cancer and the time a patient would seek care on his/her own initiative for signs and symptoms in the absence of an early detection protocol. The duration of this interval is related to the growth rate and other biological characteristics of the tumor as well as to the patient's awareness of cancer signs and symptoms. Both these factors can influence the patient's survival rate from the time of diagnosis. For example, patients with long preclinical intervals may have tumors with slower growth rates that may be less malignant. Thus, compared to control or untested patients, patients detected at scheduled examinations have longer preclinical intervals, which may imply slower growth rates, less malignancy, and longer survival. Patients with cancers missed at the early detection examination but self-detected between examinations have shorter preclinical intervals, which may mean faster growth rates and lower survival.

However, if a patient tends to delay a long time before seeking care, the preclinical interval may be lengthened by postponing its end point, without changing the cancer's rate of growth. Conversely, a patient with cancer self-detected between scheduled examinations may have a short preclinical interval due to an awareness of signs and symptoms, and may have a better than average prognosis. In the Health Insurance Plan (HIP) study,[8] patients with cancers self-detected between examinations had better case-survival rates (57.6 percent) over a seven-year period than the control group (53.3 percent). However it affects survival, the length bias—the selection of cancers with long preclinical intervals through early detection examinations—complicates the interpretation of data derived only from a screened population.

Overdiagnosis. Another problem can confuse the interpretation of case-survival rates and data related to the tumor stage at the time of diagnosis in an early detection program. Early detection examinations are given to find cancers when they are tiny. Unfortunately no sharp boundary exists between nonmalignant and malignant cells, and it is possible to overdiagnose as a very early cancer a lesion that is not, and would never become, cancer. If this were to occur, not only would the number of "cancers" detected be increased, but the reported number of cancers detected in the earliest stage would be inflated. As these lesions would never become clinically significant cancer, survival statistics can, concomitantly, become inflated. To correct for this possibility, physicians must be certain that all of the patients actually have cancer, which is determined by actually counting cancer deaths.

Summary. Because of such problems as lead time, patient selection, overdiagnosis, and length biases, to draw conclusions from knowledge of a tumor stage at the time of detection or from case-survival rates can be misleading. These biases can be eliminated by randomized controlled trials (RCTs) in which mortality in the entire group offered early detection examinations is compared to mortality in the control group. But because of the high cost and long duration of RCTs, this information is rarely available, and estimates of the value of early detection usually must be based on the results of other types of studies. Fortunately the presence of the biases mentioned does not mean that information obtained from studies in which they occur is useless. Nonrandomized and uncontrolled trials can provide a great deal of evidence, as can projections from the stage of the cancer at the time of detection, case-survival rates, and other measures. Outcomes estimated from these data must be interpreted with great care, however, and the problems inherent in them must be recognized and adjusted for whenever possible.

Errors in Formal Cost-Effectiveness Analyses of Screening

Confusing Incidence and Prevalence. Confusing incidence and prevalence is a common error that occurs in formal cost-effectiveness analyses of screening programs. In analyzing the value of a screening program, one of the most important factors is the probability that an asymptomatic patient who walks into a screening clinic or a physician's office has the disease. This probability is the *prevalence* of occult cancers—cancers that can be detected by a test but have not yet caused the patient to seek care for signs and symptoms. In trying to estimate this probability, analysts may incorrectly use the annual *incidence rate*—the probability that an asymptomatic person will be diagnosed as having the disease in the coming year.

For example, in an analysis of screening for cervical cancer, an economist modeled the problem as a decision tree and stated that "an individual is in either of two states of health: he either has a disease—State D, with probability θ; or he does not—State D_0, with probability $1 - \theta$."[9] The analyst then wrote that "the probability of State D is θ (the disease incidence)," and he actually used the age-adjusted annual incidence rate as his estimate of θ. Proceeding from this he stated: "Cancer of the cervix in the United States is variously reported to occur in women at the rate of 31.5 or 30–40 per 100,000 women . . . For the present purpose, 35/100,000 is taken as the estimate of θ." What the author wanted was a statement of the probability that an asymptomatic patient has a cancer detectable at the time of screening (the prevalence of an occult cancer), not the probability that a cancer will develop in an asymptomatic patient in the coming year.

Cure and Mortality Assumptions. Another problem that occurs frequently in formal cost-effectiveness analyses is the assumption that if a cancer is detected through screening, the patient is cured, and if a patient discovers the cancer through signs and symptoms, either in the interval between screening examinations or in the absence of screening, death will occur immediately. This assumption, however, is quite contrary to clinical experience: many patients with early-stage cancers detected by screening die within a few years of diagnosis, and many patients with even regional cancer survive their disease. A similar error is to assume that if a cancer is detected through screening, it is found in an early stage and the case-survival rates for early stages can be used to estimate the prognosis. Similarly some investigators assume that if the patient is not screened but seeks care for signs and symptoms, the cancer was necessarily found in a late stage. The inappropriateness of this assumption is easily displayed by data from a randomized controlled trial of breast cancer: 30 percent of the cases detected through screening were in a late stage (had axillary lymph node involvement), while 46 percent of the unscreened cancer patients had no lymph node involvement.[8]

True-Positive Rate. Other problems arise in the use of information about the true-positive rate—the probability that the test will detect an existing cancer. Unfortunately no single value for this rate can be established. If a cancer is one cell large, the true-positive rate is zero, and if it is 14 cm large, the true-positive rate approaches 100 percent. A continuous function is needed to describe the true-positive rate as a function of the age or state of development of the cancer (for example, its size). Nonetheless, almost all analyses of screening programs use a single number for this rate. It may be appropriate to assume the true-positive rate is constant when analyzing the use of a diagnostic test in patients with signs and symptoms, because in virtually all these cases, the patients tend to have detectable (and therefore advanced) cancers. In a population of asymptomatic people, however, we are looking for tiny cancers that are, by definition, difficult to detect, and the true-positive rate of the test varies with the size of the lesion; it is lower than the rate measured when the test is used for differential diagnosis.

Another problem in the interpretation of true-positive rates occurs when the presence of cancer is uncertain, and it is necessary to use other tests to determine whether a patient actually has the disease. This can be very misleading. For example, the following statement reported the results of a very large study designed to determine the accuracy of mammography. In conducting 10,743 mammograms, it was found that, "In this combined series . . . the true-positive rate was 87.6 percent, the true-negative rate was 93.8 percent, and the agreement in radiologic and pathologic diagnosis was 91.6 percent. This represented a total of 1856 patients coming to biopsy."[10] Because the true- and false-positive rates had to be based on patients who had been definitively diagnosed, the investigators based the true-positive and true-negative rates on the results of patients who had undergone biopsy. Thus the announced rates pertained to the patients who had undergone biopsy and were not the rates in the "combined series of almost 11,000 patients," as the investigators stated. This error led the investigators to overestimate the true-positive rate and underestimate the true-negative rate. Regarding the true-positive rate: because the mammogram may have missed some tiny cancers, the x-ray was negative, and these patients did not undergo biopsy. The true-positive rate for these cases would in fact be lower than the rate measured in the cases where biopsy occurred. The true-negative rate is underestimated because most of the patients who did not have biopsies did not have cancer and had negative mammograms.

These examples are not unusual. Elsewhere in his book, the same author wrote, "In Table 24 are the diagnoses of the total number of mammograms, 3818, in the series. Mammograms had been performed for 725 malignant and 489 benign lesions coming to biopsy. In the entire series, the true-negative rate was 91.0 percent and the roentgenographic and pathologic agreement was 94.6 percent."[10] The true-negative rate is almost certainly underestimated here and is probably closer to 99 percent than 91 percent. Errors such as these can be extremely important. For example, an analyst using these published rates to estimate the benefits and the costs of a national screening program would be working with a difference of 8 percent in the true-negative rate (99 percent–91 percent), representing about three million women who would be candidates for breast biopsy. The analyst's estimates of the financial cost would be off by about $1.5 billion.

Use of Rates Observed in Diagnostic Clinics. A related problem is the use of true-positive and true-negative rates observed in diagnostic clinics. These rates, measured for patients who have signs and symptoms, are sometimes erroneously used to analyze the use of the test for screening an asymptomatic population: "It is probable that 80–90 percent of breast cancers determined by surgical biopsy can also be suspected from mammography. Therefore, the latter technique is quite useful as an additional screening device."[11] These rates of 80 percent to 90 percent were observed in patients who underwent biopsy and therefore were likely to have detectable disease, but are inappropriate for estimating the value of mammography in an asymptomatic population who would have far tinier cancers and in whom the sensitivity of the test would be lower.

Successive Applications of Tests. Another problem is the assumption that successive applications of the same test are independent. Suppose a test has a true-positive rate of 70 percent. If it is assumed that successive applications are independent, it is possible to raise the effective accuracy to virtually 100 percent simply by applying the test several times at one sitting. It is clearly not reasonable to assume that the results of successive tests are independent. A test with an overall true-positive rate of 70 percent might miss a cancer because the lesion is too small or lacks a characteristic feature of malignancy, and it would not be proper to assume that if the test were performed a second time, it would again have a 70 percent chance of finding the lesion.

Interpreting Randomized Controlled Trials (RCTs)

Most of the biases in evaluating screening techniques just discussed can be corrected for by using randomized controlled trials. With this method, however, the very beauty of the design of RCTs and the amount of work entailed can bias the use of their results. First, the fact that a trial shows a statistically significant difference in outcomes between the control and study groups does not necessarily mean that screening is appropriate and should be done. It may not be worth the costs or risks, or the resources may be used to better effect elsewhere. Second, there is a strong tendency to advocate the particular configuration (that is, the choice of tests, ages for starting and stopping screening, and frequency of examinations) that was studied in the RCT. For example, if an RCT of breast cancer evaluated annual mammograms and annual physical examinations, the screening options that can be recommended are immediately limited; dropping the mammogram or recommending that it be done biennially or at some other frequency becomes difficult. Finally, even if a randomized controlled trial does not show that screening makes a difference, it may still be desirable to screen. Because of the huge populations needed to evaluate screening programs, many trials have a very large probability of a Type II error—an error that occurs when an RCT misses a true effect. The decision to screen should take into account the costs and the risks of screening as well as the expected benefits. If an RCT fails to show a benefit, but the probability of a Type II error is high and the test has a very low cost and risk, screening might still be appropriate.

In summary, many types of errors and biases complicate the evaluation of screening programs. In making an explicit recommendation, a practicing physician, a health planner, or an analyst should exercise great care, and the reasons should be explained and evaluated by other professionals.

Endnotes

1. Wolfe, J.N. (1967). *Mammography*. Springfield, Ill.: Charles C. Thomas.
2. Wasserman, N.F. (1974). Thermogram and mammogram. *Minnesota Medicine* 57:971–974.
3. McClow, M.V., and Williams, A.C. (1973). Mammographic examinations (4030): Ten-year clinical experience in a community medical center. *Ann. Surg.* 177:616–619.
4. Rogers, J.U., Dowd, R. and Egan, R.L. (1966). Comparative mammography study. *Amer. J. Roentgenol. Rad. Ther. and Nuc. Med.* 97:748–754.
5. Rice, D., and Hodgson, T. (1978). "Social and Economic Implications of Cancer in the United States." Presented to Expert Committee on Cancer Statistics of the World Health Organization and International Agencies for Research on Cancer, Madrid, Spain, June 20–26.
6. Berlin, N. (1976). Letter to the Editor. *National Observer*, September 11.
7. Occupational Safety and Health Administration. (1975). OSHA proposed standard, exposure to coke oven emissions. *Federal Register* 40:32268–32282.
8. Shapiro, S. (1977). Evidence on screening for breast cancer from a randomized trial. *Cancer* 39:2772–2782.
9. Schweitzer, S.O. (1974). Cost effectiveness of early detection of disease. *Health Services Research* 22–32, Spring.

10. Egan, R.L. (1972). *Mammography*. 2nd ed. Springfield: Charles C. Thomas.
11. Allen, J.G. (1977). Breast. In *Surgery: Principles and Practice*, 5th ed., ed. J. E. Rhoads. Philadelphia: J. B. Lippincott.

PITFALLS AND BIASES IN EVALUATING DIAGNOSTIC TECHNOLOGIES

by S. J. Adelstein

Both the proponents and opponents of equipment-embodied, medical technology have been searching for a means of evaluating the efficacy of diagnostic procedures. In the past several years various approaches have been developed. A general schema for incorporating these various approaches is shown in Figure 1.[1] In this scheme, two decision paths are described. The first follows the course of a patient presenting to a physician with a constellation of signs and symptoms. The physician may decide: (1) that he has sufficient information to make a diagnosis and proceed to treatment or (2) that additional laboratory testing is necessary to gain more information about the state of the patient's health and also to guide his decision in terms of the choice of treatment. Both sequences result in a series of health out-

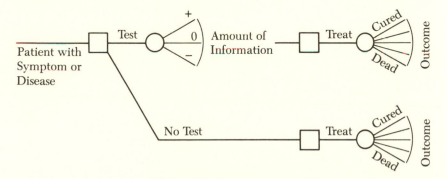

Figure 1   General Schema for Evaluating a Diagnostic Procedure. Tests may be evaluated in terms of the information they produce, how they influence physician behavior with regard to treatment, or the patient's ultimate health. (*SOURCE: McNeil, B. J., and Adelstein, S. J., Determining the value of diagnostic and screening tests*, J. Nuc. Med. *17:439, 1976. Reprinted with permission.*)

comes that can be compared in the most favorable circumstances. In addition, both sequences have financial costs that can be calculated.

In evaluating diagnostic tests a number of attributes may be examined. First, the technical merit of the test can be determined and compared with the technical merit of similar tests; this is particularly true in the field of imaging modalities where technical criteria have been carefully elaborated. Second, it is possible to compare the accuracy of diagnostic tests in terms of their sensitivity, specificity, and predictive outcome. Third, one can attempt to determine the extent to which test outcomes influence physician behavior and, indirectly, patient care; in the comparison of the test and no-test situations, one could try to compare the behavior of physicians with and without the information that the test provides. Fourth, the influence of the test on health outcome can be assessed; these outcomes are usually expressed in terms of morbidity, mortality, and quality of life. Fifth, the cost-effectiveness of the procedure can be compared to other procedures or to the no-test scenario; cost-effectiveness calculations are usually expressed in terms such as cost per case found, cost per death averted, or cost per hospital day avoided.

Each step of this process is susceptible to errors of commission, omission, and interpretation. I shall attempt to demonstrate some of the difficulties that can be found at the various stages represented in this schema. Because of the areas of my own clinical and research experience, I have used examples from nuclear medicine and radiology.

Relation between Technological Improvements and Accurate Diagnoses

Improvements in technology are not necessarily translated into more accurate diagnoses. In nuclear medicine we have seen an enormous improvement in technology, both in instrumentation and in radiopharmaceuticals, dating from 1954 to the present. In the case of liver scanning for the detection of metastatic malignancy, there has been a revolutionary change both in the instruments and radio-indicators employed. This improvement can be summarized quantitatively by a figure-of-merit (FOM) that incorporates factors relating to the fidelity of the instrument (measured in terms of resolution and sensitivity), the number of photons available from the radionuclide incorporated into the liver, and the radiation dose received by critical organs. Table 1 shows a comparison of two techniques in which the figure of merit is raised significantly (a factor of 25) by alterations in the radiopharmaceutical (^{131}I to ^{99m}Tc) and the instrument

Table 1 Effect of Changes in Figure of Merit on True Positive (TP) and False Positive (FP) Ratios

Year	Radiopharmaceutical	Instrument	Figure of Merit	Test Indices TP	Test Indices FP
1954	^{131}I-Rose Bengal	3" scanner	7	.94	.37
1972	^{99m}Tc-Sulfur Colloid	Anger camera	172	.90	.37

(3" scanner to Anger camera). Despite these changes the sensitivities and specificities are virtually the same.

A similar observation has been made in the employment of ultrasound for abdominal abscesses. Real-time scanning was introduced to improve the man/machine interface by selecting the most appropriate imaging planes for inspection. When real-time scanning was compared with B-mode scanning, no significant difference in sensitivity or specificity could be found.[2]

Relation between Data Base, Accuracy, and Predictive Value

A good data base for absolute or relative measures of accuracy and predictive value is very difficult to obtain. Very often a single investigator or institution will attempt to collect sufficient data to determine the sensitivity and specificity of a diagnostic test and then, in some instances, to obtain predictive value from these data. A number of hazards have been identified with studies of this type.[3] First, there may be biases in work-up and the establishment of truth. Second, instrumentation technique, data collection, and follow-up may be inadequate. Third, instabilities in the prevalence and spectrum of disease may flaw any attempt to establish predictive outcome.

The Liver Scan. Some of these problems can be illustrated by the study of 650 sequential liver scans taken at the Peter Bent Brigham Hospital in Boston (Figure 2).[4] Of the 429 patients with abnormal liver scans in this group, 61 percent (263) had histological diagnosis made by biopsy, surgical exploration, or autopsy; however, of the 221 normal scans, only 37 percent (81) had a histological diagnosis. Because the decision to obtain histological data was determined to some degree by the outcome of the test, the population from which the sensitivity and specificity of the liver scans were determined was not an unbiased one. This type of bias occurs frequently with retrospective analyses and is one reason for preferring prospective studies, in which the definitive standard for diagnostic end point

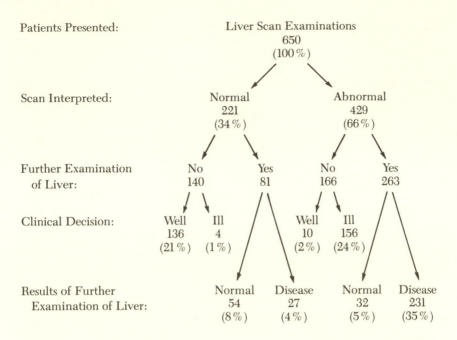

Patients Presented: Liver Scan Examinations
                                              650
                                            (100%)

Scan Interpreted: Normal   Abnormal
                                   221   429
                                  (34%)   (66%)

Further Examination   No   Yes   No   Yes
of Liver: 140   81   166   263

Clinical Decision: Well   Ill   Well   Ill
                      136   4   10   156
                     (21%)   (1%)   (2%)   (24%)

Results of Further   Normal   Disease   Normal   Disease
  Examination of Liver: 54   27   32   231
                            (8%)   (4%)   (5%)   (35%)

Figure 2 Clinical-Pathological Outcomes in 650 Sequential Liver Scans Performed in a Single Nuclear Medicine Unit. Of these, 61 percent of the abnormal scans and 35 percent of the normal scans had further histological liver examination. The prevalence of liver disease in the entire group (clinically ill or pathologically diseased) was estimated to be 64 percent.

is selected independently of test outcomes. From the histologic examinations (see Figure 2: "Results of Further Examination of the Liver"), the sensitivity of the liver scan in this group of patients was 90 percent and its specificity 63 percent (false-positive value of 37 percent).

Use of these sensitivity and specificity data to calculate the predictive value of the liver scan* in a particular patient requires knowledge of the prior probability of disease in that patient. This is notoriously difficult to obtain because the prevalence of disease in an entire population can be used only as a first approximation of the value for a particular patient. Even then, this population prevalence may be biased because it depends so heavily on the referral patterns and behaviors of referring physicians. For the liver scan data, the prevalence of disease was determined using both histologic and follow-up data at 18 months (see Figure 2, line marked "Clinical

---

* Predictive value generally means either the probability that a patient with an abnormal liver scan actually has liver disease, or, conversely, that a patient with a normal liver scan is actually free of liver disease.

Table 2 Comparison of Computed Tomography and Radionuclide Brain Scanning in the Diagnosis of Cerebral Tumors

Hospital	Total Cost	Total No. Patients	Total Patients Available for Analysis
A	$ 417,000	599	0
B	461,000	574	25
C	373,000	525	15
D	463,000	634	67
E	523,000	597	29
Total	$2,237,000	2,929	136

Decision"). With this figure of 64 percent, the probability that a patient with an abnormal scan will have liver disease is 90 percent and the probability that a patient with a normal scan will be free of disease is 86 percent.

CT of the Head. An alternative and preferable manner for collecting data to evaluate a particular test is exemplified by a prospective cooperative study involving several medical centers. Under the best of circumstances, such studies are expensive and require considerable planning and expertise to achieve unbiased results within a reasonable period. A recent study by the National Cancer Institute on computerized tomography and radionuclide scanning in the diagnosis of cerebral tumors illustrates some of the problems with prospective studies.

At the time the study was started, computerized tomography of the brain was on the border between an emerging and a new technology, and five institutions were selected for this comparative study. The institutions were comparable in terms of their technical capabilities for CT scanning, and it was *assumed* that they were comparable in terms of their radionuclide studies. In fact, though, the institutions were quite different; one of the consequences of this difference was that none of the patients in one hospital was available for comparative purposes.

Several facts emerge from a careful analysis of this study. First, the costs of the study were large, averaging about $450,000 for each hospital for a total of $2.2 million (Table 2).[5] Second, the total number of patients entered in each hospital was large, nearly 3,000, but the total number available for final analysis was small, only 136. This loss was due to a variety of factors, including incomplete follow-up, lack of histologic data, lost copies of examinations, and so forth. Fortunately the sample was sufficient to make an analysis of the comparative accuracy of computerized tomography and radionuclide scanning. Finally, the cost per patient entered into the original study was about $765, but the cost per patient used for the analysis was $16,500.

Criterion-free Measures of Accuracy

If measures of accuracy are not criterion free, confusion may result. In practically all instances lax or strict criteria can be used to interpret diagnostic data from a wide variety of sources. Most diagnostic tests are not perfect. If they were, they would easily segregate normal populations from diseased populations. In fact, there is considerable overlap and room for type 1 (false negative) and type 2

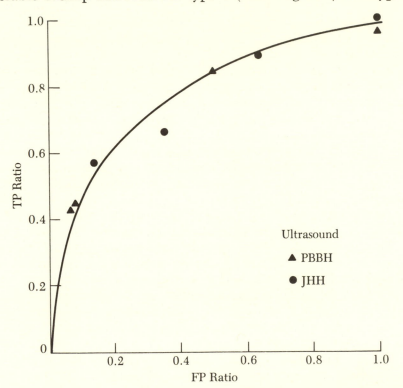

Figure 3   ROC Curve for the Detection of Focal Sources of Sepsis by Ultrasound. Were the test to be no better than a chance selection of positive or negative results, the data would fall along the 45 line. A perfect test would have a TP ratio of 1 and an FP ratio of 0. (*SOURCE: McNeil, B. J., Sanders, R., Alderson, P. O., Hessel, S. J., Finberg, H., Siegelman, S. S., Adams, D. F., and Abrams, H. L., A prospective study of computed tomography, ultrasound, and gallium imaging in patients with fever*, Radiology 139:647–653, 1981. *Reprinted with permission.*)

(false positive) errors. A descision must therefore be made as to where to set the upper (lower) limits of normal or the lower (upper) limits of abnormal. Setting this value depends upon the percentages of false positive and false negative outcomes one is willing to accept. In the imaging modalities, the biased interpretation of individual readers produces a floating criterion. The best current method for deriving a criterion-free measure for imaging purposes is the receiver operating characteristic (ROC) curve or a plot of the true-positive ratio against the false-positive ratio (Figure 3).[6,7] This method results in a continuum of inversely related sensitivities and specificities for a particular procedure. In Figure 3, for example, when the sensitivity is 80 percent, the specificity is 60 percent (100 percent minus the false-positive percent). When the sensitivity is lowered to 60 percent, the specificity is raised to about 80 percent. These curves can be very useful in the analysis of a host of problems, including the relative accuracy of two diagnostic tests, the utility of taking diagnostic tests together,[2] and the construction of disjunctive analyses using multiple variables.[8]

Relation between a Good Test and Improved Health

Cephalopelvimetry. Just as a new technology does not necessarily result in improved accuracy, so improved accuracy may not result in an improvement of health outcomes. In fact, relatively few controlled studies have been conducted in which such cause and effect relationships can be determined. Several years ago a study by Crichton[9] involved a prospective analysis of the use of cephalopelvimetry for the diagnosis of cephalopelvic disproportion and consequent improvements in morbidity and mortality in the perinatal period. In this protocol all patients referred to the radiology department for cephalopelvimetry were entered into the study on a random allocation basis. Approximately one half of the patients had x-ray examinations and the other half did not. The test apparently had an influence on physician behavior in that 43 percent of patients who had pelvimetry underwent Caesarean section, while the corresponding value for the control group was 32 percent. The combined neonatal mortality and morbidity was 16 percent in the pelvimetry group and 25 percent in the control group. These differences are marginally significant.

Screening for Renovascular Disease. A number of recent studies have shown a less clear-cut relationship between certain diagnostic tests and health outcome. A number of years ago, we examined the impact of screening for renovascular hypertension on the frequency of subsequent morbid events, especially strokes and heart attacks.[10,11]

We found that several factors determined whether or not this screening procedure was likely to have an impact on outcome. Compliance was of particular importance; if compliance was high, little was gained by screening patients for renovascular hypertension versus treating them with hypertensive medication irrespective of cause. If compliance was low, then screening made a good deal of difference; surgical treatment had a greater impact than it did otherwise.

Radionuclide Brain Scan. In another study, George and Wagner[12] tried to evaluate the impact of the introduction of brain scanning and other neurodiagnostic procedures on the time course of brain tumors for the decade 1962–1972. During this time period, radionuclide brain scanning increased by a factor of 10; an analogous increase was seen in the number of cerebral angiograms performed over the same period. In 1962 brain scanning was an emerging technology; by 1972, it was a well-established technology. The sensitivity of the scan did not, however, change perceptively over the interval, although quite an impact occurred in terms of disease discovery. Before 1962 the duration of symptoms in patients with brain tumors was about four years. After these neurodiagnostic procedures were used, the duration of symptoms to the time of disease discovery was reduced to less than one year. Despite this, survival curves for patients with brain tumors were not significantly different from the beginning of that era to its end. The survival curves after operation were in fact virtually superimposable; and if one assumes that the time to operation was decreased over the period of ten years, then the survival of patients might actually have been shortened. In terms of survival, the value of a positive brain scan is not great. Alternatively, other issues must be considered in this case; perhaps the most obvious is the importance of prognostic information which is made available as a result of a normal test in a patient without a brain tumor. Most of the patients scanned fell into this category.

Interpreting Cost-Effective Calculations

Screening for Renovascular Disease. It is often easier to make a cost-effective calculation than it is to interpret one. At times the interpretation of cost-effectiveness calculations is straightforward. This is particularly true when two procedures of equal accuracy are compared. The cost per case found or the cost per year of life extended can then be compared in a straightforward manner. However, when an absolute judgment as to the value of cost-effectiveness ratio must be made, criteria for making judgments may or may not be available. Occasionally the complete analysis can be of some

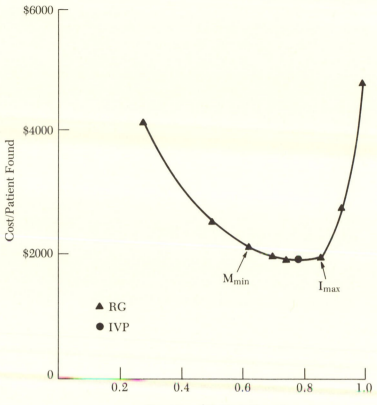

Figure 4   Cost of Case Finding in Renovascular Hypertension as a Function of Fraction of Patients Identified. M_{min} represents a minimum in the costs of mistakes in diagnoses; I_{max} maximizes information content. From a cost-effective viewpoint I_{max} produces the greatest return per unit cost invested. (*SOURCE: McNeil, B. J., Varady, P. D., Burrows, B. D., and Adelstein, S. J., Measures of clinical efficacy: Cost-effectiveness calculations in the diagnosis and treatment of hypertensive renovascular disease,* N.Engl. J. Med. *293:216, 1975. Reprinted with permission of the* New England Journal of Medicine.)

value. In screening patients for renovascular hypertension, we were able to calculate the cost per patient discovered as a function of the percentage of renovascular hypertensive patients identified (Figure 4), as well as the total cost of screening 100 patients as a function of the percentage of renovascular hypertensive patients identified (Figure 5).[11] In both instances, one point appears to provide an optimal return for invested costs. In the first instance (Figure 4), the curve is parabolic and concave upward. There is a rather broad minimum, and a point at which about 90 percent of patients discovered is the extreme before the curve turns up again. This suggests that this

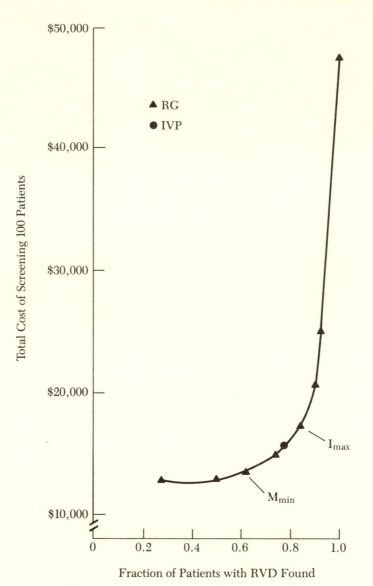

Figure 5   Total Cost of Screening for Neurovascular Hypertension as a Function of Fraction of Patients Identified. At M_{min}, 60% of patients are found and the total cost begins to ascend; at I_{max}, 90% of patients are found and the total cost begins to ascend rapidly.

Table 3   Impact of Computed Tomography on Neuroradiologic Costs and Case Finding at Two Boston Teaching Hospitals

	1974	1977
Fraction of patients with disease identified	75%	86%
Fraction of all patients tested who had disease	68%	36%
Total diagnostic costs*	$532,000	$867,000
Cost per patient found*	$ 565	$ 805

* In constant dollars relative to 1977 costs.

point (which also maximizes the information content) optimizes the relationship between the fraction of patients discovered or identified and the cost per patient. With reference to the renogram as a screening procedure, the *total* costs also seem to change significantly at this same point (Figure 5), which represents an operating point taken before the total costs rise very steeply.

CT of the Head.   The interpretation of cost-effectiveness calculations can be more problematic. As a theoretical exercise, I have attempted to evaluate the cost-effectiveness of computed tomography for neurological disorders. The data base derives from a study of Abrams and McNeil[13] in which they analyzed the number of neurodiagnostic studies and neurologic evaluations at two Boston hospitals before and after the introduction of CT of the head. If the accuracy data obtained from the study of Swets, et al.[6] are combined with these data, it is possible to tabulate the fraction of patients with disease identified, the percentage of patients found to have disease among those tested, the total costs of diagnosis, the cost per patient found, and the marginal cost (Table 3). As expected, the fraction of patients with disease who were identified rose from 75 percent to 86 percent. At the same time, the fraction of patients tested who had disease decreased from 66 percent to 36 percent; thus, physician behavior seemed to be influenced by the introduction of the test because of its perceived high degree of accuracy and noninvasive quality. Total diagnostic costs increased from $532,000 to $867,000, average costs increased from $565 to $805, and the marginal cost was $2,600 per patient discovered.

An appropriate question arising from these data concerns the meaning of the increase in the cost per patient discovered, as well as the increase in the number of patients examined by this method. Although the demonstration of a significant objective difference in health outcomes will be difficult, there undoubtedly has been a significant increase in the confidence of physicians whose patients have had this neurodiagnostic procedure and an increase in the peace of mind of those patients who have had a negative result. This analysis thus raises some critical questions. What is the value of physician confidence and patient peace of mind? Who should decide this worth—policymakers, individual patients, or society as a whole? Who should pay for it?

Conclusion

Several generalizations can be made about the evaluation of diagnostic procedures. First, at the present time we do not have a coherent epistemology for dealing with these evaluative methods. Most of the literature has a considerable anecdotal quality which will probably continue for some period of time. Second, in these early days many of our models tend to be simplistic and do not necessarily epitomize the reality of the patient-physician encounter. However, in our attempt to start somewhere, I believe that we should select examples that are both non-trivial in terms of their impact on medical care and amenable to analysis. A consequent tension has developed and will continue within the interested community. On the one hand, many will ridicule these approaches because they cannot characterize exactly the way we practice medicine on a day-to-day basis. On the other hand, others will want to use these approaches immediately for developing policy. We can only hope that ultimately we can weed out the pitfalls and biases to meet all expectations.

Endnotes

1. McNeil, B.J. and Adelstein, S.J. (1976). Determining the value of diagnostic and screening tests. *J. Nuc. Med.* 17:439.
2. McNeil, B.J., Sanders, R., Alderson, P.O., Hessel, S.J., Finberg, H., Siegelman, S.S., Adams, D.F., and Abrams, H.L. (1981). A prospective study of computed tomography, ultrasound, and gallium imaging in patients with fever. *Radiology* 139:647.
3. Ransohoff, D.F. and Feinstein, A.R. (1978). Problems of spectrum and bias in evaluating the efficacy of diagnostic tests. *N. Engl. J. Med.* 299:926.
4. Drum, D.E. and Christocopoulos, J.S. (1972). Hepatic scintigraphy in clinical decision-making. *J. Nuc. Med.* 13:908.
5. McNeil, B.J. (1979). Pitfalls in and requirements for evaluations of diagnostic technologies. In *Proceedings of the Conference on Medical Technology*, ed. J. Wagner, pp. 33–39. DHEW Pub. No. (PHS) 79–3254.
6. Swets, J.A., Pickett, R.M., Whitehead, S.F., Getty, D.J., Schur, J.A., Swets, J.B., and Freeman, B.A. (1979). Assessment of diagnostic technologies. *Science* 205:753.
7. Swets, J.A. (1979). ROC analyses applied to the evaluation of medical imaging techniques. *Invest. Radiol.* 14:109.

8. McNeil, B.J., Hessel, S.J., Branch, W.T., Bjork, L., and Adelstein, S.J. (1976). Measures of clinical efficacy III. The value of the lung scan in the evaluation of young patients with pleuritic chest pain. *J. Nuc. Med.* 17:163.

9. Crichton, D. (1962). The accuracy and value of cephalo-pelvimetry. *J. Obs. Gyn. Brit. Comm.* 69:366.

10. McNeil, B.J., Varady, P.D., Burrows, B.D., and Adelstein, S.J. (1975). Measures of clinical efficacy: Cost-effectiveness calculations in the diagnosis and treatment of hypertensive renovascular disease. *N. Engl. J. Med.* 293:216.

11. McNeil, B.J. and Adelstein, S.J. (1975). Measures of clinical efficacy: The value of case finding in hypertensive renovascular disease. *N. Engl. J. Med.* 293:221.

12. George, R.O. and Wagner, H.N., Jr. (1975). Ten years of brain tumor scanning at Johns Hopkins: 1962–1972. In *Noninvasive Brain Imaging*, eds., H.J. DeBlanc and J.A. Sorenson, pp. 3–16. New York: Society of Nuclear Medicine.

13. Abrams, H.L. and McNeil, B.J. (1978). Medical implications of computed tomography. *N. Engl. J. Med.* 298:255.

PITFALLS AND BIASES IN EVALUATING NEW THERAPEUTIC TECHNOLOGIES

by David L. Sackett

When the introduction of new therapeutic technologies has not proceeded in a series of orderly steps, we have usually paid a high price
in money, health, and, on more than one occasion, lives. This paper
will consider two major pitfalls that attend the failure to follow this
series of steps. As some technical jargon will be used that may be
new to some readers, the terms will be defined by means of the following example.

A group of clinicians might develop a machine which, when applied to a few patients with peptic ulcers, could show promise for
reducing their symptoms and healing their craters. The orderly
evaluation of this new therapeutic technology would occur in four
steps. First, the *efficacy* of this procedure would be determined by
studying whether it did more good (symptom-relief and ulcer healing) than harm (side-effects or complications) to a group of ulcer
patients who had been carefully selected for their prior cooperative-
ness and compliance with health advice. Second, if the efficacy study
gave favorable results, the *effectiveness* (or *usefulness*) of the procedure would be determined by carrying out a similar study of
whether it did more good than harm, but this time a cross-section
of ulcer patients would be used, some of whom might not agree to
undergo treatment. This evaluation would thus determine the value
of *offering* the new therapy to ulcer patients, and would thereby
provide information about the likely impact of this new therapeutic
technology on the whole population of ulcer victims.

If this new therapy were proven to be efficacious and effective, it
would come into widespread clinical use. At this point the stage
would be set for the third and fourth steps in its evaluation—the
determination of whether this new therapeutic technology was being
made *available* to all ulcer patients and whether its execution represented an *efficient* use of health resources.

In summary, the orderly evaluation of new therapeutic technology involves the stepwise determination of its efficacy, effectiveness, availability, and efficiency. Two common hazards in carrying out this evaluation will be discussed: (1) the premature evaluation of availability and efficiency, and (2) the reliance on subexperimental evidence in determining efficacy and effectiveness.

Premature Evaluation of Availability and Efficiency

Evaluation of the availability of a new therapeutic technology may be carried out according to one of five strategies: on what people want; on what experts say they need; on what people actually demand; on what people actually use; on what the health system supplies. The following description of these five strategies for assessing availability assumes that the therapeutic technology has already been shown to be efficacious and effective. It is this assumption that can lead to the errors in sound assessment.

Want. The first strategy is want, or expected technology. The public's perceptions of their own wants, expectations, and rights is emphasized. Wants can be estimated precisely by careful population surveys of self-perceived health status and expectations but are usually expressed in spontaneous acts by individuals (such as letters to the editor) or by groups (such as citizens groups that encourage the opening, or discourage the closing, of a health facility).

Need. The second strategy is the evaluation of need or technologies that "ought-to-be." Emphasis is placed on the perceptions of some "expert" groups who base their estimates of need on routine reports of births, deaths, and notifiable diseases; on the results of special surveys; or on projections of trends in population, disease risk, or use of health services.

Demand. The third strategy is the evaluation of demand, or the types and amounts of health technology requested or desired by the public once they know the costs or prices involved. Demand can be estimated by surveys which probe both what is wanted and what would be paid (or what alternatives would be foregone) to satisfy the want. Such surveys are both complex and costly.

Utilization. Evaluation of the health technology actually utilized is a fourth strategy for assessment. Such data are abundant because they are generated as a matter of course by hospitals, other institutions, and individual practitioners.

Supply. The fifth strategy is the evaluation of supply or "get-at-able" health technology. Here the focus may be either quantitative —a consideration of the numbers and distributions of facilities and health personnel relative to the populations they serve, or qualitative

—a consideration of the quality of care provided by these technologies. Quantitative data on the supply of health technology and personnel tend to be widely available, but data on the quality of health care they provide usually must be obtained through special efforts such as medical audits.

Scrutiny of the foregoing strategies reveals that none includes a test for efficacy or effectiveness; all assume that the technology under consideration does more good than harm. Evaluations of availability thus simply address the question of whether people who could benefit from this assumed efficacy have access to the health technology reputed to achieve it.

Although a number of technologies have been subjected to premature *availability*, a recent Canadian example was particularly striking. It involved the Participation Program, an intensive propaganda campaign intended to bring the assumed cardiovascular benefits of increased physical activity to all Canadians. Recently, however, a randomized effectiveness trial, supported by some of the same agencies that supported the initial propaganda campaign, noted that the risk of recurrent myocardial infarction may rise, not fall, when convalescing infarct patients engage in endurance training.[1] This campaign thus illustrated the fallacy of premature attention to issues of availability.

Premature evaluation of the *efficiency* of a health maneuver can be subjected to the same errors as are premature evaluations of the availability of technologies. For example, given that the supply of health resources cannot meet all wants and needs for health services, health providers frequently ask: How can we put our limited resources where they will do the most good? Because this question involves both costs and consequences, and because it implies a choice between alternative courses of action, an attempt to answer it is said to constitute an economic analysis. In such a comparative analysis money is usually (but not always) the unit of measurement, and the "real" cost of any health program is the sum of the effects or benefits foregone by committing resources to this program rather than to another one. In the economic context, then, the real costs of an air ambulance for a remote area is, for example, the number of nephritics who will die due to the failure to put the same resources into an expanded chronic dialysis program. This comparison of costs and effects is formalized in the tactics of cost-benefit, cost-effectiveness, and cost-utility analysis.[2]

The strategies of efficiency evaluation assume effectiveness at the outset. But for an efficiency evaluation to be rational, the effectiveness of the health technology being optimized must be established beforehand.

A useful example (even though it concerns secondary prevention) is the application of high technology, in the form of automated multiphasic health testing, to the clinical evaluation of patients admitted to hospital. Evaluations of the efficiency of various technologic strategies and tactics for admission screening have contributed to the creation and installation of automated clinical laboratory equipment of majuscular dimensions. Unfortunately, however, these evaluations of the efficiency of admission screening technologies preceded the rigorous examination of whether patients actually benefited from the technological invasion of their structure and function. The efficacy of admission screening was assumed in such efficiency analyses, or, at best, testified to by the discovery of new disease, or by the endorsements of satisfied clinicians. However, when a randomized trial of admission screening was finally carried out, no difference in health outcomes was found between experimental patients whose admissions generated innumerable screening test results and control patients whose test results were either held back or never generated.[3] An additional touch was added when the trial documented that efficiency for the total health system (as opposed to the clinical laboratory) fell when admission screening was performed.

Summary. Evaluations of the availability and efficiency of a new therapeutic technology *assume* its efficacy and effectiveness. Accordingly, such availability and efficiency evaluations are rational only *after* efficacy and effectiveness are firmly established.

Reliance on Subexperimental Evidence for Efficacy and Effectiveness

The hazard involved in subexperimental data can be avoided by an understanding of "confounding." Consider the hypothetical example of a microvascular surgeon who develops a technique for bypassing a surgically inaccessible occluded carotid artery in experimental animals and wishes to evaluate this procedure for its potential benefit to humans. Because it is a lengthy procedure and not without risk, the surgeon carries out the first few bypasses on cerebrovascular patients who are, for the most part, free of hypertension or other extraneous disorders that increase their surgical risk. The results are dramatic: all of the patients survive the procedures and their cerebrovascular symptoms remains stable or even improve. In contrast, many of the poor-risk cerebrovascular patients, who were rejected for surgery because of coexisting hypertension, either died or experienced progression of their cerebrovascular symptoms. When the results are reported, a substantial segment of the profession concludes that the bypass procedure is of obvious efficacy and ought to be performed on all good risk (and even some poor risk) cerebrovascular patients.

Three properties identify hypertension as a confounder in this hypothetical example:

1. It is extraneous to the question posed. At issue is the efficacy of bypass surgery, not the biology of hypertension.
2. It is a determinant of the outcomes of interest. Hypertensives are more likely to experience progressive cerebrovascular disease and to die than are normotensives.
3. It is unequally distributed among the treatment groups being compared. In this case the inequality is marked; very few hypertensives were bypassed because almost all were rejected for surgery.

In terms of causation, hypertension is a confounder, and the presence of such confounders has complicated the evaluation of the efficacy of almost all preventive, therapeutic, and rehabilitative maneuvers. Confounding leads to bias—the arrival at a conclusion that differs systematically from the truth—and examples abound in both experimental and subexperimental research.[4]

If confounding hampers the valid demonstration of efficacy and effectiveness, how can it be avoided? Seven strategies exist for breaking confounding; all of them attack its property of unequal distribution among treatment groups, and one is clearly superior to the rest.

First, confounding can be broken by restricting the criteria for inclusion; in the example just given, hypertensives could be excluded from either the operated or nonoperated patient groups. Second, operated and nonoperated patients could be individually matched for their hypertension status. Third, stratified sampling could be carried out, creating cohorts of operated and nonoperated patients containing identical proportions of hypertensives. Fourth, a stratified analysis could be performed on all patients, comparing outcomes among operated and nonoperated hypertensives, separate from the comparison of outcomes among operated and nonoperated normotensives. Fifth, an adjustment or standardization procedure analagous to age-standardization could be applied in the analysis. Sixth, a model could be established for risk of the outcome of interest that would include a correction factor for hypertension and could be expanded to include other possible confounders such as symptomatic coronary heart disease, diabetes, and so forth. Seventh and finally, appropriate patients could be allocated randomly to undergo or not undergo the new microvascular surgical technology.

Random allocation has a profound advantage over the other six strategies: it breaks unknown as well as known confounders, thus

acting to prevent bias from potential, as yet undiscovered confounders. This benefit to validity places the true experiment (where
allocation to the maneuvers under comparison occurs by random
allocation) above the subexperiment (where allocation occurs by
any other process) as a means for determining the efficacy or effectiveness of new technology or any other clinical maneuver. Accordingly, the true experiment has become the standard approach for
determining the efficacy and effectiveness of chemotherapeutic
agents and most other drugs, and is used increasingly in evaluating
surgical technology as well.

I believe that two circumstances exist under which it is appropriate to rely on subexperimental evidence for efficacy. First, when a
given health state has been shown to lead inevitably to death, any
therapy followed by survival is efficacious, and a randomized trial
becomes not only superfluous but, in many cultures, unethical. Examples are rare but arguably include malignant hypertension, tuber-
culous meningitis, and choriocarcinoma. Second, and recalling that
the working definition of efficacy includes considerations of harm as
well as good, some adverse effects of clinical maneuvers may be so
rare or so late as to preclude the feasibility of their experimental
verification. For example, although the relative risk of myocardial
infarction for a 30-to-39-year-old woman taking oral contraceptives
compared with nonusers is 2.7, the likelihood that she will suffer a
myocardial infarction is only about 1 per 18,000 per year.[5] As we
have described elsewhere,[6] to be 95 percent sure of observing at
least one such event in a one-year randomized trial, the number of
treated women would have to be over 50,000.

In summary, reliance upon subexperimental evidence in making
judgments about efficacy or effectiveness constitutes a second major
pitfall in the evaluation of new therapeutic technology, and the
trigger for this trap is the confounding of risk with exposure. The
true experiment, in which allocation to the technologies under comparison occurs by random assignment, breaks both known and unknown confounders and is therefore the method of choice for
determining the efficacy or effectiveness of new technology.

Random allocation does not, however, prevent all sources of bias,
nor does it ensure the generalizability of the results of an experiment.
Residual issues will now be discussed briefly.

Additional Hazards

The foregoing sections have described what I judge to be two major
hazards in the evaluation of new therapeutic technology. It is not
enough, however, simply to execute random allocation in a trial of

efficacy or effectiveness. A series of lesser pitfalls remain in carrying out these evaluations, and I list them as follows.

Nongeneralizability. The screening process through which patients must pass to become study subjects may limit the generalizability of the results of evaluative studies. This pitfall has nothing to do with random allocation, for random allocation is a device for achieving validity, not generalizability. Indeed, we have seen how one strategy for breaking confounding—exclusion of subjects with a known confounder—clearly decreases generalizability by its very execution. Thus a randomized therapeutic trial may generate a valid result that is true for its subjects but applicable to only a small portion of affected individuals outside the trial.

Since the exclusion of subjects unlikely to benefit from experimental therapy may be necessary for efficiency,[7] the appropriate response here concerns information rather than prevention. Explicit inclusion/exclusion criteria used in the trial should be reported, together with a description of potential subjects who were not entered, in order to help the reader determine the applicability of the results. Finally, to the extent that nongeneralizability is the result of centripetal or popularity bias,[4] the incorporation of all the institutions throughout a region into a trial should increase the generalizability of its results.

Ambiguous or Changing Objectives. Failure to identify whether a therapeutic trial is asking an explanatory question (*can* the therapy work under ideal circumstances) or a management question (*does* the therapy work under the usual clinical circumstances), or the attempt to shift from one type of question to the other, may be fatal to both the validity and generalizability of the trial result.[7] For example, an invalid shift from a management to an explanatory question would occur if a trial compared outcomes between compliant subjects (those who faithfully took their study medicines) and noncompliant subjects on the assumption that they differed only in the amount of therapy received. In the Coronary Drug Project placebo group, the death rate among those who faithfully took their placebos were very much lower than that observed among the noncompliant subjects.[7] This pitfall can be avoided by both identifying the nature of the question posed before the trial begins and by sticking to it once the trial is underway.

The Unstable Maneuver. The expertise, especially surgical, with which the maneuver is executed may vary sharply in time and spacing, and a biologically valuable maneuver may appear useless or even harmful when performed in a clumsy manner. This pitfall can be avoided by invoking preset specifications on clinical competency (surgical mortality, graft patency) to be met prior to joining a trial, coupled with the monitoring of this performance during the trial.

Co-intervention. The performance of additional therapeutic procedures upon the experimental group should be avoided unless these same procedures are performed with equal vigor upon the comparison group.[8] For example, if experimental patients are seen more frequently than control patients, these additional opportunities for clinical evaluation and management may spuriously inflate the estimate of the benefit of the test therapy. A major strategy for preventing co-intervention is the "blinding" of study patients and their clinicians to the experimental therapy through the use of placebo drugs and maneuvers.

Diagnostic Suspicion. A knowledge of the trial subject's prior exposure to the test maneuver may influence both the intensity and the results of the search for relevant outcomes.[8] This is especially troublesome when the outcomes of interest are "soft" events such as changing neurologic findings, self-reported exercise tolerance, and the like. This pitfall represents the second major reason for blinding both study patients and their clinicians. Other preventive strategies include explicit, objective outcome criteria and, when blinding is impossible, the use of independent outcome assessors.

Substitutions. This pitfall is the substitution of changes in a risk factor, not established as causal, for its associated outcome. For example, many therapies have been claimed to be efficacious in the treatment of coronary heart disease, not because of their effect on the coronary rate but because of their effect on risk factors such as uric acid or one of the lipid fractions. This pitfall can be avoided by not using the substitution.

Insensitive Outcome Measures. When outcome measures are incapable of detecting clinically significant changes or differences between groups, the risk of a Type II error (in which we conclude that treatment is not efficacious when, in fact, it is) rises. This is because, *inter alia*, the resulting misclassification of events among both experimental and control patients tends to spuriously decrease any real difference between the true rates of their outcomes. This pitfall can be avoided by developing and validating, prior to the trial, outcome measures that are both reproducible and sensitive to clinically significant change.

Narrow Outcome Reporting. The restriction of outcome measurements to one or a few biologic measurements may miss important dimensions of good and harm. For example, cancer therapists are increasingly concerned that the outcome measures traditionally used in their trials of new therapeutic technology (tumor size, disease-free interval, and survival) are insensitive to key elements of the quality of life of cancer patients. They are now taking steps to avoid this pitfall by generating and validating new outcome measures of physical, social, and emotional functions; peace-of-mind; and psychological well-being. These measurements are being extended beyond states of health to considerations of the value or utility of the health states produced by the new therapeutic technologies.

Competing Risks. When a major event (such as death) precludes the subsequent occurrence of a lesser event (such as stroke) in the same pathogenic sequence, the latter cannot be considered in isolation.[7] If it is, the therapy that kills patients before they have a chance to have a stroke could appear to be beneficial in the prevention of stroke. Such an interpretation has, in fact, been suggested by some researchers following two randomized trials involving clofibrate. The risk reduction for nonfatal myocardial infarction was impressive. However, there was a corresponding risk increase for total mortality.[9] This pitfall can be avoided by the use of diagnostic hierarchies in which lesser events in a sequence are not considered in isolation. When such an analysis was carried out on the clofibrate trial data and the overall rates for myocardial infarction or death compared, the lack of efficacy of the drug was unambiguously demonstrated.

Endnotes

1. Rechnitzer, P.A., Ontario Exercise-Heart Collaborative Group. (1978). A controlled prospective study of the effect of endurance training on the recurrence rate of myocardial infarction. *Ann. R. Col. Phys. Surg. Can.* 11:29–30.
2. Sackett, D.L. Evaluation of health services. In *Preventive Medicine and Public Health.* 11th ed., ed. J.M. Last. New York: Appleton-Century-Crofts (in press).
3. Durbridge, T.C., Edwards, F., Edwards, R.G., and Atkinson, M. (1976). An evaluation of multiphasic screening on admission to hospital. *Med. J. Aust.* 1:703–05.
4. Sackett, D.L. (1979). Bias in analytic research. *J. Chronic Dis.* 32:51–63.
5. Mann, J.I., Vessey, M.P., Thorogood, M., and Doll, R. (1975). Myocardial infarction in young women with special reference to oral contraceptive practice. *Br. Med. J.* 2:241–45.
6. Sackett, D.L., Haynes, R.B., Taylor, D.W., and Gent, M. (1980). Compliance. In *Monitoring for Drug Safety,* ed. W.H.W. Inman. London: MTP Press, pp. 427–438.
7. Sackett, D.L. and Gent, M. (1979). Controversy in counting and attributing events in clinical trials. *N. Engl. J. Med.* 301:1410–12.
8. Sackett, D.L. (1975). Design, measurement and analysis in clinical trials. In *Platelets, Drugs and Thrombosis,* eds. J. Hirsh, J.F. Cade, A.S. Gallus, and E. Schonbaum, pp. 219–225. Basel: S. Karger.
9. Oliver, M.F., Heady, J.A., Morris, J.N., and Cooper, J. (1978). A cooperative trial in the primary prevention of ischaemic heart disease using clofibrate. *Br. Heart J.* 40:1069–1118.

Part Four

SOCIAL AND ETHICAL IMPLICATIONS OF TECHNOLOGY

SOCIAL AND ETHICAL IMPLICATIONS IN TECHNOLOGY ASSESSMENT

by Laurence R. Tancredi

New diagnostic and treatment technologies have the greatest impact on the physician-patient relationship. These complex new technological procedures require expert knowledge for determining proper use,[1] and initially most decisions regarding their use are made by the physician or health care provider. In time, as knowledge is gained and information becomes more readily available, patients and consumers are better able to participate in any decisions regarding the diagnostic or treatment plans proposed for their care. This results in improving the effectiveness of informed consent as a protective device for the patient in the therapeutic relationship.[2]

The use of a technology should be evaluated from three perspectives: the physician as provider, the patient, and society in general. Each of these perspectives can be represented by certain standards for assessing medical technology. As a general principle, the decision-making power with regard to the use of any specific new technology shifts in the direction of the physician, based on the following considerations:

1. The degree and nature of uncertainty of information available about the efficacy of the new technology.
2. The relative inadequacy of existing diagnostic and treatment methods that would be used for the same purpose being proposed by the new technology.
3. The degree of competency of the patient that must be considered when the patient's medical condition precludes involvement in medical decisions.
4. The complex relationship between the risks and benefits of the technology.
5. The urgency of therapeutic intervention, involving some of the

same issues relative to the patient's emotional and medical competency in times of emergency.

The question of who will dominate in the decision to use a particular technology will be answered only after individual investigations of specific issues.

Perspectives for Assessment

It must be emphasized again that the three assessment perspectives —physician, patient, and society—actually represent standards that delineate specific issues affecting the principal actors involved in the decision to use a new technology. Another important consideration in terms of the power of decision making relates to timing and the question of when to introduce the technology into clinical medicine. Therefore the significance adhering to any of the following standards rests not only on the degree of complexity of a new technology but also on its stage of development.

Mechanistic Medical Efficacy Standard

One criterion for assessing technologies is the therapeutic or mechanistic medical efficacy standard, which includes a wide range of issues dealing with the mechanical aspects of the technology. These are addressed in detail earlier in this volume by Eddy and Adelstein. Basically these standards attempt to ascertain whether or not a diagnostic or therapeutic technique is successful, and whether it achieves an appropriate medical objective. These questions avoid the more complicated conceptual issues of differentiating between health and disease and what role if any, notions like "the quality of life" should have in assessing the medical success of the technology. For the most part physicians and health care providers are trained to think in terms of the purely medical effectiveness of technologies. They rely on tests such as random clinical trials to provide the information for their decisions regarding the treatment of specific pathological processes. Because their training emphasizes the more quantifiable, measurable aspects of disease processes, physicians will be inclined to preserve and apply the values inherent in that emphasis to relevant diagnostic and treatment methods. Therefore in a purely provider-based decision the mechanistic medical efficacy standard will predominate. Statements like "in the best interest of the patient" and "would choose if knowledgeable" justify the physician's or provider's desire to strongly recommend and thereby influence the final agreement, presented as consensual, between the physician and patient concerning the use of a specific technology.

Personal Efficacy Standard

The second major standard that should be used for assessing technologies is the personal efficacy standard. (These are discussed in greater detail in this volume by McNeil and Pauker and by Pauker et al.) In general this standard addresses the many concerns the patient may have when deciding on a particular diagnostic or treatment program. In addition to the purely medical efficacy of a technology, the patient will be interested in its impact on the quality of his or her life.

The first consideration regarding the quality of life is whether the patient will be able to return to a baseline of functioning, particularly in cases requiring dramatic therapeutic technologies such as renal dialysis or the proposed artificial nuclear-powered heart. The medical treatment may be effective in keeping the patient alive and in correcting the biochemical and physiological abnormalities of a disease or defect, but may at the same time create significant difficulties which change the quality of his or her life. For example, studies have shown that from 50 to 83 percent of chronic hemodialysis patients maintain a productive life despite the need to adhere to an irregular and time-consuming treatment program.[3] Other studies have suggested that factors such as the predialysis vocational functioning of the patient, work satisfaction, and basic dissatisfaction with the role of being sick are critical in contributing to any prediction about the patient's vocational adjustment while on hemodialysis.[4-8] Still others have shown that 60 percent of clinic populations have poor emotional adjustment.[9] However, in his review of studies conducted on emotional adjustment of dialysis patients, Armstrong determined that 19 of these studies showed that a median of roughly 46 percent of the adults adjusted poorly to the treatment. In contrast, he determined in nine studies involving children that a median of approximately 22 percent were considered poorly adjusted.[3,10-12] The difference may reflect the fact that pediatricians rather than psychologists were evaluating the children. Even given that factor, however, the studies certainly suggest that a large percentage of adult patients adjust poorly to hemodialysis.

The second consideration involving the quality of the patient's life would be the degree of relief from expected symptoms. The patient would be most concerned with relative freedom from pain or discomfort, including those that may be induced by the treatment itself.

The third consideration, the requirements of the treatment, would also be particularly important to those patients suffering from chronic illnesses, such as end-stage renal disease. The hemodialysis patient, for example, may have to spend as much as six hours, three times a week, attached to an artificial kidney machine in an institutional setting.

The fourth consideration would be the psychological effects of the technology—the short-term as well as the more permanent long-term changes that may ensue. In the case of patients requiring renal dialysis, psychological studies[4] have demonstrated a significant increase in the suicide rate, estimated in one study[13] to be 100 times higher than the normal rate, which is about 10 in 100,000 individuals in the United States. If deaths attributed to the failure of the patient to comply with medical regimens were included as suicide, the rate would be 400 times that of the general population.[13]

Short of suicide, patients on hemodialysis have other psychological effects that can prevent them from receiving maximum benefit from the treatment.* There is, for example, a significant increase in the percentage of patients who suffer from anxiety and depression, as well as sexual dysfunction.[14] In contrast to the normal population, these patients more often develop marital problems.[4,15] Hemodialysis patients who are being treated through a home program where the family becomes intimately involved in their care exact more than just emotional support from their spouses. The relationship can become so strained that family disharmony and even separation can result. A host of physical changes such as intermittent azotemia, chronic and severe anemia, and bone difficulties (usually due to secondary and tertiary hyperparathyroidism) can augment the psychological dysfunctions experienced by the patient.

The fifth consideration in understanding quality of life issues is the impact of the new technology on the self-image of the patient. This issue has particular relevance to mutilative surgery such as radical mastectomies, jaw resections, and other extreme surgical procedures for seriously ill cancer patients. Although radical procedures are usually employed only when dictated by severe necessity, and although they have the potential of curing the disease, they can create immense difficulties for patients by causing distortions in body image and self-evaluation. In her book *First You Cry*, Betty Rollin revealed the intense psychological impact of a partial radical mastectomy on her sense of body integrity and her function as a woman.[16]

---

* It is interesting to observe that despite these numerous psychological effects, patients on renal dialysis do not rate their state in life as low as might be expected. See, for example, Table 4 in the paper by McNeil and Pauker in this volume.

Just as issues concerning the quality of life are important during the development and implementation of a therapeutic technology, issues concerning the quality of death also deserve consideration within the purview of the personal efficacy standard. In 1972 the National Heart Institute set up a panel to review issues surrounding "the totally implantable heart."[17] The panel was concerned with the dilemma of prolonging the life of a patient with end-stage heart disease on the one hand, and of dealing with the possibility that the patient would develop a more serious, long-term disease in the future. Certainly whenever a technology extends the length of life, the possibilities of more undesirable and serious long-term and disabling illnesses that may result in a more painful death must be considered.[18]

Economic issues, especially the cost of the treatment, are also germane to considerations of the personal efficacy standard. The patient must have adequate medical insurance to deal with all costs involved with the therapy, possible complications, and follow-up treatment. Again, all of the considerations discussed under the personal efficacy standard—the degree of competency of the patient, the availability of information, and the strength of consent—will affect the criteria used in deciding on specific diagnostic and therapeutic technologies.

Societal Efficacy Standard

This standard deals with the broad societal issues concerning new technologies, the impact of these technologies on the health care system, and the implications they have on principles of fairness. By society I refer essentially to individuals in the aggregate. Traditionally societal concerns have not entered into either the physician's or the patient's decisions regarding medical care. Increasingly, however, they are becoming more important, not so much regarding individual decisions on the use of a technology, but for policymakers involved in decisions regarding implementation and development of specific technologies in the future. Three issues included in this broad category are discussed here.

The Overall Cost Effect. The economic impact of new technologies has become particularly important since the passage of the Social Security Amendment in 1973, which allowed coverage for renal dialysis and transplantations for patients suffering from end-stage renal disease. When initially passed it was projected that within ten years taxpayers would pay roughly $1.5 billion a year to cover the treatment of patients suffering from this condition. Instead, the escalating expense for this treatment now places the projected cost by 1984 for the care of these patients at well over $3.5 billion a year.[19]

The impact of these figures underlines the importance of considering cost issues when evaluating highly expensive technologies, such as the artificial kidney, that have particular application to the chronically ill. The decision to care for end-stage renal patients with these expensive treatments results in a deflection of critical resources, money, facilities, and personnel that could be channeled to patient populations suffering from other conditions.

The Population Impact.   Technological biomedical development can have crucial effects on the size and composition of the population. An increase in heart transplantation for patients suffering from end-stage cardiac disease could result in increasing the elderly population in this country and thereby creating a deflection of medical resources not only for the cardiac care of these patients but for the numerous chronic and acute conditions that are concomitants of advanced age.

Technology can also affect the composition of the population through the development of techniques to minimize the childhood fatalities of genetic diseases. If children with various genetic diseases survive the adolescent years and reach young adulthood, their procreation inevitably would bring about significant expansions of the gene pools for these conditions and a concomitant increase in the number of patients with genetic disease. Such an increase would add to the health care system's burden of caring for these patients.

Of similar concern are technologies that actively increase the length of the dying process by maintaining patients in a state of incompetence. For example, the Byrd respirator, a commonly used technology, keeps patients alive who are often beyond the point of being salvaged. As we have seen with landmark cases such as that of Karen Ann Quinlan,[20] this technology has created ethical issues pertaining to time of death and the responsibility involved with extraordinary care and dying.

Bias in the Selection of Patients.   The third factor of concern to society addresses the potential for arbitrariness and bias in the selection of patients for expensive but limited therapies. In the early days of the artificial kidney, committees such as the Seattle Committee were set up to develop criteria for selecting patients for this treatment.[21] At that time very few machines were available; at most only about 500 new patients per year could be accommodated, although an estimated 10,000 patients per year could have benefited from chronic, as opposed to acute, hemodialysis. The criteria for selection included highly value-laden information, such as age and sex; intelligence; the "likelihood of vocational rehabilitation"; a social welfare evaluation of the patient and his or her family; demonstrated social worth of the patient; and potential social contribution if rehabilitated. Even less personal criteria, such as the patient's proximity to the hemodialysis center, were considered.[22] Many of these criteria were quite arbitrary, and in the early years of artificial kidney treatment, the elderly and children were not selected for chronic hemodialysis.

Information Disclosure and Consent

A patient's competency to understand available information is important in determining which of the two major models will predominate —those dealing with the medical efficacy criteria or those involving personal considerations. The ostensible goal of informed consent is to shift, whenever possible, the decision-making power onto the patient, and to incorporate the patient's values in dealing with the social, psychological, economic, and medical issues involved in his or her particular treatment. For the patient, informed consent is a powerful tool, but often one of unrealized potential; in fact, under certain circumstances, it may have only minimal effect.

The concept of informed consent first appeared in the mid-1950s when one court ruled that a physician had the duty to disclose to patients "all the facts which mutually affect his rights and interests, and of the surgical risks, hazards and dangers, if any."[23,24] From the patient's perspective informed consent is an instrument against medical paternalism. It essentially reinforces themes of individualism, autonomy, and self-determination; and it underscores the rights of the individual to privacy. A fully realized informed consent therefore would include the right to refuse treatment—a choice being considered in a series of cases involving the rights of the mentally ill, even though involuntarily committed, to refuse treatments such as psychotropic medications.[25,26]

As conceptualized, informed consent deals with considerations relevant to therapeutic consent, such as the benefits, risks, and alternatives to treatment. Where ideal informed consent exists, the patient should be able to establish the materiality of those considerations. Other issues pertaining to experimental treatments, particularly the right to terminate experimental treatment without prejudice and to be treated by appropriate traditional methods, are also addressed.

As a theoretical notion informed consent can hardly be criticized. However, practical application presents considerable conceptual problems. The *Canterbury* case[27,28] in 1972 was a landmark decision which attempted to shore up the protection provided to patients in the therapeutic context. *Canterbury* was the first case to state that the degree of information disclosed to the patient must be information that would be deemed material for a patient in deciding whether to undergo the recommended treatment. Pragmatically, this means that when an issue of informed consent is raised, a jury must decide whether or not the information that was *not* provided was material enough from the patient's perspective to be essential in making a reasonable decision regarding consent. This position clearly is not based on the physician's concept of good medical care or on his opinion regarding information that should be disclosed about a technology. It has often been argued that this case was so vague in delineating the materiality of the information the patient should be given to make a reasonable decision that it essentially extended the notion of information disclosure to its extreme—the disclosing of *all* the benefits and risks of a therapeutic intervention.

Some subsequent judgments[29] have retrenched somewhat from that position because of the unrealistic extent of the information to be presented to the patient. These cases seem to result in expanding the application of the traditional notion that therapeutic privilege is essential; that that privilege articulates a physician's right not to disclose information that would adversely affect the patient's decision on consent, resulting in a detriment to the overall treatment. Furthermore, in some cases, the standard seems to revert to the "professional disclosure" standard of the amount and kind of information a reasonable physician under similar circumstances (type of practice, condition of patient, social supports available, etc.) might disclose to a patient about the benefits, risks and alternatives of treatment. Despite the *Canterbury* developments, the "professional disclosure" standard remains the test in the majority of jurisdictions in this country.

The issues relevant to informed consent are: (1) whether it is voluntary; (2) whether the patient is competent to understand the information; and (3) whether the information or data available about the therapy or experimental treatment reaches a threshold level of materiality. These issues will increase in importance as new technologies are introduced and the legal system continues to construct protective devices to assure the rights of patients in the health care system.

The issue of voluntariness is particularly significant when the patient is confined in a mental institution or is incarcerated, and when the patient's options are clearly limited. Competency is important, particularly in regard to mental patients, children, the mentally retarded, and prisoners. However, many issues dealing with competency apply to the average unimpaired patient.

The materiality of the information available to the patient is a concept concerned essentially with the patient's ability to make a reasonable appraisal of the information given regarding the diagnostic or treatment technology so that an informed consent can be made. The threshold level of certainty should be sufficient for this purpose;

if the uncertainty level is too great in regard to the nature of the risks involved, then this criterion has not been met.

The notion of materiality emerged from a case involving psychosurgery.[30] Amygdalotomies were to be performed to decrease violent behavior in prisoners with a history of committing violent acts. The court determined that the information regarding this psychosurgical procedure was not sufficient to justify its use, in that it could not elicit a truly informed consent. This argument would apply to those technologies that have not been rigorously studied in terms of the benefits and risks involved and for which data concerning these matters could be viewed as being insufficient.

Very little is known about risks, particularly in experimental treatments. However, a recent article by Gilbert, McPeek, and Mosteller[31] demonstrated that there may be a fallacy in the notion that patients who receive experimental treatments are facing increased risks. In this study, 46 papers were reviewed that involved at least ten patients per trial who were receiving experimental treatments in surgery and anesthesia. The results of the evaluations seem to establish that the groups of patients receiving experimental treatment, for the most part, were not doing much better or worse than those receiving the regular treatment for the disease condition.

Competency to Understand

Data are available supporting the view that the competency of patients to understand the medical information required to make a meaningful informed consent in either an emergency or non-emergency situation is not high. One study,[32] for example, involved interviewing 20 patients four to six months after heart surgery. The average score of primary recall regarding medical information about risks and complications of the surgery was about 10 percent of all the information that was given. After the patients were presented with the topics that were discussed regarding the risks and serious complications of surgery, approximately 23 percent of the information was recalled. The average recall of all the information presented to the patient, such as the diagnosis, risks, benefits and alternatives to surgery, was only 29 percent. When it was presented to the patients the second time, only 42 percent of the information was recalled. Sixteen of the patients claimed that major items had not been discussed at all. Of course, this study has the methodological disadvantage that it involved interviewing these patients after a considerable passage of time from the surgery and the session when informed consent was obtained.

A very recent study[33] dealing with the capacity of patients to understand information presented to them involved the ability of

patients suffering from cancer to recall information regarding chemotherapy, radiation therapy, or surgery which appeared on the consent
forms that they signed. The day after the forms were signed, these
patients were asked to fill out a questionnaire to determine their level
of understanding and their opinions regarding the purposes of the
consent document. Sixty percent of the 200 patients who filled out
the questionnaire indicated that they understood the nature and purpose of the treatments. However, only 55 percent of those sampled
were able to identify even one major complication from the three
broad procedures discussed. Just 30 percent of the patients actually
read the forms carefully, and most seemed to rely on information
given to them orally by the physicians regarding the nature and purpose of the treatment. At best, they only scanned the consent form.
In discussing the recall discrepancies demonstrated in their studies,
the researchers pointed to three factors they consider implicative.
First, the medical condition of the patients seemed to have a direct
effect on their ability to recall the information. Ambulatory patients
did better than those who were bedridden. Second, a relationship
seemed to exist between the ability of the patients to read and understand the material presented and their educational background.
Third, a direct relationship was suggested between the care with
which the patients read the consent forms, their level of recall, and
their overall level of perception. Even given the most ideal circumstances, however, there was some question as to whether the forms
were sufficiently readable. A recent study[34] involving the examination of five surgical consent forms showed that those sampled were
very difficult for the general population to understand.

In the case of mental patients—even those who enter into treatment voluntarily—consent becomes more problematic. Only a few
studies have dealt with this issue, and two are particularly noteworthy because they elucidate clearly the difficulties of obtaining
informed consent from this group of patients. One study[35] involved
100 patients admitted to an institution voluntarily. Only eight of
these patients were able to understand fully the terms of the voluntary admission form. A second study[36] involved 40 patients who were
questioned on their ability to understand the legal rights form, information release form, and voluntary admission form. Fewer than 50
percent could understand these forms when examined on the information.

These studies highlight the difficulties that exist in the ability of
patients to understand medical information, and show that patients
generally do not have a high level of comprehension, due very often
to the way in which medical procedures and risk information are
presented. For some time clinicians have assumed that patients will
understand the relevant issues in their treatment if given enough

information concerning the full range of benefits, risks, and alternatives, and will therefore be knowledgeable enough to make an informed consent.

Judgments of Uncertainty

In addition to the comprehension issues relevant to informed consent, questions also arise regarding judgments made under uncertain conditions. Even experts in their fields may distort the information presented to them and be so influenced by their biases that they make poor choices. Tversky and Kahneman[37] suggest that at times of choice people fall back upon a circumscribed number of heuristic principles* that essentially avert the complicated task of deciphering probabilities involved in a particular decision. This tendency suggests that even those in the best position intellectually and emotionally to understand the material presented to them will lapse into bias-laden heuristic reasoning or intuitive judgments that preclude an objective understanding of the other alternatives available. If this is the case, then informed consent is clearly more of an idealized notion than one that can be practically applied.

The need to resort to heuristic mental processes is rooted in human psychodynamics. When an outcome involves a significant degree of uncertainty, taking refuge in these intuitive mental operations provides a means by which an individual can avoid anxiety and discomfort. The result, of course, can be distortion, such as a failure to consider the human error present in the implementation of a technology, or overconfidence in the knowledge base available for that procedure.[38] Bunker and Brown[39] conducted a study of physicians' personal use of the health system with respect to surgical and anesthesiological procedures. The investigators felt that physicians would provide some insight into the untoward results of overconfidence regarding the achievements of technology. The study demonstrated a

---

* It is perhaps useful to outline the three main heuristic principles described by Tversky. *Representativeness* involves linking an event A with a class B because A and B may resemble each other. This reliance on the resemblance could result in insensitivity to prior probability of outcomes to inadequacies of sample size and to misconceptions about chance and regression. *Availability* refers to the process of arriving at a notion about probability based on the ease with which occurrences of a similar nature can be brought to the mind of the person making the decision (for example, deciding that a particular disease is a risk to a particular age group on the basis of one's exposure to individuals in that age group who have the disease). Such a process would also be insensitive to statistics and probabilities. *Adjustment and anchoring* refers to a process of arriving at estimates based on an initial value derived, perhaps, from previous studies or from one's individual experience which is adjusted to result in the final answer.

rate of 25 to 30 percent higher usage over the general population of surgical services for both the physicians and their spouses. This difference in rate was especially evident with regard to hysterectomies and appendectomies, procedures which had already come under considerable criticism in terms of their necessity and effectiveness in treating many abdominal and pelvic difficulties.

The higher rate of usage of these controversial procedures implies that the acculturation and socialization of the physicians, through their training process, is so intensive in the direction of affirming the values of intervention for virtually any condition that it even results in distortions involving choices of questionable medical benefit. Again it could be argued that the intuitive processes of these consumers have been shaped by years of comfortable retreat into the values of the medical model and that an objective assessment in any individualized case seems terribly difficult.

Placebo Effect

The placebo effect has an impact on the power of informed consent by distorting the nature of information available for competent choices.[40] In a recent study Benson and McCallie[41] evaluated five abandoned treatments for angina pectoris which demonstrated the inflated impact of the placebo effect on the assessment of medical technologies. These five treatments followed each other over a period of several years, beginning with the use of xanthines and followed by khellin, vitamin E, ligation of the internal mammary artery, and implantation of that artery. When each of these treatments was introduced into the medical care system in the 1950s and 1960s, the pattern of response was very similar. Initially the efficacy studies (which were not double-blind, random clinical trials) showed a 70 to 90 percent effectiveness. After several years a more skeptical group of investigators conducted another assessment of each treatment; the efficacy was reduced to 30 to 40 percent, which is comparable to the level of the placebo effect itself. The extent to which certain treatment inflates the results may be considerably influenced by its role in creating a placebo effect. Study designs of new technologies must therefore take the impact of the placebo into account to produce meaningful results. Particular care must also be taken because experiments that incorporate a placebo group do raise many ethical questions.[42] In fact Bok[43] argues that administering a placebo without informing the patients or subjects of an experiment is a form of deception that should not be tolerated except under very extraordinary circumstances—for example, if the individual does not want to know whether a placebo or an active agent will be used.

Technology and Information Expansion

In the discussion thus far we have emphasized the difficulties created by technologies in helping patients to arrive at well-reasoned treatment decisions. Three broad concerns have been reviewed—competency of the patient, reliance on heuristic processes or statistical biases, and the impact of the placebo effect. These issues create questions regarding the significance and validity of informed consent as a protective device for patients receiving medical care. Technology expands the degree of available information concerning medical care but initially appears to create a disequilibrium in favor of provider values based on the assumption that clinicians or physicians are in the best position to truly understand the benefits and risks of the technology. In time, however, due to an inevitable expansion of information available to the patient, the power balances in medical decision making will equalize, eventually shifting more in favor of the patient. The introduction of a new technology, such as those that treat angina pectoris, not only expands the availability of information through clinical trials but also permits the evaluation of traditional medical treatments for the same diseases when these are used for comparative purposes in controlled trials.

Well-designed clinical studies are fundamental to the expansion of information regarding new technologies.[44] Some of the studies conducted in recent years and published in medical journals have not been well designed. Two defects in these studies seem particularly conspicuous and deserve attention by researchers. First, many of the studies do not seem to have sufficient "power" to answer the questions that they ostensibly address. In a recent review of approximately 160 cancer investigations, Mosteller, Gilbert, and McPeek[45] discovered that many of these cancer studies lack sufficient sample size and differences in performance of treatment to establish the advantages of one modality over another. The researchers did acknowledge that some studies that rely on various measurements over a period of time may be useful even though the sample size is not large. What is important in their observation is that when taking into account the range of differences in performance of treatment, the sample sizes, and the measurement variability, many of the studies did not result in sufficient power to differentiate successes from failures in the technology.

Along the same line, Freiman and his associates[46,47] reviewed 71 negative trials of random clinical trials to determine the reasons for the rejection of the treatments involved. They also demonstrated insufficient power, for example, 50 of the trials involved could have missed a 50 percent improvement and 67 had a 10 percent chance of missing a 25 percent improvement. Here again the investigators worked with sample sizes that were too small to result in statistically accurate data.

Sackett and Gent[48] address a second major problem that many of these studies demonstrate. They describe difficulties regarding the counting and conceptualizing of the ends to be evaluated. They showed that the ways in which one conceptualizes the criteria for technological effectiveness will influence the validity of the results. For example, one study they describe included only those people who were available for follow-up; essentially ignored in the evaluation were subjects who might have died from the procedures being evaluated or who might have had other problems in the interim period following the application of the technology. Inevitably the results were skewed and may have resulted in characterizing the technology as more effective than it actually was.

Fletcher and Fletcher[49] provided further evidence of the conceptual weaknesses in clinical studies. They studied 600 articles appearing between 1946 and 1976 in the *New England Journal of Medicine*, *JAMA*, and *The Lancet*, demonstrating an increasing frequency of studies with weak research designs. Consequently, the information being obtained from studies in 1976, for example, will not be as substantial for informing patients as previous studies. The validity of the experimental design for studies of new technologies is crucial for obtaining adequate information to inform patients so that they can have a greater role in decisions to accept or reject treatment options.

Extending this one step further, the technology assessment must also include the individualized success rate in the use of the technology in specific health care systems. Linn's study of burn care units and regular care facilities for burned patients illustrates this point. Another interesting study, conducted by Luft and associates,[50] revealed differences in the success rate for certain surgical procedures such as heart surgery, vascular surgery, transurethral prostate resection, and coronary by-pass surgery among various hospitals with varying numbers of procedures performed. For example, they found that hospitals that engage in 200 or more operations of a specific type have a significantly higher success rate than those that engage in less. Therefore, as with the ethical considerations of conducting well-designed clinical studies to determine the efficacy of technology, it is also essential that good studies be conducted to show relative and contrasting success rates in the use of these technologies in various health care systems. Furthermore, patients should be made aware of an institution's success rate in the application of certain technologies, particularly as compared with other institutions. Explication of the risks and benefits of the technology to patients is not sufficient. These studies show that similar evaluations must be conducted across a range of procedures, and that the information on relative risks must be presented to patients making treatment decisions.

Ethical Issues

Information relevant to the psychosocial impact of technology on the patient and the patient's family is important for an informed consent. When a patient enters the health care system, information is provided on the medical effects, benefits, risks, and alternatives of treatment for specific conditions. Little attention, if any, is paid to the long-term effects of these technologies on the patient's quality of life—a matter of major concern to the patient. Studies on the psychosocial impact of hemodialysis, for example, reveal a significantly high increase in suicide rate, serious depression and anxiety, sexual dysfunction, and serious difficulties with relationships, particularly with spouses and other members of the family, among hemodialytic patients. These effects, augmented by the onerous need to spend perhaps six hours a day, three days a week on a hemodialysis machine, greatly affect the patient's psychological state while undergoing continuous treatment.[3,5] Information regarding the psychosocial impact of these technologies is never really explained to the patient when treatment is being considered, and in terms of the quality of life, this information must be provided when treatment decisions are made.

Two remaining issues of ethical concern should be considered. The first pertains to the conflicts of interest for the physician and provider in implementing technological development, and the range of information to be considered by the physician in making decisions about the use of technologies is an important aspect of this issue. For example, a physician may have a patient who might benefit from the use of an existing but highly expensive technology. Although part of the expense can be borne by the patient, much of the cost that goes into the development and implementation of the technology is paid for by the public. In addition, because expensive technologies create a deflection of resources that might be used for the treatment of other disease conditions, questions of distributive justice inevitably must be raised when they are used.

Should the physician be expected to evaluate all of these broad economic and societal concerns when making a decision concerning an individual patient? The physician is placed in a position of conflict between traditional expectations—providing the best possible treatment for the patients—and society's expectations—making decisions concerning underlying socioeconomic issues. Fried[51] would argue that socioeconomic considerations should not enter into the

individual physician-patient decision. Instead he feels strongly that
the emphasis should be on the patient's needs. Fried is adamant that
the physician should not withhold treatments that are available because they might be considered wasteful for a particular patient. The
primary incentive should be what is necessary and best for the patient's condition.

From another perspective, however, having the patient as a primary focus does not excuse the physician and provider from addressing socioeconomic concerns in the developmental stages of new
technologies. The medical profession should be expected to provide
information to policymakers about what technologies can and cannot do and their relative costs and benefits. In addition physicians
and providers should inform patients and consumers of health services about the relative merits of technologies. While the economic
and societal implications of these technologies should not enter into
decisions between physician and patient regarding treatment plans,
physicians and providers in the aggregate have an ethical responsibility to inform the public of cost factors. Eisenberg and Rosoff[52]
argue that that is part of the informed consent requirement. Physicians should also be required to inform patients not only of the costs
of the treatment but the extent of coverage by third-party payers
of various treatments. Proponents of this position feel that economic
information about the technology is as important as the medical and
psychosocial impact and should be incorporated in an informed consent obligation.

Although it can be argued that the primary focus of the physician
should be the care of patients, it would be naive to assume that socioeconomic considerations do not enter, if only subconsciously, into
the physician's decision regarding the treatment of patients. Crane[53]
addresses some aspects of this problem in a study she conducted of
practitioners in four specialties—neurosurgery, internal medicine,
pediatrics, and residents in pediatrics and medicine. She sent questionnaires to over 3,000 of these practitioners to determine what their
triage patterns were in regard to seriously ill patients. Focusing her
study on one small group of terminally ill and essentially non-relating
patients, she discovered that the nature of the psychological and
interpersonal interaction figured prominently in the physicians' decisions to spend less time with these patients and to engage less in the
medical treatment of their condition. Hence, the patient in a withdrawn or terminal state received minimal attention by the physicians
so they could spend more time with those more likely to survive.
This study shows that broader societal concerns do enter into physician decisions—more, perhaps, as a result of social learning than of
an explicit, well-defined decision on the part of the practitioners to
minimize intervention.

Finally, the issue of the patient's responsibilities and relation to the broad societal considerations of technology development and implementation should not be overlooked. This issue concerns the control of individual risk-taking behaviors that ostensibly have no direct effect on others in society. These behaviors may, however, be indirectly responsible for causing patients to become physically disabled and to require highly expensive treatment. The ethical and legal considerations here concern society's right to control behaviors such as cigarette smoking, alcohol consumption, and noncompliance with medication regimens for the purpose of forestalling and even precluding the need for highly expensive technologies in chronic disease states. Patient responsibility in these matters has not been carefully addressed, yet individual patient values ultimately interact with the use of technologies and will have major secondary economic impacts on the entire health care system. Because discussion of societal controls of personal health behavior involves complex legal, psychosocial, ethical, and political issues carefully guarded in our Bill of Rights, articulating the problem is extremely difficult. However, in light of increasing health care costs and the effect of immediate deflections of costs in various areas on treatments available for other conditions, the importance of understanding the impact of individual risk-taking behaviors on others in society is strengthened. Patient values regarding these risk-taking behaviors play a major role in determining which treatment technologies are given priority for development and implementation.

Conclusion

This discussion examined technologies from three broad perspectives: (1) therapeutic or purely medical standards; (2) individual or personal values; and (3) societal concerns. Particular emphasis was placed on the strengths and weaknesses of informed consent as a device for allowing greater patient input into medical care decisions involving diagnosis and treatment, and the complex informational issues entailed in medical decision making were discussed as a barrier to realizing its full potential. The social and economic dimension[54] of biomedical technologies will continue to provide an inescapable backdrop against which physicians and providers make choices and decisions regarding their use.

Endnotes

1. Tancredi, L.R., and Barsky, A.J. (1974). Technology and health care decision making—Conceptualizing the process for societal informed consent. *Medical Care* 12:845–858.

2. Waitzkin, H., and Stoeckle, J.D. (1972). The communication of information about illness. *Advances in Psychosomatic Medicine* 8:180–215.
3. Armstrong, S.H. (1978). Psychological maladjustment in renal dialysis patients. *Psychosomatics* 19:169–173.
4. Levy, N.B. (1977). Psychological studies at Downstate Medical Center of patients on hemodialysis. *Med. Clin. N. Amer.* 61:759–762.
5. Levy, N.D. (1979). Psychological problems of the patient on hemodialysis and their treatment. *Psychother. Psychosom.* 32:260–266.
6. Sullivan, M.F. (1973). The dialysis patient and attitudes toward work. *Psychiatry in Medicine* 4:213–219.
7. Goldberg, R.T., Satow, B.L., and Bigwood, W.A. (1973). Vocational adjustment, work interests, work values, and rehabilitation outlook of women on long-term hemodialysis. *Rehabil. Psychol.* 20:94–101.
8. Meldrum, M., Wolfram, J., and Rubini, M. (1968). The impact of chronic hemodialysis upon the socio-economics of a veteran patient group. *J. Chronic Dis.* 21:37–52.
9. Reichsman, F., and Levy, N.B. (1972). Problems in adaptation to maintenance hemodialysis: A four year study of 25 patients. *Arch. Intern. Med.* 130:859–865.
10. Fine, R.N., Korsch, R.M., Grushkin, C.M., et al. (1970). Hemodialysis in Children. *Am. J. Dis. Child.* 119:498–504.
11. Haffke, E.A., Rajendran, S., and Egan, J.D. (1978). Dialysis, depression and antidepressants. *J. Clin. Psychiatry* 39:759–760.
12. Levy, N.B., and Wynbrandt, G.D. (1975). The quality of life on maintenance hemodialysis. *Lancet* June 14:1328.
13. Abram, H.S., Moore, G.L., and Westervelt, F.A. (1971). Suicidal behavior in chronic dialysis patients. *Amer. J. Psychiatry* 127:1199–1204.
14. Speidel, J., Cock, U., Balck, S., et al. (1979). Problems in interaction between patients undergoing long-term hemodialysis and their partners. *Psychother. Psychosom.* 31:235–242.
15. Finkelstein, F.A., Finkelstein, S.H., and Steele, T.E. (1976). Assessment of marital relationships of hemodialysis patients. *Amer. J. Med. Sci.* 271:21–23.
16. Rollin, B. (1976). *First You Cry.* New York: Signet.
17. The Artificial Heart Assessment Panel. (1973). The totally implantable artificial heart: Economic, ethical, legal, medical, psychiatric, and social implications. (NIH) 74:191, June.
18. Calabresi, G. and Bobbitt, P. (1978). *Tragic Choices.* New York: W.W. Norton and Co.
19. Salvatierra, O., Feduska, N.J., Vincent, F., et al. (1979). Analysis of costs and outcomes of renal transplantation at one center. *JAMA* 241:1469–1472.
20. *In the Matter of Karen Ann Quinlan,* 355 A.2d 647 (1976).
21. Ramsey, P. (1970). *The Patient as Person.* New Haven: Yale University Press, pp. 239–275.
22. Katz, A.A., and Procter, D.M. (1969). Social psychological characteristics of patients receiving hemodialysis treatment for chronic renal failure. Kidney Disease Control Program, U.S. Dept. HEW K. No. PH–108–66–95 (July).

23. *Salgo* v. *Leland Stanford, Jr., University Board of Trustees*, 154 Cal. App. 2d 560, 317 P.2d 1970 (1957).
24. *Natanson* v. *Kline*, 186 Kan. 393, 350 P.2d 1093, opinion on denial of motion for reviewing, 187 Kan. 186, 354 P.2d 670 (1960).
25. *Rogers* v. *Okin*, 478 F. Supp. 1343 (1979).
26. *Rennie* v. *Klein*, 462 F. Supp. 1131 (1979).
27. *Canterbury* v. *Spence*, 464 F.2d 772 (D.C., 1972).
28. *Cobbs* v. *Grant*, 8 Cal. 3rd 229, 104 Cal. RPTR 505, 502 P.2d 1 (1972).
29. See: *Karp* v. *Cooley*, 349 F. Supp. 827 (1972), *aff'd* 493 F.2d 408 (1974); *Ficklin* v. *MacFarlane*, 550 P.2d 1295 Utah (1976); *Bly* v. *Rhoads*, 222 S.E.2d 783 (1976); *Butler* v. *Berkeley*, 213 S.E.2d 571 (1975); *McMullen* v. *Vaughan*, 227 S.E.2d 440 (1976).
30. *Kaimowitz* v. *Michigan Department of Mental Hygiene.* Unreported, Cir. Ct. Wayne County, Michigan (1973).
31. Gilbert, J.P., McPeek, B., and Mosteller, F. (1977). Statistics and ethics in surgery and anesthesia. *Science* 198:684–689.
32. Robinson, G., and Merav, A. (1979). Informed consent: Recall by patients tested postoperatively. *The Annals of Thoracic Surgery* 22:209.
33. Cassilith, B.R., Zupkis, R.V., and Sutton-Smith, K. (1980). Informed consent forms. *N. Engl. J. Med.* 302:896–899.
34. Grundner, T.M. (1980). On the readability of surgical consent forms. *N. Engl. J. Med.* 302:900–902.
35. Olin, G.B., and Olin, H.S. (1975). Informed consent in voluntary mental hospital admissions. *Amer. J. Psychiatry* 132:938–941.
36. Palmer, A.B., and Wohl, J. (1972). Voluntary admission forms: Does a patient know what he is signing? *Hosp. and Comm. Psychiatry* 23:250–252.
37. Tversky, A., and Kahneman, D. (1974). Judgment under uncertainty: Heuristics and biases. *Science* 185:1124–1136.
38. Starr, C., and Whippe, C. (1980). Risks of risk decisions. *Science* 208:1114–1119.
39. Bunker, J.P., and Brown, B.W. (1974). The physician-patient as an informed consumer of surgical services. *N. Engl. J. Med.* 290:1051–1055.
40. Benson, H., and Epstein, N.D. (1975). The placebo effect—A neglected asset in the care of patients. *JAMA* 282:1225–1226.
41. Benson, H., and McCallie, D.P., Jr. (1979). Angina pectoris and the placebo effect. *N. Engl. J. Med.* 300:1424–1429.
42. Simmons, B. (1978). Problems in deceptive medical procedures: An ethical and legal analysis of the administration of placebo. *J. Med. Ethics* 4:175–181.
43. Bok, S. (1974). The ethics of giving placebos. *Scientific American* 321:17–23.
44. Dyar, D.P., Simon, R.M., Friedewald, W.T., Schlesselman, J.J., DeMets, D.L., Ellenberg, J.H., Gall, M.H., and Ware, J.H. (1976). Randomized clinical trials. *N. Engl. J. Med.* 295:74–79.
45. Mosteller, F., Gilbert, J.P., and McPeek, B. Reporting standards and Research strategies for controlled trials. Agenda for the Editor. *Controlled Clinical Trials*, in press.
46. Freiman, J.A., Chalmers, T.C., and Smith, H., Jr. (1978). The importance

of beta, the type two error and sample size in the design and interpretation of the randomized clinical trial. *N. Engl. J. Med.* 299:690–694.

47. Tukey, J.W. (1977). Some thoughts on clinical trials, especially problems of multiplicity. *Science* 198:679–683.

48. Sackett, D.L., and Gent, M. (1979). Controversy in counting and attributing events in clinical trials. *N. Engl. J. Med.* 301:1410–1412. See: Koran, L.M. (1974). The reliability of clinical methods, data, and judgments. *N. Engl. J. Med.* 293:645–646.

49. Fletcher, R.H. and Fletcher, S.W. (1979). Clinical research in general medical journals—A thirty-year perspective. *N. Engl. J. Med.* 301:180–183.

50. Luft, H.S., Bunker, J.P., and Enthoven, A.C. (1979). Should operations be regionalized? *N. Engl. J. Med.* 301:1364–1369.

51. Fried, C. (1975). Rights in health care—Beyond equity and efficiency. *N. Engl. J. Med.* 293:241–244.

52. Eisenberg, J.M. and Rosoff, A.J. (1978). Physical responsibility for the cost of unnecessary medical services. *N. Engl. J. Med.* 299:76–78.

53. Crane, D. (1975). *The Sanctity of Social Life: Physicians' Treatment of Critically Ill Patients.* New York: Russell Sage Foundation.

54. Banta, H.D. and Sanes, J.R. (1978). Assessing the social impacts of medical technologies. *J. Comm. Health* 3:245–258.

INCORPORATION OF PATIENT VALUES IN MEDICAL DECISION MAKING

by Barbara J. McNeil and Stephen G. Pauker

> *. . . in the face of bureaucratic and economic pressures, it may never have been more important than it is now to stress the necessity of a patient-centered ethic; it is perhaps as important to stress other values that physicians ought to serve as well—social, family, and communal values.*[1]

Over the past five years many members of the medical community have become concerned with the need to incorporate patient values into medical decisions and societal values into social and public policy decisions. A number of attempts have been made to use the tools of economics, business, cognitive psychology, and preference theory to derive information about these individual and social values. This review will summarize some of the general issues involved in these areas as they apply to medicine and will briefly indicate how some of the techniques developed outside of medicine have been applied to a few problems in medical decision making.

Three basic questions concern us. First, what is the relative importance to a patient of immediate vs. long-term survival; that is, what are his or her time preferences? Second, how do individuals value varying states of health? Third, how do preferences for survival at varying times interact with preferences for different health states?

Time Preferences and Choice of Therapy

One prototypical study involving an assessment of time preferences focused on the treatment of patients with "operable" lung cancer.[2] Such patients have a choice between two therapies: (1) surgery with a five-year survival rate of 33 percent and a surgical mortality rate ranging from 5 to 20 percent, but averaging 10 percent; and (2) radiation therapy with a five-year survival rate of only 22 percent, but no peri-treatment mortality rate (see Figure 1). For the first two years after treatment, the chance of being alive is higher with radiation therapy than with surgery, but after two years, the reverse is true. Thus the question becomes: How do we determine the relative importance of long- and short-term survival?

The approach used to assess this decision was derived from utility theory and required that individuals respond to a hypothetical situation involving the choice between a fixed period of certain survival or a gamble on longer survival. For example, a 48-year-old man might be asked to consider two options: (1) a 50/50 chance between living out his full life expectancy (25 years) or dying within a month or two; or (2) a guaranteed but intermediate survival of "a" years (Figure 2). The number of years designated by "a" measures the importance of the near term versus the far term for the patient—the smaller "a" is, that is, the smaller the guaranteed survival he would be willing to settle for to avoid the risk of a gamble, the more important is the present compared to the future." Conversely, an increase in the number for "a" indicates the importance of the future compared with the present. These gambles can generate a "utility curve" in which each time/utility pair corresponds to the relative value of living up to, but not beyond, that time (Figure 3). For example, if the patient just described chose five years of guaranteed survival rather than risking near-term death to obtain full life expectancy, considerable value would be placed on the near term. Instead of obtaining 50 percent of the value of living for 50 percent of his life expectancy (0.5×25 years = 12.5 years), he would achieve 50 percent of this value by living five years, a considerably shorter time. Such an individual is considered to be risk averse, that is, one who values the near term more than the far term. This person is in contrast to the risk-neutral individual who values each additional year of life equally and whose utilities vary linearly with survival (see Figure 3, diagonal line).

With data from the utility curve and the varying periods of probable survival after surgery and radiation therapy, an index of expected utility can be derived, and the efficacy of these alternative treatments can be compared. Expected utility is calculated by add-

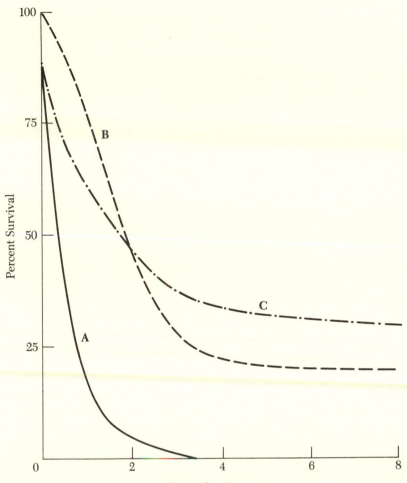

Figure 1 Survival Data for Patients with Lung Cancer. The ordinate shows the cumulative percentage of patients surviving and the abscissa indicates time since diagnosis. Curve A represents survival data for patients with metastatic disease regardless of mode of therapy (2). Curve B represents survival data for patients with presumably operable bronchogenic carcinoma who are treated with radiation therapy. Curve C represents survival data for patients with presumably operable bronchogenic carcinoma who are treated with surgery. This curve includes a 10 percent perioperative mortality rate. *(SOURCE: Reprinted with permission from* Radiology *132:606, 1970).*

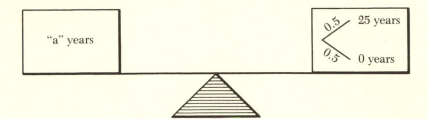

Figure 2 Hypothetical Gamble for Assessment of Utilities. An individual is given a choice between a 50/50 gamble (indicated on the right) between two periods of time and a guaranteed intermediate period of certain survival. The gamble displayed here represents a 50/50 chance of dying shortly (0 years of life) or living out a full life expectancy (25 years). The period of fixed intermediate survival, designated "a" years, is used as a means of assessing the importance of near-term versus far-term years of life for the patient.

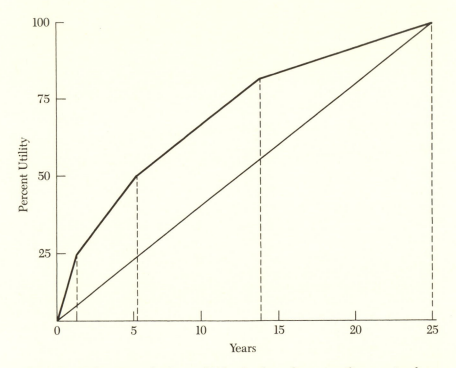

Figure 3 Utility Curve for Years of Life. On the ordinate is utility, ranging from 0 to 100 percent and on the abscissa is years of survival. The light diagonal line indicates the utility structure for risk-neutral individuals. The curved line represents a risk-averse individual. The dotted vertical lines correspond to points obtained by a series of gambles as described in Figure 2. For example, the dotted line representing a utility of 50 percent at five years would result from an answer of "a" = 5 years in Figure 3.

ing the products of utility and probability for all periods of survival up to the maximum, expressed symbolically as:

$$\sum_i U_i \times P_i$$

U_i equals the relative value or utility of living up to, but not beyond, age i (obtained from the patient's utility curve), and P_i equals the probability of living up to, but not beyond, age i (obtained from published survival curves). The best treatment for a particular patient is the one with the highest expected utility. For example, the patient with lung cancer who is faced with a decision between surgery and radiation therapy should select the treatment with the higher expected utility. Because surgery, in effect, trades short-term for long-term survival (that is, increased long-term survival is obtained at the calculated risk of the surgical mortality rate), the patient's attitude toward survival in these different time frames will influence his or her choice of treatment.

Patients with lung cancer were interviewed and their responses were used to calculate the expected utilities for the two treatments. Taking into consideration a 10 percent surgical mortality rate, the results suggested that 21 percent of 60-year-old men and 43 percent of 70-year-old men should have radiation therapy instead of surgery (Table 1). (These figures drop at lower surgical mortality rates and increase at higher rates.) The findings are particularly interesting in that therapeutic choices are not generally made on the basis of expected utility, as indicated here, but on the basis of absolute five-year survival rates. As Figure 1 suggests, radiation therapy would never be chosen over surgery with this criterion of therapeutic efficacy.

In summary, this prototypical investigation was primarily concerned with the choice between two therapies—surgery, which offered an increased chance for prolonged survival but the risk of early death, and radiation, which offered a smaller chance for pro-

Table 1   Optimum Choices of Therapy in Operable Lung Cancer

Age	% Who Should Receive Radiation Therapy Rather than Operation, with Operative Mortality Rates of:			
	5%	*10%*	*15%*	*20%*
60 years	7	21	43	64
70 years	14	43	50	71

SOURCE: Based on data in McNeil, B.J., Weichselbaum, R., and Pauker, S.G., Fallacy of the five-year survival rate in lung cancer, *N. Engl. J. Med.* 299:1397–1401, 1978.

longed survival but little risk of early death. The results indicate that measures of therapeutic efficacy which ignore patient preferences for surviving over the near term compared with over the far term may lead to a suboptimum choice of therapy (see Table 1).

Time Preferences and the Value of Diagnostic Tests

Patients' time preferences should influence not only choices in therapy but also choices in diagnosis, particularly when the decision to perform diagnostic tests might influence the near term survival. A prototypical question might be: Given the sensitivity and specificity of various diagnostic staging examinations in patients with apparently operable lung cancer, which patients should receive these tests?[3] The decision tree shown in Figure 4 offers one approach to this problem.* To simplify matters, we have assumed that surgery is the conventional treatment for operable disease and that few, if any, diagnostic staging tests are done on apparently operable patients ("No Test" strategy, upper half of Figure 4). However, despite surgery, we know that 20 percent of these apparently operable patients have occult metastatic disease from which they will ultimately die. Surgery is therefore ineffective in such patients and exposes them to a 10 percent perioperative risk of death. The possibility thus arises of using preoperative tests to search for occult disease ("test" strategy, lower half of Figure 4). If tests offered infallible information about the presence of such disease, their performance would always be in the patients' best interests. However, false-negative results would lead to unnecessary surgery for some patients, while false-positive results would rule out surgical benefits for other operable patients. Patients' attitudes toward the importance of near-term versus far-term survival are thus important because of these errors and their resulting effects on survival.

Incorporating patient attitudes into an evaluation of the usefulness of preoperative staging tests requires specific information: survival data on patients with lung cancer as a function of both the mode of therapy and the presence or absence of metastatic disease; knowledge about patient attitudes toward survival for varying time periods and the risk of perioperative death; and data on the accuracy of diagnostic tests in staging oncologic disease. Using these data, we

---

* This analysis was predicated on surgery as the primary treatment for operable disease. Theoretically it would be possible in a more complex analysis to relax this assumption and evaluate the role of diagnostic testing for surgery *or* radiation therapy as the primary modality.

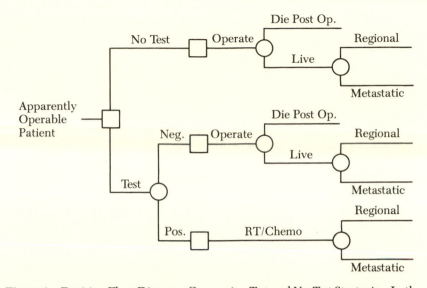

Figure 4 Decision Flow Diagram Comparing Test and No-Test Strategies. In the no-test strategy (upper branch), patients with presumably operable bronchogenic carcinoma undergo surgery; 10 percent die perioperatively. Of the remaining patients, 80 percent have regional disease and are cured, while 20 percent have occult metastatic disease and are not cured. In the test strategy (lower branch), patients with presumably operable bronchogenic carcinoma are examined pre-operatively to identify occult metastatic disease. Those patients with negative tests are treated surgically. Length of survival after operation depends on whether they have regional (true-negative test) or metastatic (false-negative test) disease. Those patients with positive tests are treated palliatively with radiation therapy or chemotherapy. Their length of survival also depends on whether they have regional (false-positive test) or metastatic (true-positive test) disease. (*SOURCE: Reprinted with permission from* Radiology *132:605, 1979*).

determined that preoperative testing for occult disease should never be performed if the goal were to maximize the five-year survival rate or life expectancy (Table 2). However, if near-term survival were an important goal, testing should be performed frequently (Table 3). For example, 47 percent of patients with operable lung cancer should undergo a typical diagnostic staging test with a sensitivity and a specificity of 80 percent (Table 3).

Wilfred Card[4] also analyzed the role that patients' preferences play in determining the value of a diagnostic test. To estimate the relative gain associated with performing gastroscopy he required for patients about to undergo surgery for suspected cancer an assessment of their preferences for survival over the far term versus the near term. Taking into account the survival rates for gastric cancer and benign ulcers, the utilities for life (see Figure 3), and the diagnostic information attainable for gastroscopy, Card concluded that gastroscopy,

Table 2   Evaluation of Preoperative Testing in Patients with "Operable" Lung Cancer: Traditional Objective Approaches

Strategy		Five-Year Survival (%)	Life Expectancy (yrs.)	Optimal Decision
No preoperative testing		32.5	6.42	
Preoperative testing where test has:				
Sensitivity (%)	Specificity (%)			
100	100	32.5	6.43	Test
90	95	32.0	6.36	No test
90	90	31.5	6.29	No test
80	95	32.0	6.35	No test
80	90	31.5	6.29	No test
80	80	30.5	6.15	No test
50	90	31.5	6.28	No test

SOURCE: Reprinted with permission from McNeil, B.J., and Pauker, S.G., The patient's role in assessing the value of diagnostic tests, *Radiology* 132:608, 1979.

Table 3   Evaluation of Preoperative Testing in Patients with "Operable" Lung Cancer: Incorporation of Patient Attitudes

Test Characteristics		Patients Who Should be Tested (%)
Sensitivity (%)	Specificity (%)	
100	100	100
90	95	68
90	90	60
80	95	66
80	90	56
80	80	47
50	90	50

SOURCE: Based on data in McNeil, B.J., and Pauker, S.G., The patient's role in assessing the value of diagnostic tests, *Radiology* 132:609, 1979.

in general, does not result in an increase in expected utility. However, in a derivative analysis, he showed that if a putative ulcer failed to heal after six weeks of medical therapy, gastroscopy did increase expected utility.

The study of testing in patients with lung cancer indicates that patients' attitudes can have a marked influence on the value placed on diagnostic tests. Specifically, failure to consider these attitudes can lead to an underestimation of a test's true value. The study of gastroscopy indicates the importance of considering patients' attitudes, not only in the first stage of a medical decision (that is, operate immediately following the first test) but in all stages of decision making (that is, try one treatment first, then reevaluate).

Thus far this review of patients' preferences has looked only at the relative value of near years versus far years and has thus implied that quality of life is constant. Basically, we have assumed that only survival is important. For lung cancer patients undergoing treatment, this implication is approximately true. In the case of many other diseases and therapies, however, it is not. In fact most therapies are more likely to change quality of life than quantity of life. We must therefore examine patients' preferences for particular states of health at a given point in time and determine the value that is placed on a healthy versus an unhealthy state.

Preferences for Varying States of Health

The first and most commonly used approach to obtaining preferences for alternative states of health involves a time trade-off technique in which an individual is asked to estimate the number of years of ill health he or she would be willing to forego to live in a state of good health.[5,6] This technique assumes that perfect health is assigned a value of 1.00 and death a value of 0.00. Thus, for example, if an ill state of health involved a lifetime of intermittent chest pain, and if a 48-year-old man with a life expectancy of 25 years were willing to have 20 pain-free years rather than 25 years of life with pain, the relative value of a life with chest pain to that individual would be 20/25 or 0.80.

Sackett and Torrance[6] described different states of health under various circumstances and noted a range of values dependent upon three variables: (1) the state itself, (2) the length of time in that state, and (3) the experience of the individual with that state (Table 4). For example, some states of health were valued considerably lower than others (for example, depression vs. tuberculosis). States lasting over a long period of time were disvalued greatly. For example, a life involving only three months of hospital dialysis was rated at about 0.62 in contrast to a lifetime of dialysis at 0.32. Finally, ill states actually experienced by the individual questioned were not disvalued as highly as those by individuals asked to imagine such a state: for example, hospital renal dialysis was valued as 0.52 by patients compared to 0.32 by the general population.

Another approach to patient state of health preference was originally suggested by Schelling[7] and later used by Card and others. An individual is asked to specify the maximum surgical mortality he or

Table 4   Mean Utilities for Several States of Health for a General Population and Dialysis Patients

State of Health	Utility
Perfect health	1.00
Tuberculosis	0.68
Mastectomy for breast cancer	0.48
Depression for 3 months	0.44
Home dialysis for life:	
by general population	0.39
by dialysis patients	0.56
Hospital dialysis for life:	
by general population	0.32
by dialysis patients	0.52
Death	0.00

she would be willing to accept to avoid a particular state of ill health. Again, with this technique perfect health is assigned a relative value of 1.00, death a value of 0.00, and the value of the intermediate states is calculated from $1.00 \times (1 - \text{maximum operative mortality})$. For example, if an individual were willing to undergo a 20 percent surgical risk to avoid a life of chest pain, the value of a life with chest pain would be $1 \times (1 - 0.20) = 0.80$. Using a slight variation of this technique, Card and his colleagues[8] assessed the importance of various levels of visual acuity to different subjects (Figure 5). These levels were described in terms of the occupations and recreations possible with them, and then a series of gambles was presented to determine the relative value of one level of vision compared to another. These gambles were based on the premise that perfect acuity had a utility of 1.00 and no acuity (blindness) had a value of 0.00. In another study, Card[4] determined that for a series of volunteer physicians and other medical personnel, the average utility for complete blindness was about 0.80, indicating that a linear change in the ordinate of Figure 5 is probably necessary.

A variation of the gamble technique has been used to evaluate different consequences of a pregnancy and different results from plastic surgery for cleft palate. In the first case, Pauker and his colleagues[9] used this approach to estimate the consequences to a parent of having a deformed fetus (a child with Down's Syndrome, for example), an elective abortion to avoid delivering such a fetus, or an accidental miscarriage incurred during testing to discover the presence or absence of a deformed fetus. Knowledge of the relative values placed on these outcomes has been used to determine whether prospective parents should undergo prenatal testing. This approach is discussed

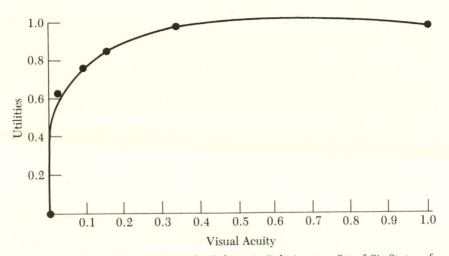

Figure 5 The Utility Function of a Subject in Relation to a Set of Six States of Visual Acuity. On the ordinate is the utility and on the abscissa is the visual acuity. Data from one patient are recorded. [SOURCE: *Reprinted with permission from "Outcome of Severe Damage to the Central Nervous System," CIBA Foundation Symposium 34 (New Series).*]

elsewhere in this volume by Pauker and his colleagues, as well as by Fletcher and Swint.

In the second case, Krischer[10] assessed attitudes toward several sequelae associated with surgery for cleft palate. His technique involved a choice between a gamble that offered either a state worse than the presurgical state (p) or the optimum result possible ($1 - p$), and a guaranteed result of intermediate quality. The sequelae included monetary costs for surgery, speech quality, and cosmetic appearance. He found significantly different attitudes among clinicians in different specialities and among different family members.

Both the time trade-off and surgical mortality techniques described above have several drawbacks that limit their ability to assess the value of various health states. The time trade-off technique may overestimate disutility because it asks about shortening life in the very distant future (such as at the end of one's normal life expectancy). For example, an individual with a life expectancy of 25 years may regard as trivial the loss of five years at the end; hence, a utility resulting from that estimate (that is, $20/25 = 0.8$) may be too low. Questions that ask about shortening life at different points (some involving only brief periods of survival) may be a more accurate means of assessing various health states.

The surgical mortality technique errs in the reverse direction. Uneasiness about the immediate risk of death may cause the respondent to underestimate disutility. Moreover, this technique runs into operational difficulties when a gamble to assess a particular state of health is either medically unfeasible or too hypothetical to be treated seriously by the respondent.

Despite such difficulties two potential uses for these two techniques are immediately obvious, assuming that the information provided is reasonably accurate. First, such information can be coupled with time preference data to obtain an integrated evaluation of treatments that affect both the quality and quantity of life. Second, these techniques can indicate which states of health are least acceptable to the general public and can suggest areas for additional research or for the development of alternative forms of health care delivery. For example, the findings of Sackett and Torrance[6] might suggest that if women who have had mastectomies value their life at only 48 percent of normal life, greater thought should be given to alternative nonsurgical forms of therapy for breast cancer.

Integration of Time and State-of-Health Preferences

For a particular patient the quantity and quality of life might have to be weighted before a choice can be made between alternative therapies for the same disease. For society in general, the same weighting process may be important to develop a rank order of benefits achieved from various health interventions. For example, many benefits offered by all medical interventions or within interventions for a particular disease process may be ranked. Such an ordering would aid in an optimum allocation of health care resources and would require the integration of time and health-state preferences—no easy matter in many areas of medicine, where quality of life and length of life are differentially affected by treatments. For example, surgery for coronary artery disease may or may not prolong life, depending upon its associated operative mortality rate; under appropriate circumstances, however, this surgery nearly always improves quality of life.

Two approaches to the integration of time and state-of-health preferences have been suggested. Neither has had extensive clinical application nor verification and both suffer from operational difficulties. The first approach was originally suggested by Zeckhauser and Bush and their colleagues[11,12] and involves the concept of "quality adjusted life years" (QALYS). In brief, this index assumes that an individual is risk-neutral in preferences for time.[13] It also assumes that an individual, regardless of life expectancy, will always accept the same proportionate decrease in healthy years to avoid an unhealthy state. Thus, for example, if a patient were willing to settle for 20 healthy years rather than 25 unhealthy years, we could expect to see a choice of 12 healthy years instead of 15 unhealthy years, 8 healthy years instead of 10 unhealthy years, 3.2 healthy years instead of 4 unhealthy years, and so forth. The major limitations of this index are two-fold: first, older patients are not likely to be risk-neutral for years of life; and second, most individuals are not likely to exhibit a constant proportional trade-off. In fact, their trade-off values for short periods of survival may be very different from those for longer periods.

An alternative approach, although cumbersome, eliminates the difficulties associated with the use of QALYS. Basically, this technique involves obtaining preferences over time by the lottery technique (as described in Figure 3). Then, at sequential points in time along that curve, a series of time trade-off questions are asked that generate a series of new points, thus creating a new utility curve for incorporating quality of life.[14] This curve is displaced downward from the curve describing time preferences only.

Summary

Patient and societal involvement in decisions that affect patient welfare is important although the means for obtaining such involvement are unclear. The early investigations summarized have applied techniques from other disciplines to medical decisions that involve actual patients. In the case of therapeutic decisions, the investigations have led to proscriptive results—what the patient *should* want, given his values or preferences. The results, therefore, suggest the better of two alternative actions. Other techniques of direct (contrasted to proscriptive) assessment of preferences for alternative actions have not yet been applied extensively in medicine. These would be based on recent work in cognitive psychology and would ask patients to choose explicitly among various options *after* they had been given certain abbreviated information about such options. Research involving both proscriptive and direct techniques is necessary to improve future studies involving patient preferences. (Other contributions in this volume address the need for appropriate techniques to determine the relative value of societal programs.) Such efforts would provide more effective health care on an individual level while simultaneously helping to establish priorities on a societal level.

Endnotes

1. Callahan, D. (1980). Shattuck Lecture: Contemporary biomedical ethics. *N. Engl. J. Med.* 302:1228–1233.
2. McNeil, B.J., Weichselbaum, R., and Pauker, S.G. (1978). Fallacy of the five-year survival rate in lung cancer. *N. Engl. J. Med.* 299:1397–1401.
3. McNeil, B.J., and Pauker, S.G. (1979). The patient's role in assessing the value of diagnostic tests. *Radiology* 132:605–610.
4. Card, W.I. (1980). Rational justification of therapeutic decisions. *Meta-medicine* 1:11–28.
5. Torrance, G.W. (1976). Social preference for health states: An empirical evaluation of three measurement techniques. *Socio-Econ. Plan Sci.* 10: 129–136.
6. Sackett, D.L., and Torrance, G.W. (1978). The utility of different health states as perceived by the general public. *J. Chronic Dis.* 31:697–704.
7. Schelling, T.C. (1966). The life you save may be your own. In *Problems in Public Expenditure Analysis*, ed., S.B. Chase, Jr., pp. 127–176. Washington, D.C.: The Brookings Institution.
8. Card, W.I. (1976). Estimation of the utilities of states of health with different visual acuities using a wagering technique. In *Decision Making and Medical Care*, eds., F.T. deDombal and F. Gremy, pp. 239–247. Amsterdam: North Holland Publishing Company.
9. Pauker, S.G., and Pauker, S.P. (1977). Prenatal diagnosis: A directive approach to genetic counseling using decision analysis. *Yale J. Biol. Med.* 50:275–289.
10. Krischer, J.P. (1976). Utility structure of a medical decision-making problem. *Operations Research* 24:951–972.
11. Zeckhauser, R.J., and Shepard, D.S. (1976). What's new for saving lives? *Law and Contemporary Problems* 40:5–45.
12. Chen, M.M., and Bush, J.W. (1976). Maximizing health system output with political and administrative constraints. *Inquiry* XIII:215–227.
13. Pliskin, J.S., Shepard, D.S., and Weinstein, M.C. (1980). Utility functions for life years and health status. *Operations Research* 28:206–224.
14. McNeil, B.J., Weichselbaum, R., and Pauker, S.G. (1981). Speech and survival: Tradeoffs between quality and quantity of life in laryngeal cancer, forthcoming.

Part Five

MEDICAL TECHNOLOGY: COSTS AND PUBLIC POLICY

THE ROLE OF TECHNOLOGY ASSESSMENT IN COST CONTROL

by Louise B. Russell

"Technology assessment" is a relatively new concept, and its application to the field of medicine is even more recent, a consequence of the growing concern over medical care expenditures. In general, the term refers to an evaluation of the costs and effects of a medical technology. The traditional clinical trial is at the heart of such an evaluation since the most important questions about any technology are: Does it work and, if so, how well? But technology assessment goes well beyond clinical trials to include estimates of the cost of providing the technology to individuals and to society as a whole. An assessment may even attempt to describe how the technology affects people other than the patient, and its emotional and philosophical consequences beyond the usual medically defined outcomes.

Many people hope that technology assessment can help to control the growth in expenditures that continues to strain the budgets of individuals, their employers, and their governments. In this chapter I will discuss the role of technology assessment in cost control, and will particularly emphasize its uses and limitations in this role.

The Growth of Medical Care Expenditures

I will begin by describing the nature of the cost problem in medical care from the perspective of hospital costs. The problem has been most acute in this area. In addition, I believe that the principles that apply here correctly describe the trends and tendencies in medical care generally.

The growth in hospital expenditures over the last several decades is often attributed incorrectly to inflation. Although hospital care is

I am grateful to Barbara McNeil for her comments on an earlier draft of this paper. The opinions expressed in the paper are my own and should not be attributed to the officers, trustees, or other staff members of the Brookings Institution.

subject to the same inflation as the rest of the economy, that is not the reason it has claimed a growing share of the gross national product—2.0 percent in 1965 and 3.6 percent in 1978.[1] (Over the same period, total spending on medical care rose from 5.9 percent of the gross national product to 9.1 percent.) Instead, most of the growth in costs over and above the effects of general inflation has been due to the increasing amounts of resources used by hospitals. Some of the new resources have provided more days of care, especially for the elderly, and more outpatient visits. But most have been used to provide more and better services in the course of a hospital day— such as intensive care, open-heart surgery, respiratory therapy procedures, CT scans, and the like—all of which require new staff, space, and equipment.

Since 1965 the cost of a day in the hospital has risen, on average, more than 12 percent per year. About half of this growth has been due to inflation—higher prices and wages. The other half, or about 6 percent per year on average, has been due to increases in the resources used per day.[2] The accumulated growth since 1965 means that the level of resources used per hospital day in 1980 is double the level of 1965.

These facts provide the dimensions of the growth in resources, but they do not explain why resources, and thus costs, have grown so much, or why—until the industry imposed voluntary restraints in an attempt to avoid federal cost legislation—that growth showed no signs of slowing down. In the simplest and strongest terms, the answer to these questions is that society designed the system to work this way using the mechanism of third-party payment. With the growth of private insurance and the introduction of Medicare and Medicaid, third parties now pay more than 90 percent of all hospital costs.[3] The principle behind third-party payment is that people should not have to worry about cost; when they need care they should be able to get it.

Need for a service does not, however, mean that the benefit is greater than the cost. It means that the benefit—on average—is greater than zero. When need is the sole criterion, the number of worthwhile services that can be provided in medical care is virtually limitless. The rapid growth in costs attests to this fact.

In terms of specific technologies, the goal of meeting needs requires the investment of large amounts of resources to provide services that bring smaller and smaller benefits. The benefits may be small for each person (relief from a cold, for example), or they may be small because the procedure is seldom successful (emergency surgery that saves only one person in 100). As long as a need remains unfilled, the system will try to fill it, no matter how small or costly the gain.

Dialysis for people with end-stage renal disease is a particularly good example because so much of its history is a matter of public record and the issues are not obscured by arguments about its effectiveness.[4] Dialysis first became possible for people with chronic kidney failure in the early 1960s. The pressure to make it widely available developed quickly. An article published in 1968[5] estimated that, at that time, 1,000 patients were receiving dialysis but that another 6,000 died for lack of treatment. These numbers imply an estimate of about 35 new patients per million population per year.

Financing for dialysis continued to expand, and in 1973 Medicare assumed responsibility for the costs for most patients. When Congress reviewed the new program in 1977, the most frequent estimate it was given was that new patients would soon level off at 60 per million population per year. The enormous difference in the two estimates, made only ten years apart, stems from changes in the criteria for selecting patients. In the late 1960s, when there were not enough facilities to treat all patients with end-stage renal disease, priority for treatment was frequently given to those who would gain the most in length and quality of life—young or middle-aged adults who had no serious disease other than kidney failure. Younger and older people, and people with other serious diseases, do not benefit as much from dialysis, but they do benefit. Today they too receive dialysis, and the new estimate is based on this broader group of patients.

Congress was also concerned about the shift from home to center dialysis. In 1972, 40 percent of dialysis patients dialyzed at home; in 1977, fewer than 20 percent did. The benefit of center dialysis is its convenience for the patient's family, who may find it difficult to work, raise a family, or go to school if they must also help with the time-consuming dialysis treatments. But this benefit is costly: Medicare estimates that the first year of home dialysis, when the patient is trained in the procedure, costs $21,000, and subsequent years cost about $12,000, compared to $22,000 for every year of center dialysis.[6]

The pattern appears repeatedly. Intensive care accounts for close to 20 percent of all hospital costs and continues to grow.[7] Yet many studies find only modest benefits, such as fewer complications, and studies that concentrate on only one measure of outcome—mortality rates—often find no benefit at all.[8] Open-heart surgery is performed for some patients with no overt symptoms, although it has yet to be shown that surgery prevents heart attacks or prolongs life in these cases.[9] Expensive new forms of radiotherapy are introduced in the hope of modest gains in survival rates.

I am not trying to argue that these technologies are worthless. This is true in some cases, but in many others future research may prove their value in uses that are now controversial. The point is that the costs are high, while the benefits are modest at best. Again, the crux of the cost problem is that the opportunities for investments of this sort are effectively unlimited in modern medicine. The decision to limit the further growth of costs is thus implicitly a decision that, even though the resources can be put to beneficial use in medical care, their use for other purposes is valued more highly.

The caring side of medicine offers the same unlimited scope for spending that the curing side does. Nursing home care is an excellent example. Expenditures on nursing homes are still much smaller than hospital expenditures, but they are growing more rapidly.[10] The number of nursing home residents per 1,000 people 65 years of age or older doubled between 1963 and 1973,[11] and nursing home payments are the largest single component in the costs of the Medicaid program (almost 40 percent).

The United States is not alone in its predicament. In 1975 the United States spent 8.4 percent of its gross national product on medical care; West Germany spent 9.7 percent; Sweden spent 8.7 percent and the Netherlands 8.6 percent; France spent 8.1 percent.[12] The problems are international and other countries are considering many of the same proposals as possible solutions.

The United Kingdom, at 5.6 percent of GNP, illustrates the problems that arise when the problem of costs is solved—deciding which of many worthwhile services should be provided, and which should not. It is not hard to find evidence of the difficult decisions that have been made. Per million population, fewer people receive dialysis in Britain than in the United States or a number of European countries; this fact has been the subject of recurring debate in that country and has stimulated an unusual degree of intervention by the national government in an attempt to get more of the medical budget allocated to dialysis.[13] The waiting list for hospital care is long—500,000 in a country of 50 million people; and people with low-priority conditions, such as those waiting for a hip replacement, often turn to the private sector for treatment.[14]

Alternative Objectives and Cost Control

Given this description of the cost problem, how can technology assessment help to control costs? The answer depends on which of two objectives the nation chooses for the medical care sector. The first possibility is to continue, at least temporarily, to try to provide all medical care of benefit, that is, all services that are needed. In this case, technology assessment would be restricted to helping identify inefficient or valueless procedures. The second possibility is to decide not to let costs grow as rapidly as in the past and to limit the resources available for medical care to less than would be required if need alone were the guiding principle. In this second case, once the limits were set, technology assessment could help society and individual practitioners choose from among the many worthwhile possibilities.

Dialysis illustrates the difference between the first and second objectives. If need were the sole criterion, dialysis would be provided to the same people in much the same way as it is now. But if competing needs had to be weighed against the limited resources made available for medical care, the decision might be to restrict the use of dialysis, perhaps returning part-way to the sort of guidelines often used in the United States during the 1960s and currently applied in the United Kingdom.

I would like to discuss the role of technology assessment in each of these situations in more detail. Technology assessments provide certain kinds of information, information which is subject to various technical and philosophical limitations. The case study sections in this volume will try to explore these matters more fully. But some limitations are more important for one objective than the other.

Consider the first objective—providing services based solely on need. Under this objective, the only question that matters is whether the technology works. If the answer is yes, for certain conditions, then the technology will be adopted and used for those conditions. Any technology that is not totally valueless or actively harmful will be adopted. Clinical trials are clearly the critical and perhaps the only necessary part of the assessment in this case. Information about costs is secondary, even tangential. In those relatively few cases in which several procedures produce the same result, cost information is needed to determine which one does the job most cheaply. In all other cases costs are useful only as an indicator of the expenditures necessary to provide the service.

It is not easy to prove that something is without value, or that it produces exactly the same results as some more expensive alternative. The problem is underlined in a study by Freiman and her associates that reviewed 71 "negative" clinical trials.[15] Each trial tested a new therapy against an alternative and concluded that the new therapy was not an improvement. After reviewing the statistical basis for each trial the authors concluded: ". . . 67 of the trials had a greater than 10 percent risk of missing a true 25 percent therapeutic improvement, and with the same risk, 50 of the trials could have missed a 50 percent improvement." In other words, substantial benefits might have been missed. The authors argued that most of the 71 therapies had not been given a fair trial.

The difficulty was that, in most of the trials, the sample was rather small. It therefore ran a substantial risk of being dominated by observations from the "no benefit" end of a distribution that, in fact, had an average benefit quite different from zero. Each of the trials could have produced firmer, clearer results had it been based on a larger sample. Undoubtedly the results would also have been stronger if the trials had covered the widest possible range of outcomes—mortality, morbidity, disability, emotional effects, and so on—and if they had controlled for all relevant variations among patients, practitioners, and treatment settings. Studies that include more than one institution need to be particularly careful about controlling for these differences because they are likely to be greater among institutions than within the same institution.

In sum, the trials could have been more rigorously conducted. However, large samples, carefully tested protocols, multicenter trials, and longitudinal surveys are more expensive than an informal trial that uses the few patients available in a single center. If technology assessment is to help control costs, it is fair to ask under what conditions it is likely to save more than it costs. And with respect to the costs of medical care itself, it is important to recognize that, when the objective is to fill all needs, technology assessment may succeed in eliminating some procedures without affecting the growth of total costs. Technology assessment might help us spend the money a little better, but without helping us spend any less.

If we are serious about controlling medical care costs, we must adopt the second objective and limit the claim of the medical care sector on the economy's resources. Implicitly or explicitly, we must then accept the principle that not all needs will be met and that there is some level of benefit, or of benefit relative to cost, below which the value of the service is not sufficient to justify the resources required to provide it. This need not be a formal determination. It can be the result of a series of informal, intuitive judgments in answer to the question: Is this activity worth the cost?

If this second objective is adopted, practitioners and patients must decide which services make the best use of the resources available. Technology assessment provides information that can help in deciding what should be done and for whom. For example, an assessment of kidney dialysis would indicate which people receive the greatest benefit from dialysis in terms of length and quality of life, and the costs. As another example, an early cost-effectiveness study[5] compared the two major methods for treating patients with end-stage renal disease—dialysis and transplantation—and concluded that transplants were preferable whenever they were possible. Assessments such as these offer one basis for allocating limited resources among competing uses and for setting treatment priorities.

Technology assessment can supply the information for these decisions, but it cannot make them. They depend unavoidably on human judgments about the value of achieving certain results. For example, assessment can show that intensive care is more costly than care in regular wards, and that for stroke patients in the early 1970s, intensive care reduced complications but did not affect mortality rates (Drake, et al. in Endnote 8). Another assessment can indicate that screening for phenylketonuria shortly after birth costs $9,000 per case found, and that rescreening a few weeks later, to pick up cases that were not clearly developed at birth or that were simply missed, costs $260,000 per case found.[16] Further detail can clarify the magnitude and nature of the costs and benefits. But technical information alone will not tell us whether we value the results enough to commit some of our limited resources to provide those services.

The decisions are more difficult than simply deciding when a benefit is large enough, because choices must be made between services that help the same people in different ways, or, still more difficult, that help different people in different ways. The value of kidney dialysis for the elderly must be weighed against neonatal intensive care; hip replacements must be weighed against open-heart surgery. These problems do not arise when the objective is simply to meet all needs, which is part of what makes need such an appealing objective. Here again technology assessment offers very useful, albeit imperfect, information about the costs and benefits of each kind of care for patients with different initial conditions, but it does not automatically produce the answer at the end.

If medical services must meet certain standards before they are provided, then similar standards must apply to technology assessment. It is not enough that the value of assessment be positive, on average. Like the technologies themselves, technology assessment must produce a benefit that is "large enough" relative to its costs. The funds available should be directed to those areas where more information is likely to produce the greatest improvements in the allocation of resources, and where good decisions are most difficult to make based only on the information generated by the experience of individual practitioners.

Conclusion

At the present time the United States seems to be in a state of transition between the first and second objectives. Particularly since the 1960s we have tried to pursue the first objective—providing care whenever it is needed, without regard to the size or cost of the benefit. We came closest to the ideal in hospital care. But third-party payment has not eliminated the economic problem of unlimited wants and limited means, it has only recast the problem in more aggregate terms. As individual patients and doctors, freed from the restraint of costs, have pursued the many worthwhile developments in medical care, costs have risen rapidly, and the claim on the nation's resources has become increasingly difficult to meet.

No change in policy has been announced. Public statements still maintain that the goal in medical care is to provide all the care that is needed, but the programs for cost control passed in the early 1970s and the programs that have been proposed more recently are quite different. Professional Standards Review Organizations (PSROs), created by the Social Security Amendments of 1972, are supposed to eliminate *unnecessary* hospital admissions and *unnecessarily* long hospital stays for Medicare and Medicaid patients, but not care that is beneficial to the patient. Similarly, Certificate of Need (CON) programs, first passed at the initiative of individual states and now required for all states by the National Health Planning Law of 1974, are charged with preventing *unnecessary* investment and duplication of services. These two mechanisms require careful investigation of individual circumstances before a decision is made regarding the need for an admission, a longer stay, or a new facility.

The more recent proposals—the Carter Administration's hospital cost containment bill, which is included in the Administration's National Health Plan as well, and similar but broader proposals in Senator Edward Kennedy's national health insurance bill—do not include any provision for linking cost control to judgments about the need for the service. The Carter Administration's bill would limit the growth in a hospital's budget to inflation plus or minus an "efficiency" factor based on the hospital's routine costs compared to those of similar hospitals. The Kennedy bill would limit the rate of growth in expenditures for covered services to the rate of growth in the gross national product (averaged over three years), effectively fixing their share of GNP. Neither proposal recognizes the possibilities for further spending or the benefits that might result from those possibilities, and both would cap the rate of growth in costs at levels well below those of the last fifteen years.

Although their proponents do not say so, these proposals implicitly make a judgment that the costs of new services have outweighed their benefits in recent years, and that a much lower rate of growth is to be preferred. No matter whether one of these programs, or some other, is finally adopted, cost control requires this kind of judgment. The benefits of medical services must continually be weighed against the costs. By formal or informal means, people at all levels in the system must frequently ask whether the benefits of a particular activity justify the costs.

Technology assessment can help in the decision to move from the first objective to the second by making clear the small benefits of some services that are provided now, the high costs, or, in some cases, both. When choices between services must be made, assessment can indicate the costs and benefits for different groups of patients and can suggest the criteria for choosing among patients and services. But technology assessment is not a magic solution to the cost problem. In some cases, the technical and ethical obstacles to assessments will mean that decisions must be based on hunch and intuition, at least temporarily. Furthermore, technology assessment costs money and it does not make economic sense to assess every aspect of every technology, no matter what the expense. In all cases, technology assessment can do no more than supply pertinent information. The nation as a whole, individual practitioners, and patients, must make the final decisions.

Endnotes

1. Gibson, R.M. (1979). National health expenditures, 1978. *Health Care Financing Review*, vol. 1, no. 1 (Summer), Tables 1 and 3. Cooper, B.S., Worthington, N.L., and McGee, M. (1976). *Compendium of National Health Expenditures Data*, Social Security Administration, DHEW Pub. No. (SSA) 76–11927, January, Tables 1 and 6.
2. Feldstein, M., and Taylor, A. (1977). *The Rapid Rise of Hospital Costs*. Council on Wage and Price Stability, Staff Report, January, especially Table 7.
3. Gibson, *op. cit.*, Table 4.
4. Except as noted, the material in the next three paragraphs is from Russell, L.B. (1979). *Technology in Hospitals: Medical Advances and Their Diffusion*. Washington, D.C.: Brookings Institution.
5. Klarman, H.E., Francis, J., and Rosenthal, G.D. (1968). Cost-effectiveness analysis applied to the treatment of chronic renal disease. *Med. Care* 6: 48–54.
6. End-Stage Renal Disease Program, "Quarterly Statistical Summary: Quarters ending 12/31/78 and 3/31/79."
7. This estimate is based on more recent data than the estimate presented in Russell, *Technology in Hospitals*. It uses the technique discussed there (p. 48), and data on beds in mixed, cardiac, and neonatal intensive care units from *Hospital Statistics* (1979 ed.). Chicago: American Hospital Association, Table 12a.
8. See, for example: Drake, W.E., Jr., Hamilton, M.J., Carlsson, M., and Blumenkrantz, J. (1973). Acute stroke management and patient outcome: The value of neurovascular care units. *Stroke* 4:933–45. Griner, P.F. (1972). Treatment of acute pulmonary edema: Conventional or intensive care? *Ann. Intern. Med.* 77:501–06. Astvad, K., Fabricius-Bjerre, N., Kjaerulff, J., and Lindholm, J. (1974). Mortality from acute myocardial infarction before and after establishment of a coronary care unit. *Br. Med. J.* 1:567–569. Mather, H.G., Morgan, D.C., Pearson, N.G., Read, K.L.O., Shaw, D.B.,

Steed, G.R., Thorne, M.G., Lawrence, C.J., and Riley, I.S. (1976). Myocardial infarction: A comparison between home and hospital care for patients. *Br. Med. J.* 1:925–929.

9. Meyer, L. "Heart Operation Called Questionable," *Washington Post*, 26 July, 1977. Brody, J.E. "Personal Health," *The New York Times*, 8 February, 1978. Brody reports that patients without overt symptoms are usually discovered "through a routine physical exam that includes an exercise stress test."

10. Gibson, *op. cit.*, Tables 3 and 5.

11. Data on residents from: National Center for Health Statistics, *Characteristics of Residents in Institutions for the Aged and Chronically Ill: United States—April–June 1963*, Series 12, No. 2, Table 5; and *Utilization of Nursing Homes, United States: National Nursing Home Survey, August 1973–April 1974*, Series 13, No. 28, Table 1.

12. Simanis, J.G., and Coleman, J.R., Health Expenditures in nine industrialized countries, 1960–76, *Social Security Bulletin*, vol. 43, no. 1 (January 1980), pages 3–8.

13. Stocking, B. (1980). The management of medical technology in the United Kingdom. In *The Implications of Cost-Effectiveness Analysis of Medical Technology, Background Paper #4: The Management of Health Care Technology in Ten Countries*, Office of Technology Assessment, U.S. Congress. *Renal Failure: A Priority in Health?* Office of Health Economics, London, April 1978.

14. See, for example: Waiting-Lists Lengthen. *Lancet* 1 (January 15, 1977), reviewed in *Med. Care Review* 34:188–190. Lister, J. (1978). By the London Post. *N. Engl. J. Med.* 299:1454–1455.

15. Freiman, J.A., Chalmers, J.C., Smith, H., Jr., and Kuebler, R.R. (1978). The importance of beta, the type II error and sample size in the design and interpretation of the randomized control trial: Survey of 71 "negative" trials. *N. Engl. J. Med.* 299:690–694.

16. Sepe, S.J., Levy, H.L., and Mount, F.W. (1979). An evaluation of routine follow-up blood screening of infants for phenylketonuria. *N. Engl. J. Med.* 300:606–609.

INFLUENCES OF REIMBURSEMENT POLICIES ON TECHNOLOGY

by Robert A. Derzon

Reimbursement and payment policies to finance personal health care services have joined other major forces in accelerating technological innovation, application, and diffusion in American medicine. Fee-for-service payment and cost-based institutional reimbursement are imperfect mechanisms. However, the propensity for technological development and transfer is so great that the elimination or improvement of these finance methods may retard but would not eliminate much of the current technology now accepted and desired by consumers and providers.

In my discussion two indisputable facts are evident: (1) the American medical care system, for better or worse, has become massively dependent upon the discoveries and advancements in science; and (2) few "free lunches" are available to the American medical care consumer. We, the American medical care system, have deployed an uncomplicated, unimaginative, and often indiscriminate "pay as we go" program to bring these two truths together and, in the process, to launch what many believe is the leading growth industry of the 20th century. The manner in which these two aspects of the American health scene intersect, mutually support, and influence each other is the subject of this chapter.

Policies of health financing have not been the singular driving force in technology development, use, and diffusion; but the impact of reimbursement and payment practices as a key factor in shaping the health system and the behavior of its components should not be ignored. The following comments are intended to present a balance between the technological imperatives in health and the role of financing. The paper is organized to:

1. Describe those technological imperatives.

2. Illustrate certain critical characteristics of our current reimbursement practices.
3. Explain how these characteristics have influenced medical technology adaptation.
4. Assess, within limits, the influence of these payment policies on technology.
5. Suggest alternative strategies that could weaken the demand for ever-increasing dependence upon technology.

The Technological Imperative

Technological emphasis on medical care in the United States is not the result of happenstance; rather, it is the national extension of producers' and consumers' efforts to harness science and technology—in this case for greater certainty of diagnosis and more effective therapeutic interventions. As Americans, we have been disposed toward technology in every facet of our lives and in our livelihoods. To expect the discipline of medicine to be immune from this cultural legacy is to disregard totally the mores of a sophisticated society.

Our national priorities have reflected our infatuation with technology. The constant support of a powerful bioscientific research enterprise is one example. Strong state and federal subsidizing of research grants and educational programs in the health field is another. Proliferating subspecialization in medicine is the sign of another need to fractionate further the expanding knowledge base of medicine. Patent protection and tax incentives are other means of encouraging the development of technologies in all fields.

The mass media has been a force in alerting the American public to new discoveries, often in advance of their appearance in professional journals and invariably ahead of the results of controlled clinical studies and proven effectiveness. Public demand, when coupled with the anxieties of illness or discomfort, spur the providers to active responses.

Other strong influences on technological development have been competition among physicians, prestigious faculties of aggressive universities and their hospitals, and community hospitals for the affections of their physicians. Local pride, voluntarism in health care, and the marketing practices of hospitals are high on the list of technological inducers.

The American jurisprudence system and its approach to liability has been another important force in the technological imperative. Tort law has been a direct stimulus to defensive medicine and the unfortunate trade off between legal security and cautious, conservative, moderate interventionist practice. In the United States a preoccupation with safety, quality, and self-regulation has been the forerunner of peer review and medical audit with their apparent zest for measurable physical and process standards, many of which seem to have technological underpinnings.

These examples are cited to make a simple case. The environment in health care is rife with powerful forces that spur our propensity for a type of scientific medicine highly reliant on technology. If health insurance dollars had never been circulated, there is still strong reason to believe we would have been presented with too many coronary care units, intensive care beds, and overused fetal monitoring. However, in the above insurance coverage the differences in the utilization of these technologies would be marked. Without a doubt our financing system has been, and is, a complementary factor in the rampant introduction and use of new technologies in medicine.

Certain characteristics of our present payment system do, in fact, operate to reinforce technology use and change. Once understood, features of the reimbursement system could be modified to induce countervailing pressures on the adoption and use of new and existing technologies.

The Special Character of Our Payment and Reimbursement System

In spite of the various generic inducers of medical technology, reimbursement of hospitals and payment to physicians has also promoted and reinforced the allure of machine-age medicine. Several features of the American health financing payment system need to be highlighted. An understanding of the following payment and prepayment characteristics is an essential precursor to that analysis:

1. Payment certainty.
2. Consumer insurability.
3. Government assumption of risk.
4. Benefits based on "medical" necessity.

Certain keystones of the health insurance system go to the heart of the rationale for insuring health expenses.

Assurance of Provider Payment for Services

Slightly more than seventy years ago hospitals were imbued with almost as much magic as medicine. Since the value of their services was quite meager, the rewards for their outputs were slim and often inconsequential. After the turn of the century, hospitals began to attract patients of means, and physicians gradually began to use more expensive equipment of moderate scientific value but of emerging monetary value. Paradoxically it could be argued that the x-ray machine, asepsis, surgical amphitheatres, and reconditioned medical education programs preceded payment and, in fact, brought compensation matters "kicking and screaming" into this century. Science and technology drove payments, rather than the reverse.

The economic downturn of the 1930s did not reduce the relatively inelastic demand for essential services. Not unexpectedly the provider was the first to be concerned about payment certainty. Providers needed stable, continuous financing—the assurance of continuity in financing services. Hospitals were the first to organize prepayment and were quickly followed by physicians who formed medical insurance organizations. Our present payment methods are imbedded in the early activities of providers who were the architects of protective plans. Protection was primarily for the physicians and secondarily for their patients. At the outset, these developments could have been labeled "second" party financing rather than "third" party.

Insurance was set up to assure physicians of their "usual or prevailing" fees. The concept had predictable upward mobility, and it raised intriguing problems, particularly where innovation brought new procedures to the market. Without usual, customary, and prevailing experience, the physician had to form a new profile for each new procedure; invariably it was a high profile encouraged by this highly inflationary payment principle.

Hospital prepayment history is less complicated. Most hospitals were voluntary, community, nonprofit enterprises. They were integral elements of the local social fabric, and were limited in their main financing objective to the recovery of operating costs. Blue Cross was formed to "make the hospital whole," paying only its incurred costs. Blue Cross plans marketed a service benefit and believed in community rating—two ideas which spread cost and risk evenly. Cost reimbursement was a sound device for averaging the impacts of high and low users, high and low technologies, and short and lengthy stays. Although the focus on Blue Cross has been modified to cope with experience rating competition, its service benefit ideal persists and has been duly copied by Medicare and Medicaid. This is not a complaint but an explanation that reflects the next characteristic.

Consumer Insurability

Health insurance has become as popular for the consumer community as it has been for the provider community. The fundamental expectation of a consumer of health services at the point of purchase is that insurance premiums, once paid, will buy coverage. Benefits should be broad, not exclusionary, and should pay the complete cost of most health services. Health service buyers and sellers have less difficulty agreeing on that issue than do economists. In fact, the public deliberately buys secondary or supplementary coverage to fill the gaps in a primary health insurance policy. Working spouses often provide double coverage for their family, and Medicare beneficiaries often purchase useless, supplemental policies to offset Medicare copayments.

This quirk in public buying habits suggests that the consumer expects the insurer to pay for all medical necessities. For all practical purposes, the beneficiary believes that all services ordered by the physician, podiatrist, or chiropractor are medically necessary. As the purveyor of protection, the insurer is in a poor position to monitor the use and pace of technological medicine.

Governmental Assumption of the High Risk and Vulnerable Citizen

The third underlying characteristic of payment in health that relates to technology is the peculiar role of the government in financing health care services. There are two key aspects:

1. The government at all levels now finances 38.7 percent of all health expenditures, 53.8 percent of all hospital care, and 53.1 percent of all nursing home care. If government were to accept its reasonable responsibility for the entire population in need, these governmental shares would have to be raised by another 5 to 10 percentage points.
2. The public for which government is responsible is a skewed sample—chronically ill, poor, aged, or all three. Special problems arise when government underpays, requires copayment, or fails to pay because a particular surgical procedure is less effective than a new one recently developed.

On the one hand, government, more than other insurers, has a credibility problem—a problem that stems from its commitments to consumers to cover and its commitments to providers to pay. The government is not powerless in defining benefits or in attempting to influence improved and more responsible health care. However, without public beneficiary and provider support and understanding, government program managers are not likely to be successful in moderating medical abuse and overuse of medical technologies. One characteristic of the current payment system—namely, the government's extraordinary share of health care financing—is the support of technological applications.

Benefits Based on Medical Necessity

The last characteristic of American payment practice that relates to the technology issue is the determination by payers of covered services. Who makes those decisions and on what basis are they made? Defining the term "coverage"—the health care benefits to be insured —is difficult. Virtually all programs with which I am familiar rely on language similar to that of the Social Security Act. This law and its implementing regulations mandate that Medicare shall pay only for services which are "reasonable and necessary for diagnosis, treatment, or improved functioning." In practice, with the exception of a few statutorily exempted services, Medicare pays for physicians' ordered services unless they are deemed experimental and not generally accepted in conventional practice. Coverage decisions also may be based on safety and efficacy criteria, but little evidence exists at this time that cost effectiveness measures are limiting coverage.

Thus the law that guides the single most important purchaser of care affords a rather generous guideline for the range of beneficiary services and, indeed, gives the primary responsibility for making the determination of a covered benefit to the practicing physician. Moreover, when the Department of Health and Human Services, rather than physicians, attempts to restrict the range of benefits, Congress often participates in solving the coverage question.

Most coverage questions are subtle and indeed difficult. The Public Health Service (PHS) has a shared responsibility with the Health Care Financing Administration (HCFA) to determine Medicare coverage. Most significant technologies, however, are introduced and reimbursed before PHS considers the issue. Customarily, new hospital technologies resulting from equipment purchases under $150,000 are added to a hospital's line of services and are subsequently absorbed into its cost report. New diagnostic and therapeutic procedures originating through physician billing are usually uncontested. The Medicare carrier simply announces that a new number and value has been assigned to the procedure. Moreover, no effective screen exists for the high volume, low cost technologies unless an intermediary or carrier is willing to argue with a prescriber. The cost of monitoring is not insignificant and the burden of denial rests with the insurer, who is trying to maintain a satisfied group of beneficiaries. If payment is denied, the patient is an innocent victim of excessive treatment and usually must pay for his own services.

Coverage policy and its related payment terms can also distort the application of medical care technology. The end-stage renal disease (ESRD) program, with financing policies heavily weighed in favor of dialysis at a center, has not only discouraged home dialysis but, in some locations, has virtually crippled the development of transplant surgery as well. The ERSD program is a perfect laboratory model of a single-source national health insurance plan that offers benefits for catastrophic illnesses and, in my view, its financing incentives have had predictably catastrophic results. The program exemplifies the influence reimbursement has had on the diffusion of high technology as well as the political vulnerability that a monolithic public payment structure can impose in forging an effective, balanced reimbursement policy.

In sum, these four characteristics of payment policy are philosophically imbedded in our social fabric, and each has propelled sound and perhaps unsound technology in medicine. If we did not have health insurance programs and propensities to provide financing, would we have intensive care units (ICUs) and coronary care units (CCUs), open heart surgery, and end-stage renal disease care in 1981? My answer is yes—though not as costly, not as widely distributed, and not as broadly available to the poor. Without extensive insurance, physicians would likely have applied more stringent indications to their judgments, and the public probably would have demanded another financing vehicle to obtain access to the expensive new products of science. Perhaps such alternatives as tax credits, a more favorable treatment of deductions for medical care costs, or a more extensive public hospital system would have developed with less prepayment. Certainly the dependence of Americans on the judgment of physicians and their mutual confidence in technology transcend the financing issues and will continue to do so in the foreseeable future.

Can We Measure the Impact of Reimbursement on Technology?

As an intellectual exercise this question challenges both the economist and health policy researchers. If such a measurement were possible, a model of various reimbursement options could be developed that could predict with reliability the consequences of each option on technology infusion. Due to multiple factors involved in technology flow, however, I am doubtful that this could be done. Further I believe we can and should identify features of the Medicare payment system that, when changed, will force more thoughtful choices in the selection and use of technology.

With this in mind, I would like to discuss briefly payment and reimbursement strategies that may have a measureable impact in terms of decreasing the use and slowing the diffusion of medical technologies. These strategies assume that barriers are needed to distribute expensive medical technologies more wisely and prudently; I am willing to make this assumption, as I believe that highly increased costs with only marginal benefits have resulted from the overuse of many technologies.

Alternative Financing Strategies for Lessening Technology Demand

Recent cost increases in health care are known to be disproportionate to the rise in the government's capacity to raise revenues. This inflation has been the single greatest threat to distributive equity in access to basic health services. Public support of essential but unaffordable health care services could collapse in the 1980s, in several regions of the nation. I believe that certain strategies can evolve to ease that threat. We do have payment policy choices.

Hospital Reimbursement

Reform in hospital reimbursement is overdue. We ought to move from hospital retrospective cost reimbursement to prospective revenue "per stay." Targeting hospital revenues should be complemented by strategies to set a reasonable limit on capital indebtedness; thus, physicians and hospital administrators, by necessity, would examine overlapping and marginally beneficial procedures. Current Medicare payment limitations on routine costs only (223 rules) should be eliminated since they, more than any other factor, have probably influenced the growth of special care units and nonroutine cost activities (60 percent of total hospital costs) such as lab, x-ray, and other ancillary services. Excess profits from ancillary hospital operations should be contained by requiring hospitals to price all major revenue center income within 10 percent of their costs. If case mix analysis can be perfected, incentives for reducing unnecessary tests can also be introduced into hospital reimbursement.

Physician Payment

Fee-for-service medicine will continue in the 1980s to be the dominant method for compensating physicians. The main concerns will be determining how to establish fees and determining whether insurers will pay all or part of the charges regardless of the fee levels. As professionals and independent practitioners, physicians cherish the principle of determining the value of their services. For the past twenty years, physicians have argued that government should pay at fee levels consistent with those fees charged to nongovernment beneficiaries.

As noted earlier, a fee standard considering "usual, customary, and reasonable" (UCR) physician charge practices has led to high fees for newly developed technologies and has rewarded procedural medicine more favorably than undramatic office practice and conservative clinical judgment. Ideally the physician payment system should be neutral with respect to the application of technology; in fact, it is not. Fees for new procedures, in my view, often bear little resemblance to the skill and experience of a physician, the relative diagnostic or therapeutic value of the service, or the time and costs incurred by a professional in performing the work. Separate studies by Schroeder[1] and Blumberg[2,3] in recent years describe the powerful economics of procedural medicine.

Substitutes for UCR should be tried in an attempt to moderate differentials in physician payment. The possibility that changes will occur in the absence of public pressure, however, is remote, due in part to the lack of options to current fee setting. Perhaps the most effective pressures will be generated from within organized medicine, as groups of primary physicians heighten their own awareness of the gross inequities evident in the current medical marketplace. Negotiated fee schedules are possible on a statewide or regional basis. We may see greater acceptance of capitation payments by the public and by physicians. Capitation and the subsequent pooling of premium revenues allow physician groups to revamp the distribution of the medical care dollars locally, a process that can moderate gross income differentials between competing specialists and primary care physicians.

Consumer Education

The public can and must be educated about the values and the limitations of medical science and technology. The attitudes of an informed public toward health care are quite valuable, particularly in the ensuing decade when a greater portion of the population will wish to take more responsibility for health care decisions. For example, consumers are now somewhat more cautious about undergoing frequent chest x-rays; women are more knowledgeable about the potential side effects of the birth control pill; patients are better informed about surgical risks; and many individuals facing terminal illness are rejecting technological care that, in their view, impedes death with dignity.

The public has become more sophisticated in regard to personal health services. In the final analysis, public awareness of the limitations and benefits of medical technology may be the best tool to moderate questionable utilization. Health professionals may argue that consumers are neither qualified nor motivated to make these decisions; certainly the prescribing physician will always have an advantage. That advantage is narrowing, however, and with increased public awareness, the support of medical technology will become more selective.

Consumer Incentives

Moderation of technology application can be achieved through the adoption of health financing programs that create financial benefits for those who use less expensive services. Health insurance can be tailored to those people who are willing to forego unlimited choices of care. Health insurance that provides second dollar coverage with substantial deductibles may deter casual use of medical technologies. Medicaid beneficiaries may have their care channeled into less costly alternatives. An experimental insurance program in Mendocino, California* strongly discouraged the use of technologies by creating a cash bank for the health care deductible of $500; this money belongs to the employee unless it is used for health services.

Other Financing Ideas

Other strategies for constraining costs, thereby limiting the financial inducements for more intensive services, include:

1. Limiting the choice of providers. Perhaps Medicare and Medicaid should purchase care from a limited number of high quality, lower cost coronary by-pass surgical groups.
2. Allowing buyer coalitions. Perhaps government could join with private insurers in negotiating tougher purchase agreements with hospitals.
3. Allocating patient care funds for selected research. Perhaps Medicare should be permitted to finance biomedical research to eliminate certain expensive half technologies.

Ineffective Financing Strategies

If appropriate use of technology in medicine is the main objective of financing policies, the following strategies will not be effective:

---

* Further information on this program can be obtained from the publication *Proposition 13*, Information Service, Vol. 2, No. 13, California School Boards Association, Sacramento, California or by contacting the Superintendent of Schools in Mendocino County.

1. Decreasing support of research and development. This negative public policy strategy would have an extremely deleterious economic impact. Such a policy suggests that we know how to select and control the development of new information; there is no evidence that we do.
2. Promoting catastrophic health insurance. Most experts conclude that catastrophic awards will directly support the instances in which high technologies play the greatest role. Catastrophic insurance will in fact be a strong inducer for hospitals to spare no expense in providing elaborate services for major illnesses, since there will be one very certain payment dollar for every cost dollar incurred for those serious illnesses.
3. Relying on technology assessment and cost effectiveness tests. This strategy is interesting, ponderous, and costly. We should not plan to resolve technology control through these assessment processes except in a limited number of well-chosen instances. The health care financing apparatus cannot wait the desired length of time needed to assess many technologies.
4. Being indiscriminate in the finance of training. The current policies of financing house officers in specialties which are in apparent oversupply simply adds to the momentum for more medical technology. That output should be tempered to make it more consistent with other prevailing concerns.

Conclusion

Costs have risen rapidly as expenditures for health services have increased. Medical technology has contributed to these costs and has been well financed by all payment programs. Public values stimulated the introduction and dispersion of highly effective, as well as marginally effective, technology. The fundamental characteristics of the payment system need to be understood before the limited range of available options can be analyzed. In my view proposed changes in payment and financing could have a tempering effect on the utilization of medical technology; in the final analysis, however, these changes will not eliminate altogether the public's faith in technological developments in American medicine.

Endnotes

1. Schroeder, S.A., and Showstack, J.A. (1978). Financial incentives to perform medical procedures and laboratory tests: Illustrative models of office practice. *Med. Care* 16:289–298.

2. Blumberg, M. (1979). Rational provider pricing: An incentive for improved health delivery. In *Health Handbook*, ed. G.K. Chacko, pp. 1049–1101. New York: Elsevier-North Holland.

3. Blumberg, M. (1979). "Physician Fees as Incentives," address presented at the Proceedings of the 21st Annual Symposium on Hospital Affairs, University of Chicago Center for Health Administration Studies, June.

ROLE OF THE HCFA IN THE REGULATION OF NEW MEDICAL TECHNOLOGIES

by Leonard D. Schaeffer

The Health Care Financing Administration (HCFA) administers Medicare, Medicaid, and other programs promoting the well-being of the nation's aged, poor, and disabled. Every day about 800,000 Medicare and Medicaid beneficiaries see a physician; 300,000 receive care as hospital inpatients; and about 90,000 receive services in nursing homes. The costs of paying for these services continue to climb. In 1980 HCFA paid about 25 percent of the nation's health bills and 40 percent of its hospital bills, spending some $50 billion.

To meet our obligations to our beneficiaries, choices need to be made in regard to expenditures on their behalf. All effective procedures must be available to beneficiaries but, at the same time, expenditures for unnecessary or ineffective services must be avoided. It is important that we act immediately: if expenditures continue to rise at current rates, this country may not be able to afford to purchase the care that will be needed by the poor, aged, and disabled in the future. Moreover, our program dollars, which affect the entire system, must be used to improve productivity in the health delivery system generally.

Intervention in administrative and clinical affairs on a day-to-day basis should be limited and regulation employed only where necessary to: (1) limit overutilization and unnecessary utilization of resources; (2) structure competition for revenues and capital; and (3) encourage strong and prudent management. In the absence of global controls, however, medical practice patterns and economic reality dictate that we make specific decisions now. The problem of medical technology needs to be addressed.

Issues Raised by the Use of New Technology

What issues does the use of new technology raise insofar as HCFA is concerned? In my view it is not necessarily that we have too much or too little medical technology, or even that the current mix is wrong. The fact is that we have no certain analytical or clinical standards with which to make such across-the-board judgments, nor do I think we will have them soon. Nevertheless we know that health costs continue to rise rapidly, and that capital investment, including technology, is a major contributor to these escalating costs.

Recent studies[1-4] have shown that initial purchase prices for new equipment represent only a very small percentage of operating costs and that the collective costs of thousands of small tests and procedures play an important role in the growth of overall health expenditures. We know of instances in which these costs of new equipment are not justified by the benefits patients receive or by the number of patients able to receive those benefits.

The issue raised by new technology for health financing programs is not that we have too much or too little, but that we lack an adequate process and knowledge base for making decisions about what to purchase with limited funds. Equally important is the assurance that new technology will contribute to both improved health outcomes and operating efficiency. We will pay for new technology and procedures if they are proven to be safe, efficacious, and cost effective, and are delivered in an efficient manner.

The Need for a Government Role

Many would agree that the costs associated with new technology pose important problems, but I continue to believe strongly that government should play a role. In most industries new technology is introduced because it is expected to enhance profitability. There is little desire to invest in new technology in the absence of the promise of improved operating efficiency or some other advantage affecting profits. Indeed, there is currently a growing concern that various forms of regulation have reduced the potential profit contribution of new technology in industry, resulting in a national rate of innovation lower than necessary to maintain acceptable levels of productivity.

The health industry, however, operates under a different set of incentives. Financial profits are not the main concern of most institutional providers, and efficiency improvements are only rarely the purpose of technological innovation. The availability of capital financing for purchasing new technology and third-party reimbursements for its clinical use, combined with the receptivity of physicians, hospitals, and patients to new and more sophisticated diagnostic and therapeutic techniques, guarantee a ready market for new medical technology. Physicians are committed to doing all that is possible to seek a diagnosis and a cure; hospitals wish to provide state-of-the-art medicine to attract physicians and to expand; and patients take advantage of all the information available. The role of physicians is particularly important; studies[5,6] show that they control over 70 percent of the resource allocation decisions made in health.

In short, the structure of the medical system is such that market forces are not likely to act as an efficient allocator of medical technology. Self-restraint seems unnecessary when the benefits of new technology are experienced locally but the costs are hidden and distributed over a much larger population. In the American system the primary responsibility is between provider and patient; neither a built-in sense of responsibility to the general economy nor an institutionalized sense of obligation or accountability for total costs is evident. Hospitals, doctors, and patients optimize locally and suboptimize nationally.

The weaknesses of the market as an allocator of medical technology have been recognized for some time, and government has already taken a number of steps to address them. I will review some of these efforts briefly.

Government Programs to Date

The health planning and Professional Standards Review Organization (PSRO) programs are the most notable among the programs already enacted to influence the distribution and use of medical technology. The health planning program has addressed the problem of health care technology chiefly through Certificate of Need (CON) review, which was made mandatory under the law, and through the development of state and regional health plans, some of which are specific as to the distribution and utilization of high cost medical technology. The Health Systems Agency (HSA) and state agencies are now beginning to implement a third authority, appropriateness review, which has the potential to influence the distribution of medical technology. At present, however, HSAs and state agencies are not required to perform institution and service-specific appropriateness reviews, and only do so at their own option. A fourth review function unique to the HSAs, the proposed use of federal funds review, or "PUFF," could conceivably influence the distribution of technology in the future.

Among the above, Certificate of Need (CON) has the most potential for influencing the distribution of medical technology. Recent studies[7-9] have not shown CON to be effective in containing growth in beds or in other capital assets, including technology. None of these studies, however, has been able to measure the deterrent impact of CON; they are unable to document what the distribution of expensive technology would have been in the absence of CON controls. For example, we now have roughly 1,200 CT scanners in the United States, more per capita by a wide margin than any other developed nation. How many would we have if planning agencies had not taken such a close look at scanner technology, imposing virtual moratoria in some states until efficacy could be established?

Planning agency performance across the country has varied considerably however. Planners have attempted to regionalize advanced technology, while providers, acting individually, have pressured for greater decentralization. The success achieved by planners has depended on a number of factors, including the point at which they intervene in the diffusion process, the match between planning goals, and the inclinations of the hospital industry in particular states, and the frequency with which individual providers have been able to circumvent CON programs.

Two years before passage of the health planning program in 1974, Congress enacted the PSRO program and in doing so sought to make the medical profession self-regulating. In their capacity as locally organized physician monitoring systems, PSROs incorporate features of earlier utilization review systems governing the delivery of care in hospitals. However, insofar as they are more highly formalized, are externally validated and mandated, and have their decisions backed by financial as well as professional sanctions, PSROs represent a new form of influence over medical practice patterns.

Each of the three PSRO review functions—concurrent review of admissions and length of stay; in-depth, retrospective medical evaluation studies; and profile analysis—have some capacity to influence the use of medical technology. Moreover, PSROs, unlike CON programs, have the capacity to analyze and influence the use of "little" technologies as well as very expensive and sophisticated technologies.

HCFA evaluations of PSRO performance, conducted for 1977, 1978, and 1979,[10-12] have shown increasing savings for the program, mainly due to reduction in hospital days. However, these studies also indicated variation in PSRO effectiveness across the country. The 1979 evaluation shows that PSROs have improved compliance with accepted quality standards and suggests that PSROs have had an impact on selected diagnoses and procedures.

Taken together, however, the planning and PSRO programs do not represent a sufficiently coordinated or completed structure to enable the government to deal with the introduction of new technology. To be effective persuaders, PSROs in particular need a source of consensus information about technological advancements. In response to the need for coordination and the growing concern for the issue of technology, Congress in 1977 established the National Center for Health Care Technology (NCHCT). Among the items in the Center's charter is the charge to administer "a program of assessments of health care technology which takes into account their safety, effectiveness, and cost effectiveness and social, ethical, and economic impacts . . . ." The Center has a variety of other charges, including the dissemination of its findings.

HCFA Efforts

The Health Care Financing Administration (HCFA) has a special concern for the diffusion of new technology since it funds 40 percent of hospital costs which typically reflect the impact of technology changes. To date HCFA's primary involvement with issues of advanced technology has been through the Medicare program, which is administered nationally by contractors using policies developed by HCFA. There are two generic types of policy decisions which HCFA must make:

1. Coverage decisions: Whether or not an item or service is one for which the program can pay.
2. Reimbursement decisions: How much is appropriate to pay for a covered item or service.

Coverage

The major authority for dealing with coverage of technology in the Medicare law is an exclusion which prohibits payment for items and services which are not "reasonable and necessary for the diagnosis or treatment of an illness or injury." The provision has been in the law since its inception and has been used to deny payment of claims for items and services which are not safe and effective (either in general or for specific indications), not reasonable or medically necessary in a particular case, and not furnished in an appropriate setting.

The provision has been used as the authority under which national coverage policy determinations are made in regard to procedures or items of questionable value. In the past the criteria employed to make these determinations permitted payment whenever an item or service had gained general acceptance in the medical community. Payment has only been denied under three circumstances: when general acceptance has not occurred; when convincing scientific evidence states that an item or service is unsafe or ineffective; or, in the case of new, unusual, or experimental items and services, when real scientific evidence does not exist.

Medicaid law permits the individual states to make similar coverage policies, and HCFA intends to step up its efforts to make its findings available to the states and to encourage them to act. The "reasonable and necessary" exclusion under Medicare is applied to individual claims by program intermediaries who have particular responsibilities; processing individual claims in accordance with program regulations and guidelines, including coverage rules; and applying the general "reasonable and necessary" test to all services, whether or not a specific coverage rule refers to them.

In the past HCFA's response to coverage issues has been primarily reactive. Questions have come from carriers processing physician claims because they have not had the resources to make determinations about certain items and services. Usually these have been new, little known, or unproven procedures about which the contractors' medical consultants are unable to draw conclusions. Current procedures for making coverage decisions are as follows: HCFA evaluates a question after it has been received and seeks advice internally as to whether payment should be made. If no conclusive answer can be developed by program staff, the matter is referred to a panel of HCFA physicians and dentists. The panel determines whether the issue is one of national medical and scientific significance that warrants development of a national policy. If it is not, the contractor is advised of the HCFA's medical opinion and instructed to deal with the claim on the basis of this opinion. If there is a national medical and scientific issue involved, the issue is referred to the National Center for Health Care Technology for a finding as to safety, efficacy, and cost effectiveness.

The Center then researches the literature, consults within the public health service and with outside groups as necessary, and provides HCFA with a finding as to whether, and for what indications, the item or service is reasonable and necessary. HCFA then makes a coverage determination and promulgates a coverage rule which is incorporated into the program manuals used by the contractors.

The HCFA intends to improve this process so that coverage questions can be handled more effectively, and a number of important steps are being taken in this connection.

1. A closer working relationship between HCFA and the National Center for Health Care Technology (NCHCT) has been established, and joint procedures and respective responsibilities are being specified in great detail.

2. The HCFA is establishing a medical professional coverage office, headed by a physician, within the agency's bureau of program policy. This office will serve as HCFA's principal advisor on Medicare and Medicaid coverage issues. One of its primary tasks will be to set priorities for coverage issues so that we can deal more effectively with our contractors and with the national center by focusing efforts on issues with the greatest potential impact on health status or cost.
3. Steps have also been taken to assure that HCFA's coverage instructions to contractors are current and medically accurate. A review of existing coverage instructions is underway, and we plan to subject coverage instructions to a thorough medical review before they are made final.
4. For the first time the HCFA is developing regulations which will set out formal criteria that HCFA will use to make "reasonable and necessary" determinations. These criteria will incorporate cost-effectiveness concerns, and we will establish a formal process under which proposed "reasonable and necessary" determinations will be published for public and professional comment before they are adopted.

In my opinion the increasingly controversial and complex nature of many coverage issues (such as CT scanners, transsexual surgery, and organ transplants) makes development of formal criteria and a procedure essential.

Limited Coverage Determinations

In addition to making general coverage determinations of this kind, the pace of technology change and the complexity and difficulty of new procedures may require that limited coverage determinations be made which allow for reimbursement under specified conditions, in certain circumstances, and for specific purposes. For example, coverage for certain new and costly procedures may have to be limited to technologies that have been demonstrated to perform safely and effectively. This is particularly important in the light of a recent study[13] demonstrating that certain surgical procedures can be performed safely and efficiently only when performed with some frequency. This study would suggest that certain advanced procedures should be performed only on a regional basis.

In the future the HCFA plans to tie coverage determinations to conformance with quality and effectiveness standards more generally. In doing this, the agency will rely heavily on NCHCT for the development of the criteria necessary to limit coverage. When criteria are developed, they will be published in a *Federal Register* notice and comments will be solicited from the public and the profession. We want to provide coverage where new procedures can be performed safely, efficaciously, and efficiently.

The HCFA also has been asked to consider a second kind of limited coverage determination applicable in situations where it is necessary to determine what conditions would be necessary to ensure safety and effectiveness. We have been asked to consider a limited investigational coverage determination that would enable Medicare payment to be made under specified constraints while we learn more.

The issue of limited investigational coverage determinations grew from efforts to grapple with another new procedure, percutaneous transluminal coronary angioplasty (PTCA). PTCA is being used to treat blockage of coronary arteries. The procedure involves the use of a balloon-tipped catheter to open blocked arteries. Proponents of the procedure assert that it is much less drastic than coronary bypass surgery and achieves a similar result.

HCFA's current position is that it cannot cover PTCA because no medical evidence is available to prove safety and efficacy. HCFA requested a medical finding from the NCHCT and learned that current data on the procedure simply do not exist. The Center has asked HCFA to provide investigational trials to learn more about the safety, efficacy, and cost effectiveness of the procedure. Medicare has never made such a determination in the past.

The issues raised by PTCA are important. If HCFA demands clinical proof of efficacy before extending coverage, does it have a responsibility to support efforts to produce that proof if it is not already available? Is it appropriate for a financing agency to invest in limited investigational trials? Does HCFA have the authority to pay for a procedure at a time when its safety and efficacy have yet to be established? An ancillary issue is whether, as a condition of payment, Medicare could require submission of additional clinical data for use in medical research.

Our consideration of these issues takes us far beyond our stated purpose. We would be moving away from asking whether there is evidence that something is reasonable and necessary to actually funding research to find out whether it is. This would be a major step and we do not know whether we should take it or even if we can take it legally.

In addition to taking steps to make better coverage decisions, we are also working to ensure that existing technology and new technology that comes into use will be used efficiently.

Reimbursement

The payment provisions of both the hospital insurance and medical insurance programs of Medicare specifically include a test of reasonableness. Part A mandates payments to hospitals on the basis of reasonable cost, and part B requires payment to physicians on the basis of a reasonable charge. Thus HCFA has the authority to limit payment when charges or costs are not reasonable. Normally this is accomplished by giving the contractors guidance in the form of recommended reimbursement screens.

HCFA is developing a regulation that would state and clarify its authority to set special national or regional reasonable charge limitations for payments to physicians under part B of the Medicare program. Under this regulation HCFA would establish payment screens or levels above which payment will not be made. These screens would be set so that payment would be sufficient only if procedures were performed in an efficient manner. This regulation will affect both the distribution and use of advanced technology and encourage only those providers with sufficient volume to operate safely and efficiently and to undertake major capital investments for new technology.

In addition to these steps other options for ensuring efficient distribution and utilization of technology are being considered. For example, reimbursement should be linked to appropriateness review when institution and service-specific appropriateness review by planning agencies becomes a reality. In addition regulations are being developed to encourage hospitals to act as "prudent buyers" and to encourage group purchasing.

Obstacles to Effective Use of Technology

Many formidable obstacles impede our efforts to come to grips with the impact of new technology. One problem is whether the existing claims-based payment system will permit us to apply fully the principles of coverage and reimbursement already developed. Under part B, new procedures and services are usually identified on the physician's or surgeon's claim. Under part A, however, the potential is much greater for the use of new technology to go unnoticed on the hospital bill. A CT scan, for example, might be entered on the hospital bill as radiology, and charges for computer-read EKGs might be listed as laboratory. If the existing system is to work effectively, or the relationship between specific technologies and health care costs is to be understood, we will need more detailed information on hospital bills. It is questionable however whether a system for gathering the information we need could be developed at a reasonable cost.

Another problem is simply keeping up with the pace of change and the growing number of coverage and reimbursement decisions that need to be made. Our list of currently pending issues includes PTCA, heart transplantation, psychosurgery, intra-ocular lenses, home use of oxygen, trans-sexual surgery, heart transplant, bilateral carotid body resection, and many others. Also, we are aware of a number of new technologies in the development stages. One such technology is the positron emission transverse tomography or the PETT scanner. It creates a three-dimensional image that reflects physiology rather than anatomy, and like the CT scanner, is complex and expensive. To date there are only a few PETT scanners in this country, all supported under grants from the National Institutes of Health.

The most fundamental obstacle is that we operate mainly at the end of the pipeline. Technology is researched, developed, and marketed before HCFA gets involved. Reimbursement considerations do not directly influence the flow of research funds, nor can we predict innovations adapted to health from other areas or imported from other countries. Given the current reimbursement system, there is a ready market among physicians, hospitals, and patients for new technology. In the absence of intervention of some kind, diffusion is virtually guaranteed.

Conclusion

HCFA prefers limited intervention in day-to-day clinical and administrative decisions. Our goal is to use regulation only where necessary to structure incentives for efficient behavior. The realities of the health sector, however, dictate that we make decisions regarding specific medical technologies. We have an obligation to our beneficiaries to make these decisions. Beyond that, we recognize an obligation to use our program dollars to improve the health delivery system generally. We are concerned with developing the mechanisms necessary to ensure that new technology is safe, efficacious, and cost effective, and that it is used as efficiently as possible.

Endnotes

1. A Feasibility Study of the Influence of Capital Expenditures on Hospital Operating Costs. Final Report, Research Demonstration Series, Report No. 6. Washington, D.C.: Health Care Financing Administration, 1978.

2. Brown, J.B., and Marks, H.M. Buying the future: The relationship between the purchase of physical capital and total expenditures growth in U.S. hospitals. In *Health Capital Issues.* Health Resources Administration, May 1981.
3. Scitovsky, A.A. Changes in the use of ancillary services for 'common illness.' In *Medical Technology: The Culprit Behind Health Care Costs?* eds. S. Altman and R. Blendon. U.S. Public Health Service.
4. Fineberg, H.V. Clinical chemistries: The high cost of low-cost diagnostic tests. In *Medical Technology: The Culprit Behind Health Care Costs?* eds. S. Altman and R. Blendon. U.S. Public Health Service.
5. Redisch, M.A. (1978). Physician involvement in hospital decision making. In *Hospital Cost Containment: Selected Notes for Future Policy,* eds. M. Zuboff, I. Raskin, and R. Hanft.
6. Gabel, J.R., and Redisch, M.A. (1979). Alternative physician payment methods: Incentives, efficiency, and national health insurance. *Milbank Memorial Fund Quarterly/Health and Society,* 57, 1 (Winter).
7. Evaluation of the Effects of CON Programs. Final Report. Washington, D.C.: Health Resources Administration, 1980.
8. Sloan, F.A., and Steinwald, B. (1980). Effects of regulation on hospital costs and input use. *The Journal of Law and Economics* 23:81–109 (April).
9. Coelen, C., and Sullivan, D. (1981). An analysis of the effects of prospective reimbursement programs on hospital expenditures. *Health Care Financing Review,* pp. 1–40 (Winter).
10. U.S. Department of Health, Education and Welfare, Health Care Financing Administration. *Professional Standards Review Organization 1978 Program Evaluation,* January, 1979.
11. U.S. Department of Health and Human Services, Health Care Financing Administration. *Professional Standards Review Organization 1979 Program Evaluation,* January, 1980.
12. U.S. Department of Health, Education and Welfare, Health Service Administration, Office of Planning, Evaluation and Legislation. *Executive Summary,* Vol. I of *PSRO: An Initial Evaluation of the Professional Standards Review Organization,* February, 1978.
13. Luft, H.S., Bunker, J.P., and Enthoven, A.C. (1979). Should operations be regionalized? The empirical relation between surgical volume and mortality. *N. Engl. J. Med.* 301, 25 (Dec. 20):1364–1369.

MISUSE OF TECHNOLOGY: A SYMPTOM, NOT THE DISEASE

by David Blumenthal, Penny Feldman, and Richard Zeckhauser

The inappropriate use of health care technology is a major problem in the American health care system. Costly medical practices, procedures, and devices allegedly often fail to yield commensurate health benefits. Health care technologies have been accused of decreasing the overall quality of patient life and making medical care less humane and personal. As the convening of this conference attests, commentators who are frequently at odds with one another— private physicians, hospital and government administrators, technology assessors, and social scientists studying the health care system —agree that the technology problem deserves policy attention.

Some Areas of Agreement

Policy toward health care technology should be based on a clear understanding of the nature of the problem, its causes, and the full consequences of proposed solutions. Though these issues have not yet been definitely resolved, we see considerable agreement on a number of issues of fact and value relevant to the technology problem.

Cost Control Essential to Policies Affecting
Health Care Technology

The principal objective in any policy relating to health care technology must be the control of costs. There are two reasons for this emphasis. The first is pragmatic and political. Most people (and certainly most policymakers) focus on the issue of health care technology because they are worried about the excessive (and therefore often inappropriate) use of medical resources of all varieties. Put another way, the driving force behind current interest in health care technology is concern about the cost of medical care.

That interest in health care technology really represents concern about health services generally is apparent in common definitions of "technology." As a working definition of the term, we have adopted the formulation used in the only federal legislation on the topic, the bill that established the National Center for Health Care Technology (Public Law 95-623, Section 309 of the Public Health Service Act). This statute defines health care technology as any "discrete and identifiable regimen or modality used to diagnose or treat illness, prevent disease, monitor patient well-being, or facilitate the provision of health care services." Although this is clearly a very broad definition, encompassing virtually all health care services, most discussions of medical technology are no more restrictive in their use of the term. This characteristic usage tends to confirm our point that concern about medical technology should be reinterpreted as concern about health care in general.

The importance of the cost problem in current debates over medical technology is reflected in statements by prominent political figures and health care experts. For example, in introducing legislation addressing the technology problem, Senator Edward Kennedy commented:[1]

> *It has been estimated that one half of the annual increase in the cost of a day of hospital care can be attributed to the use of more technology in medical practice. That means that between 1966 and 1976 expenditures for medical technology added $8 to $12 billion to our national hospital bill, which totaled $55 billion in 1977.*
>
> *What are we buying with this outlay of funds for additional technology, Mr. President? . . . The truth is that in many cases, we simply do not know.*

A major study of the technology issue, conducted by the National Academy of Sciences, noted in introducing its conclusions: "Most important, new technology is accused of raising the cost of health care."[2] Indeed, one of the seminal conferences on the technology problem was entitled, "Medical Technology: The Culprit Behind Health Care Costs?"[3]

This concern about health care costs and the overutilization of health care services reflects a widespread belief among policymakers and health care experts that we are spending more on health care in general, and on technology in particular, than is warranted by improvements in our nation's health. In economic terms, we have continued to invest in health resources until their benefits at the margin are exceedingly small, nonexistent, or even negative in some cases.

No empirical evidence proves that health care spending is excessive; however, there are findings that suggest this conclusion. Contiguous areas of Maine with similar populations spend vastly different amounts per capita on health care, yet the health of their inhabitants is apparently comparable. Health maintenance organizations (HMOs) spend between 15 and 40 percent less in caring for patients than do insurance plans in the fee-for-service sector, yet HMO populations of a given age, sex, and ethnic composition seem to be as healthy as their fee-for-service counterparts. England spends less than half as much per capita on health care as does the United States, yet measurable health indices show no ill effects for the English.[4] Finally, assessments of various health practices based on randomized controlled trials have frequently failed to demonstrate health benefits commensurate with costs.

Though not conclusive, the data are at least sufficient to justify shifting the burden of proof to those who advocate allocating more resources to the health care sector. The current outpouring of concern over health care technology suggests that a shift in that direction may already have taken place.

Inappropriate Use of Technology is a Symptom of Malincentives

A second reason why technology policy must control the cost of medical care is that the cost problem and the technology problem are largely symptoms of common maladies in the health care system. The most important of these disorders is a series of inappropriate incentives—what we call malincentives—in the way health care is financed in this country. These malincentives encourage the use of health care services with little regard to cost. They have increased the overall level of resources devoted to health care and have promoted excessive reliance on expensive and sophisticated medical practices, procedures, and devices.[3]

Third-party reimbursement programs, both private and public, cover more than two thirds of personal health expenditures; 92 percent of hospital costs are covered this way.[5] Thus many Americans pay very little for health care at the time of purchase. In addition, since private health insurance costs are subsidized through the (limited) deductibility of premiums and nontaxability of employer payments, and since most employers assume at least part of the cost of employee health protection programs, most Americans do not pay the full cost of their insurance coverage either. To what extent government should shelter the public from the costs of illness is a debatable issue; beyond dispute, however, is the fact that our extensive third-party coverage encourages the consumption of health care resources.

The tendency toward overconsumption is exacerbated by methods of payment to those who have predominant control over the use of most health resources: physicians and other health care providers. Blumberg[6] finds that physicians control at least 61 percent of the decisions concerning office visits, about 80 percent concerning patient days in the hospital, and about 90 percent concerning routine nursing service costs. Nevertheless, prevailing free-for-service modes of reimbursement give them no financial incentive to conserve resources. One more test or procedure will cost the patient very little, yet the physician will get paid for administering it.

Biases in fee schedules have the additional effect of favoring the dissemination and use of equipment-embodied technologies and other highly technical practices, procedures, and devices. This not only raises overall cost, but shifts the mix of services available in the health care system toward more sophisticated forms of care and away from primary care services.

Technology Policy Must Respond to Underlying Causes

Because we view the increasing use of technology in our health care system as a symptom of a larger problem rather than as a cause, we believe therapy should focus on underlying etiologies rather than surface manifestations. To the extent that cost control is an aim of technology policy, the fundamental causes of the escalation of health care costs—namely malincentives—should be addressed.

Possible approaches to controlling the cost of medical care might include: (1) reform in the tax treatment of health insurance; (2) elimination of restrictive regulatory programs; (3) promotion of HMOs or other modes of prospective reimbursement; (4) encouragement of competition in the delivery of health service; (5) removal of handicaps to for-profit institutions in the health care market; (6) prospective budgeting or rate regulation for the hospital sector; and (7) imposition of national caps on health care spending. Any reforms that succeed in controlling costs will probably address the concerns generally expressed in discussions of technology policy.

We do not mean to imply that cost is the only important or legitimate concern among participants in the health care technology discussion. For example, some critics of the excessive use of health care technologies, especially of sophisticated medical devices, have argued that the overutilization of particular services impairs the quality of health care; others have expressed concern over the depersonalization of care caused by the increasing use of medical hardware. We emphasize the cost issue because we feel that: (1) interest in the cost effects of the misuses of medical technologies is nearly universal; (2) cost concerns are politically most important; and (3) the factors responsible for the increasing cost of care must be addressed before progress can be made toward addressing other legitimate concerns.

Policy Implications: Principles

Health care policy should control costs by assuring that marginal expenditures produce commensurate benefits. In addition, it must have acceptable effects on other variables, such as the quality of care received by most Americans. Unfortunately there is substantial disagreement about which set of health care reforms is most likely to succeed. We can, however, agree that two points merit particular consideration in relation to technology policy: the importance of incentives and the dangers of direct government control or regulation.

The Importance of Incentives

Policies to reduce the excessive and inappropriate use of technologies, such as policies to control cost, should rely as much as possible on the establishment of incentives to use services in a cost-effective fashion. Methods should be devised to induce physicians and other providers to consider costs as well as potential health benefits in any decisions regarding the use of health care technologies and services. The performance of HMOs demonstrates that a substantial impact can be achieved through reforms in the organization and financing of health care services that impose more appropriate incentives on producers. Experiments with altered financial incentives are under way in other settings as well.

Obviously these experiments must be carefully evaluated for their effects on the cost and quality of care. As we shall discuss, changing payment methods does not, in itself, guarantee that the mix of services provided will be optimal. Nevertheless, the theoretical advantages of using incentives, in effect, to internalize the cost of medical decisions are clear. This approach ensures that the necessary but difficult tradeoff between cost and benefit will be made by those who are directly responsible for employing particular technologies.

The Difficulties of Regulation: Coverage Decisions, a Case in Point

We believe that direct governmental control or regulation of the development, dissemination, or use of health care technologies is not an effective way to control cost or to achieve other valued objectives, such as improved quality of care or increased equity in the distribution of health benefits.

Federal or local authorities could attempt to influence the development and use of technologies in several ways. For example, federal health research funds could be diverted from projects likely to produce very costly technologies, or direct regulation could prohibit the use of a technology until it had been proven efficacious, safe, and cost effective. We do not propose to review all such possible strategies here. To illustrate some of their disadvantages, however, we will discuss one quasi-regulatory approach to controlling the use of health care technology: the use of technology assessment, as discussed by Schaeffer, to make decisions concerning which technologies, both old and new, should be provided under public financing programs.

Cost control will be difficult to achieve using this approach, for several reasons. First, no commonly accepted or proven methodology exists for deciding whether a particular technology merits coverage under Medicare and Medicaid, especially when cost control is a major objective of such decisions. So-called technology assessments suffer from numerous technical problems. As Russell points out in her chapter on technology assessment, even attempts to assess the efficacy of technologies through randomized controlled trials are subject to error. These difficulties are compounded when assessments try to measure the cost-effectiveness of practices and procedures, or their relative costs and benefits. Disagreements arise over the evaluation of health outcomes, such as left-years saved or alterations in quality of life, and the assessment of health benefits realized in future periods. Cost estimates are frequently difficult to derive and are readily challenged.

Not surprisingly the Office of Technology Assessment (OTA) has concluded that cost-effectiveness analysis and cost-benefit analysis can help to structure health care decisions but cannot be used as definitive standards.[7] Russell makes the additional point (see the first paper in this section) that cost-effectiveness in particular is difficult to use in the absence of a fixed constraint on health expenditures (for example, in the context of Medicare, with its unlimited budget).

The nature of the technologies that must be reviewed further complicates coverage decisions. The technologies are numerous, varied, difficult to define, and subject to frequent modification. No commonly accepted manual of medical practices, procedures, and devices exists. Medical practices can be combined in a seemingly infinite number of permutations, depending on physician opinion and patient need; yet, each such regimen constitutes a unique medical technology. Technologies enter the market and diffuse through numerous routes that are extremely difficult to anticipate, detect, and monitor. For example, a relatively simple change in surgical technique can alter the outcome of a procedure significantly. (Examples include the by-passing of several coronary arteries during a by-pass procedure, rather than just one, or the recent development of the so-called "selective vagotomy" procedure as a variant on surgery for peptic ulcer disease.) Only the operating physicians and their support staff would know for certain which surgical procedures were undertaken in any particular case. Thus enforcement of a coverage policy could prove difficult, particularly if physician dissatisfaction resulted in substantial misreporting of actual practice.

Moreover, many worthwhile technologies could be held up during screening assessments, while some ineffective and costly procedures could escape evaluation. For example, the potential dangers of delaying the dissemination and use of particular technologies are apparent in the case of the CAT scanner. This innovation has often been offered as an illustration of the need for technology assessment before permitting the widespread utilization of a new device. In 1979 the National Academy of Sciences report on equipment-embodied technology commented that the CAT scanner had been adopted widely before sufficient information existed on its effectiveness as a diagnostic tool. In retrospect, however, the extensive early use of the CAT scanner appears to have conferred (some) medical benefits. The Office of Technology Assessment (OTA) recently commented: "There seems to be little doubt that CAT scanning has been a remarkably useful addition to the array of medical technology." The debate about computed axial tomography has moved beyond discussions of efficacy —widely accepted but not definitively proven—to such issues as whether the technology is equitably distributed among hospitals serving the poor and nonpoor. Had the dissemination of the CAT scanner been stopped, pending its evaluation, some money would have been saved, but some health benefits might have been sacrificed as well.[2,8]

Another problem with government attempts to influence directly the development and use of technologies is the persistent inability of the federal government to manage the coverage process. While the current system for making coverage determinations is a considerable improvement, it represents the fourth such system implemented by the federal government in the last six years and is already being revamped by the Reagan Administration. This institutional instability is destructive to effective coverage decision making, demoralizes existing staff, and makes it more difficult to recruit professionals with the skills needed to undertake technology assessments. (The government is also handicapped by a lack of tools. For example, Medicare lacks an information system that would allow the detection of new technologies paid for in a hospital setting.)

Finally, the attempt to implement a coverage process faces major political difficulties, which seem likely to grow worse before they improve. To the extent that they are truly binding on practitioners and health care institutions, coverage decisions are likely to be fiercely challenged in the courts and before the Congress. It is not clear that existing Medicare statutes, which require the federal government to pay for all "reasonable and necessary" practices, will allow the denial of coverage on the basis of cost. Lengthy court battles on this question would undoubtedly stall the implementation of an aggressive, cost-conscious coverage process for some years. Even if the authority of the Secretary of Health and Human Services to make such decisions is upheld, current anti-government and anti-regulatory sentiments seem likely to undermine the political support for such a process. Ongoing efforts to disband the National Center for Health Care Technology, which has been advising the Secretary on coverage issues, hint at the political storm likely to erupt if the federal government tries to preempt the judgment of physicians in deciding what is "reasonable and necessary" medical practice.

Taken together, these difficulties may undermine significantly attempts to use the coverage process for cost control. Methodological problems in determining the cost effectiveness of alternative technologies will tend to make coverage decisions controversial and subject to legal challenge. These technical difficulties also may lead to determinations that will subsequently be regarded as incorrect. Because of the number and nature of technologies, lengthy administrative delays may occur in processing new practices, procedures, and devices through a coverage decision process. This, in turn, will delay the introduction of some valuable and cost-effective techniques. Similarly, monitoring problems will tend to frustrate enforcement of decisions once made, and administrative and political difficulties will undermine the credibility, effectiveness, and viability of the coverage process over the long run.

Without actually implementing and evaluating such a system, the potential of a coverage decision process to control costs—without significant sacrifice in health benefits—cannot be discounted. However, the problems we have outlined, which are typical of many regulatory ventures, argue for caution in attempts at direct control or regulation of the use of medical technologies.

Policy Implications: Practical Approaches

Whatever the merits of the coverage decision making process for influencing the use of medical technology, the law requires that federal administrators decide which practices, procedures and devices qualify as "reasonable and necessary" for reimbursement purposes. The policy problem is how best to discharge this responsibility with the imperfect tools available.

Two extreme approaches are conceivable. On the one hand, the Health Care Financing Administration (HCFA), which administers Medicare and Medicaid, could require that any technology be proven safe and efficacious before reimbursement by the federal government can be made. HCFA might also require that technologies meet some standard of cost-effectiveness. On the other hand, HCFA could define as "reasonable and necessary" any service that providers and physicians offer. This strategy, essentially the method used until very recently, simply defines away the problem and is not inconsistent with the spirit of the original Medicare legislation.

Each of these extreme approaches has major disadvantages. The prescreening strategy suffers from all the difficulties outlined previously. However, a policy that makes no effort to define "reasonable and necessary" places the federal government in the position of approving the use of technologies that are ineffective and even dangerous. That such technologies have been and continue to be in common use is well documented.[9] Blue Cross and Blue Shield, for example, recently announced that they would no longer routinely pay for 68 commonly used practices and procedures because they were ineffective.[10]

Finding an acceptable path between these polar approaches will be a difficult and frustrating task. Choices must nonetheless be made, and some possible options can be formulated.

Limited Government Role in Technology Assessment

The federal government could support and encourage the assessment of technologies without directly linking that assessment process to reimbursement decisions. A number of arguments support such a strategy.

First, the federal government is a major consumer of health technologies, both through its third-party payments and in some instances, such as veterans' hospitals, through direct delivery. It has a fiscal responsibility—just as when commissioning fighter aircraft or hydroelectric dams—to determine whether the services it purchases are effective, appropriately priced, and worth the cost.

Second, the federal government has an obligation to protect the clients of its financing programs. Even if HCFA does not adopt the more restrictive approach of prescreening technologies, the agency at least is responsible for informing its beneficiaries of the risks and benefits of the services it purchases on their behalf.

A third argument for federal support of technology assessment is that information concerning the appropriate uses of medical technologies is otherwise unlikely to become available in sufficient quantity. Information provision on a new technology represents a classic case of market failure: Accumulating that knowledge is costly but confers few private benefits on the individuals or firms who invest in its acquisition. Moreover, once such information is acquired, there is no practicable means to charge others for its use in an effort to cover initial costs. (Even if such charges could be made, it would be inefficient to do so since the cost of supplying an additional information user is zero.)

Individual physicians concerned about their patients may attempt to make judgments on the efficacy and safety of particular technologies, but most physicians lack the training, the resources, and the data to make scientifically valid judgments.[11] The manufacturers of medical devices have some incentive to assess the technologies they develop, and the Medical Devices Law requires limited tests of efficacy and safety before certain devices are marketed. The law does not apply, however, to surgical procedures, combinations of devices and procedures, or other medical regimens.

It might be argued that once incentives in current financing systems are modified to encourage more cost-effective behavior on the part of physicians and patients, the private sector will assume the burden of assessing technologies. Health maintenance organizations, for example, have a clear stake in delivering services that are safe, efficacious, and cost effective. To help achieve this goal, large plans, such as Kaiser and the Harvard Community Health Plan, have already instituted programs in research on medical technologies.

However, it will be years, perhaps decades, before the malincentives in current financing systems are remedied. In the meantime, too little technology assessment is likely to be undertaken by the private sector. Moreover, even if those malincentives were corrected, providers would underinvest in health care research in the hope that they could be free riders and benefit, at no private cost, from the investigations of others.

Critics of technology assessment might argue that the methodological problems we have outlined make such assessments unworthy of public support. Although no conclusive test of the value of technology assessments has ever been conducted, properly conducted assessments have been successful in identifying practices, procedures, and devices that are totally inefficacious. (Examples include prolonged hospital stays for myocardial infarction, strict bed rest for hepatitis, and bland diets for ulcer patients.[9]) Clearly this information has some value. Moreover, cost-effectiveness and cost-benefit analysis, with all their limitations, provide the only rational basis for allocating resources in health care settings where funds are constrained. Properly applied by local physicians and providers, such as HMOs, technology assessments have the potential to significantly improve the effectiveness of health care delivery and the quality of care. Inadequate as such evaluations may be, they are the best tools available.

Follow-up Surveillance

A more activist approach to making coverage decisions might involve the kind of postmarketing surveillance that was proposed for drugs during recent attempts to reform drug legislation. Under this strategy, federal authorities would reimburse for certain new technologies only on the condition that physicians and providers agree to participate in whatever studies are necessary to assess those medical practices. This coverage policy might involve limiting the settings in which a technology could be used until evaluation was completed, or simply requiring that physicians provide certain data necessary to the evaluation of the procedure.

Such a strategy has several attractive features. It allows new technologies to be transferred rapidly into practice, but limits public exposure to unproven innovations. It also promotes the rapid evaluation of new practices, procedures, and devices.

The postmarketing surveillance approach also encounters potential problems. First, it is applicable primarily to new technologies and does not help with the more formidable task of ascertaining which accepted practices are inefficacious, unsafe, or excessively costly. Second, it encounters many of the practical difficulties associated with any regulatory approach to controlling the use of new and existing technologies—that is, problems defining what is a genuinely "new" practice, procedure, or device, and anticipating and detecting the emergence of these new services. Third, the postmarketing surveillance approach leaves open the question of what can or should be done when testing shows a new technology to be inefficacious, unsafe, or excessively costly.

There are no easy solutions to these problems, and whether satisfactory answers can be found remains unclear. On balance we see some justification for experimenting with a postmarketing surveillance scheme if the tasks and purposes are narrowly defined. It would be particularly important that the process do the following: (1) set a limited agenda; (2) evaluate a few technologies thoroughly rather than many inadequately; and (3) emphasize the identification of new practices and procedures that are totally ineffective or grossly unsafe. Given existing health system incentives, the information generated by postmarketing surveillance can be used to improve the quality of services provided and the rationality of medical decision making. This approach, however, is unlikely to succeed in controlling the cost of care, an objective better addressed through correcting systemic malincentives.

Conclusion

The inappropriate use of health care technologies is merely one of several symptoms of our national overconsumption of health services. Given the severity of the associated problems—high costs, excessive hospitalization, decline of personalized practice—attention to the technology problem in isolation can be a weak palliative at best. A cure will require a more radical attack on the malincentives inherent in the present American health care system.

Endnotes

1. *Congressional Record.* January 31, 1978, p. S890.
2. Committee on Technology and Health Care, Assembly of Engineering, National Research Council and Institute of Medicine (1979). *Medical Technology and the Health Care System.* Washington, D.C.: National Academy of Sciences.
3. Altman, S.H., and Blenden, R., eds. (1979). *Medical Technology: The Culprit Behind Health Care Costs?* Washington, D.C.: U.S. Government Printing Office.
4. Blanpain, J., Delesie, L., and Nys, H. (1978). *National Health Insurance and Health Resources: The European Experience.* Cambridge, Mass.: Harvard University Press.
5. Gibson, R.M. (1980). "National Health Expenditures: 1979." *Health Care Financing Review,* 2:1.
6. Blumberg, M.S. (1979). "Provider Price Changes for Improved Health Care Use." In *Health Handbook,* G.K. Chacko, ed. New York: Elsevier-North Holland.
7. United States Congress, Office of Technology Assessment (1980). *The Implications of Cost-effectiveness Analysis of Medical Technology.* Washington, D.C.: U.S. Government Printing Office.
8. U.S. Congress, Office of Technology Assessment (1981). *Policy Implications of the Computed Tomography Scanner: An Update.* Washington, D.C.: U.S. Government Printing Office. See also Banta, D. (1980). "Computed Tomography: Cost Containment Misdirected." *Am. J. Public Health* 70: 215.
9. Fineberg, H.V., and Hiatt, H.H. (1979). "Evaluation of Medical Practices." *N. Engl. J. Med.* 301:1086.
10. Greenberg, B., and Derzon, R.A. "Determining Health Insurance Coverage: Problems and Options." *Med. Care,* in press.
11. Temin, P. (1980). *Taking Your Medicine: Drug Regulation in the United States.* Cambridge, Mass.: Harvard University Press.

TECHNOLOGY VERSUS REGULATION: WHICH IS THE LOWER COST ALTERNATIVE?

by Harvey W. Freishtat

In the past, it has been almost a truism that the costs of health care technology are increasing and that regulation can play a significant role in keeping those costs down. However, within the legal forum where some of these technology/regulatory battles occur, I sense a slightly different theme. To many courts, the cost of health care regulation is becoming unacceptably high, whereas technology is viewed as offering at least the possibility of a more effective cost-benefit ratio. I would like to demonstrate through a series of court cases how the trade-offs between private technology and public regulation can be perceived so differently from the judicial perspective.*

Technology Assessment in the Context of Medical Malpractice

A good example of the way in which a technology case reaches the courts is reflected in the case of *Helling* v. *Carey*.[1] This case involved a 23-year-old woman who developed problems with nearsightedness. She consulted her physicians, Board-certified ophthalmologists, who

---

* Since the time this article was submitted for publication, a new Administration has taken over in Washington with a distinctly anti-regulatory philosophy in health care and other sectors of the economy. Whether the Administration is actively pro-technology or simply more laissez-faire in its attitudes towards technology's role in health care remains to be seen. Predictably, the role of the judiciary in resolving disputes over appropriate uses of technology in health care will remain significant.

fitted her with contact lenses. She left their offices apparently satisfied. Four years later, the woman returned complaining of eye irritation. The physicians refitted her lenses; she again departed and was not seen for four more years. When she did return complaining of more serious irritation, she was given a pressure test for glaucoma. At this point, some nine years after her initial visit, she was found to have glaucoma, with the loss of her entire peripheral vision and much of her central vision.

The patient took the case to court. Expert witnesses testifying for both the patient and the physicians agreed that the incidence of glaucoma in a patient under 40 years of age was less than one in 25,000. The question before the court was therefore whether the physicians were negligent in failing to administer the pressure test in a more timely manner, despite the fact that the patient had only a negligible chance of having the disease. The court was faced with a choice. On the one hand, the physicians could be exonerated since, in not administering the pressure test until other measures had been tried without success, they had acted in conformance with the general ophthalmological standard of care for treating younger patients. Alternatively the court could find that the standard of care established by the ophthalmological profession was simply not adequate for patient protection, and that the harm to this woman would have to be redressed even if it meant establishing a new standard of care in the process. The court reasoned as follows:

> *The incidence of glaucoma in one out of 25,000 persons under the age of 40 may appear quite minimal. However, that one person, the plaintiff in this instance, is entitled to the same protection as afforded persons over 40 [when it is] essential for timely detection of the evidence of glaucoma where it can be arrested to avoid the grave and devastating results of this disease. The test is a simple pressure test, relatively inexpensive. There is no judgment factor involved and there is no doubt by giving the test the evidence of glaucoma can be detected . . . The precaution of giving this test . . . to detect the incidence of glaucoma in patients under 40 years of age is so imperative that, irrespective of its disregard by the standards of the ophthalmology profession, it's the duty of the courts to say what is required to protect patients under 40 from the damaging results of glaucoma.*[2] *[Emphasis supplied]*

On this basis, the court found the physicians liable for medical malpractice and thereby effectively established a new standard of care requiring routine administration of a pressure test in all cases. What this court had in fact enunciated was a standard of care requiring 24,999 unnecessary tests to detect one case of disease.

CT Scanning

A similar concern about the quality of care, sometimes irrespective of cost, has led other courts to rule similarly when medical technology is at issue. CT scanning is a case in point. A review of CT scanner cases across the country reveals judicial reversal of an unusually high number of administrative decisions denying Certificates of Need (CON) for CT scanners. These reversals have occurred despite the deference courts traditionally show for administrative action, and in the face of highly restrictive health planning guidelines, widespread availability of scanners in many geographic areas, and inconclusive results from the various official CT scanning technology assessments.

A CT scanner case decided in Florida several years ago[3] illustrates just how far the courts will go to ensure that the benefits of technology are brought to the public. In that case, a Dade County community hospital applied for a CON to purchase a $600,000 body scanner. Nine scanners were already in use in Dade County, including four very close to the hospital, and several of these scanners were functioning at levels of utilization well below the 2,400 procedures per year then mandated by federal health planning guidelines. In addition, three new scanners had been approved for Dade County and were awaiting installation. Beyond that, expert testimony, including a report from the Office of Technology Assessment (OTA), suggested that there was no need for additional scanners in the Dade County area.

In the face of this evidence of oversupply and underdemand, the hospital's application was denied by both the State Health Planning and Development Agency (SHPDA) and the hearing officer on review. However, the hospital appealed directly to the court, which examined the technology of CT scanning and found as follows:[4]

> . . . (N)ext to the discovery of x-ray itself, the CT scanner is the most revolutionary development in radiology ever devised . . . [T]he use of this scanner to diagnose a patient's illness is safer, less harmful, more accurate and often less costly to use than the more traditional diagnostic procedures it replaces.

With that framework the court then addressed the SHPDA's argument about the availability of CT scanning in neighboring facilities:[5]

> The record discloses that at the time of [the Hospital's] application it generally transported its inpatients to [another hospital] some 12 to 15 miles away for CT scans, and that the process of transporting a patient is slow and its cost high. Often nurses and oxygen therapists must accompany the patient. Often three or four days transpire before a written report is received. Often the Hospital must keep a patient

in its hospital for two or three days to await the availability of the other scanner. In some cases, some patients who should receive the scans do not, because they are too ill to transport. There is also testimony that every hospital with over 300 beds should have its own CT scanner to meet community needs [and] that there are only two CT scan facilities near the Hospital which do not meet the utilization rate of 2,400 scans per year. The Hospital does not refer its patients to them because it questions the accuracy and reliability of their scan results. The (SHPDA) did not investigate the reasons for their low utilization rate, nor take into account any factors which might explain it.

Thus despite the quantitative guidelines, or sheer numbers presented, the court was not impressed. Concerning the fact that three more scanners were expected in addition to the nine existing scanners, the court found that the SHPDA had denied the hospital the equal protection of the laws. As the court phrased it, if there were unique factors that led to the granting of three CONs in the face of quantitative criteria to the contrary, then there were similarly unique factors in the case before this court.

How did the court deal with the OTA report and the other expert testimony recommending the 2,400-procedures-per-year guideline? Venerable legal doctrines of evidence known as "lack of authentication" and "inadmissibility by reason of hearsay" came to the fore:[6]

The only testimony before the Hearing Officer to support [the 2,400 procedure per year figure] was the hearsay statements of two physicians and two exhibits consisting of two unauthenticated reports, the first of which was purportedly prepared by the Office of Technology Assessment, a congressional board, for use by US Senate committee, and referred to by one of [the SHPDA's] own witnesses as a "second draft" of the report, with no indication that it was a final report prepared by that Board as its final assessment and recommendation as to the use of CT scanners or to their utilization rate. The other exhibit is a report purportedly prepared by J. Lloyd Johnson Associates in 1976, likewise unauthenticated as to its veracity or trustworthiness. Both of these exhibits were hearsay, properly objected to by counsel [for the hospital]. Certainly such evidence does not meet the test of "substantial competent evidence" sufficient to sustain [SHPDA's] order herein.

On this basis the court overturned SHPDA's denial and remanded the application for positive administrative action.

What can be seen in this opinion, and it is by no means singular, is that technology from a judicial perspective is often viewed positively, and that arguments to the contrary based on cost, duplication, lack of relative efficacy, or the like are viewed as interesting but largely irrelevant. In July, 1981, a jury in the District of Columbia declared the D.C. government liable for the failure of the D.C. General Hospital to arrange a CT scan in a neighboring hospital for a patient with severe headaches. The patient died, and the widower got a $240,000 award. This decision may have considerable significance because, as editorialized in the *Wall Street Journal*, "now . . . it might start costing states money [to deny CON's for CT Scanners], which we hope will prove the straw that breaks the back of the whole notion that the way to cut medical costs is to second-guess the hardware in hospitals." (July 5, 1981)

Certificate of Need—The Regulatory Response to Technology

Courts also are becoming increasingly impatient when lengthy and tedious regulatory procedures cost more money to implement than is saved by the underlying regulatory program. An example is a case concerning a community hospital in Michigan which sought a CON to bring its antiquated structure up to code.[7] The project cost $22 million, due not only to the extensiveness of the repairs and renovations but also to the multiplicity of codes: the Occupational Safety and Health Administration (OSHA) code, the Building Officials and Code Administrators (BOCA) code, the National Fire Protection Code, and the state and local building codes. Each code was different from the others and the only way to comply with all of the codes at the same time was through the highest common denominator approach—hence the $22 million cost.

The hospital's CON application was first reviewed by the Health Systems Agency (HSA) which responded that since the hospital had 200 beds and the area was 200 beds overbedded, the hospital should talk to the neighboring facilities about consolidation, merger, becoming a nursing home, or even going out of business. From the hospital's viewpoint, these alternatives were not only undesirable but nonsensical; the institution was an efficient 200-bed facility with a day rate lower than any of its more tertiary neighbors with whom it was being asked to merge. Even after factoring in the costs of the new project, the rates still would be the lowest in the area. Moreover, the hospital unquestionably provided quality services which were as high as those of its tertiary neighbors. The services it did not provide, such as cardiac surgery and neurosurgery, could easily be arranged through neighboring facilities.

The hospital sought to convince the HSA that closing the facility would succeed only in forcing health care to be provided in a series of larger institutions which, for the vast majority of services, provided no better care but at considerably greater cost. The hospital also sought to show the fundamental inaccuracy of the HSA's view that mergers produced economies of scale. Despite the hospital's arguments, the HSA recommended denial of the CON application.

At the SHPDA, however, the hospital's CON was approved—not because the SHPDA necessarily believed the hospital's position in terms of maintaining lower cost-efficient units within the health care system, but because the SHPDA did not believe it had the statutory authority to question the need for the hospital's ultimate existence. Therefore, the SHPDA's inquiry was limited to the needs of the hospital and, since the hospital faced delicensure if it did not conform to code, there was little question as to need in this more limited sense.

The HSA took exception to the SHPDA's approval and appealed. One of the apparent purposes of the appeal was to delay the effective date of the CON, with the hope that construction costs and interest rates would escalate to the point where construction would be effectively stymied, whatever the legal outcome on appeal. Administrative appellate proceedings were convened, but the hospital realized that no matter what the outcome of the battle, the war would be lost through mere passage of time. So the hospital decided to take the case to court for speedy resolution.

In court each side briefed and argued its position. The HSA focused on the costs of overbedding and duplication of services, while the hospital focused on the cost effectiveness of community hospitals and the legal requirements of due process of law. With the passage of each trial day, project costs increased another $5,000; each month represented an escalation of $150,000. At the conclusion of the hearing the court announced its decision from the bench. It found that the HSA had no standing even to lodge the appeal, since the SHPDA had approved the CON. On this basis both the administrative and judicial appeals were dismissed, and construction of the hospital proceeded quickly.

The court's decision was based on a technical reading of the language of the state CON statute, which recently has been amended specifically to authorize HSA standing in such cases. Behind the technical legalities of the case, however, lay a clear and fundamental reality: the court was extremely concerned about the costs of the regulatory maze involving HSA review, SHPDA review, administrative appeal, and finally court review. On several occasions, the court questioned HSA representatives as to the underlying purpose of their appeal, inquiring as to whether, in the name of cost consciousness, the appeal had not been the costliest route of all. In short, the court was concerned that a full-blown regulatory process, acting without limit or restraint, could stand in the way of health care innovation or, in this case, renovation.

The entire CON process is subject to these same concerns and criticisms. The purposes of CON are unquestionably positive as some effective means is necessary to control the introduction of technology into the health care sector. However, the dilemma is how to administer such a process without compounding the cost problem sought to be addressed.

A prime example of this problem is reflected in the ten-taxpayer mechanism* incorporated into the recently enacted amendments to the National Health Planning and Development Act (P.L. 96–79). The ten-taxpayer mechanism seeks to permit virtually anybody with an interest in a CON application to become a party to the process, with the full panoply of due process rights that automatically accompany party status. Prior to its incorporation into federal law, this same mechanism had been used by several states in administering their own CON programs. For example, it has been part of the Massachusetts CON statute since 1972.

Based upon my experience, the ten-taxpayer mechanism appears to serve a useful purpose when it accurately represents the voice of the community speaking out in favor of or against a particular CON application. Frequently, however, ten taxpayers have appeared on behalf of their own narrow and often anticompetitive interests, which generally results in lengthy hearings, costly procedures, large legal fees, and the like. For example, landlords have appeared in opposition to CON applications when their physician-tenants were seeking to construct a physicians' office building. A proprietary hospital has used ten taxpayers to oppose a CON application filed by its community hospital neighbor to force a merger or purchase of its own institution. Neighborhood residents have used ten taxpayers to force hospitals to set up day care services, to stop performing abortions, or to establish preferential minority hiring procedures within the hospital. Each of these attempts to use the CON process for outside leverage has little to do with rational, orderly planning of health care services and has nothing to do with quality of care. Yet each has added demonstrably to the costs of health care. Because of these and similar experiences with the ten-taxpayer mechanism, the federal government's move to adopt it with so minimal an attempt to assure its proper use seems highly questionable.

---

* The ten-taxpayer mechanism refers to any group of ten citizens of a state acting together who wish to make some claim.

Summary

We can learn several lessons from these cases. From the judicial perspective, the costs of regulation are increasingly perceived as too high, whereas the costs of technology seem to represent a better societal investment. Courts are displaying a growing impatience with CON programs, HSA programs, Professional Standards Review Organization (PSRO) programs, and arbitrary reimbursement programs. Each program is designed to keep costs down but, in reality, costs are not reduced where quality is undermined. In the trade-off between cost and quality, courts often tend to favor the quality side of the equation.

Courts are not particularly known for their own cost effectiveness. The American legal system is founded on the belief: "Better that a hundred guilty men go free than one innocent man be jailed." This belief is the very embodiment of constitutional due process of law, but it would not withstand even primitive OTA analysis. Technological analogues of such a belief—"Better that a hundred unnecessary scans be performed, than one necessary scan be omitted," or "Better that a hundred terminal patients be maintained indefinitely on life-saving equipment than one nonterminal patient be erroneously removed from such equipment"—would certainly not inspire unanimous agreement.

Legal cost-benefit analysis is far different from economic or social cost-benefit analysis. Many courts will not subscribe to the rationing of health care services or resources, no matter what the projected cost savings may be. Courts are accustomed to hearing cases when the projected cost savings or benefits have not been borne out. Therefore the concrete issue they face is not a "macro" socioeconomic question but a "micro" issue of providing relief to an individual person or institution.

The judicial perspective is vastly different from the regulatory or policymaking perspective. Judges do not set policy; they interpret it in the context of individual cases and people. While cost-benefit ratios may be essential in setting general policy, they often play less of a role when it is the cost side of the equation—the one in 25,000th case—that happens to be before the court. By necessity, judges often bring their personal health care concerns and family histories to the cases before them. They can often empathize with the one case that went wrong.

Courts are not predictable institutions once a case is brought before them. Patients' emotions and personal chemistry are factors which play a large role in the outcome of litigation. For this reason alone, regulators and the regulated would be well advised to avoid reaching the courts when the issue is appropriate use of health care technology. Both sides should try to establish mutually acceptable standards for the particular technology scheduled for regulation. Once a case reaches the courts, the results may not please either side.

I have presented my perspective on the interface between technology and regulation. If it bears little resemblance to technology assessment from a policymaking perspective, that may be because the courts can be a ballast and counterweight to the typical assessment process. Rational policymaking seeks to define the greatest good for the greatest number. Courts, by contrast, are often asked to provide relief in that minority of cases that would otherwise receive little consideration.

Endnotes

1. *Helling* v. *Carey*, 519 P. 2nd 981, 83 Wash. 2d 514 (1974).
2. *Ibid.*, p. 983.
3. *North Miami General Hospital* v. *Office of Community Medical Facilities of the Department of Health and Rehabilitative Services of the State of Florida*, 355 So. 2d 1272 (1978).
4. *Ibid.*, p. 1275.
5. *Ibid.*, p. 1275.
6. *Ibid.*, p. 1276.
7. *Saratoga General Hospital* v. *Department of Public Health and CHPA-SEM*, Ingham County Superior Court, Case No. 79–23944–CZ, Opinion filed November 2, 1979.

THE ROLE OF COST-BENEFIT AND COST-EFFECTIVENESS ANALYSES IN CONTROLLING HEALTH CARE COSTS

*by Joyce C. Lashof, Clyde Behney, David Banta,
and Bryan Luce*

The health policy community has been paying more attention to cost effectiveness/cost-benefit analysis as potentially significant aids in controlling the rapid rise in health care costs. The health care literature has exhibited a growing interest in cost-effectiveness analysis, and a variety of organizations have expressed an interest in a more rational method for making resource allocation decisions. For these reasons the Office of Technology Assessment (OTA) undertook a study of the methodology of cost-effectiveness analysis and its usefulness as a tool in making decisions in health care policy.

The primary distinction between cost-benefit analysis (CBA) and cost-effectiveness analysis (CEA) is that all costs and benefits in CBA are valued in monetary terms, while in CEA the desired program outcomes can be measured in nonmonetary terms such as years of life saved, decrease in disability days, or a combination of mortality and morbidity measures which may be expressed as quality-adjusted life years (QALYs). The reason for a nonmonetary measure of program effectiveness is either the impossibility or undesirability of evaluating certain outcomes in terms of dollars. Thus, unlike CBA, CEA provides a ratio expressed as dollars per life saved, dollars per QALY gained, or similar ratio. Its major usefulness lies in comparing program alternatives with the same goal or final outcome; comparison of programs with vastly different objectives cannot be done. Furthermore, knowledge that the cost per outcome, such as $50,000 per year of life saved, does not tell us whether this is an acceptable cost. Although CEA cannot make social and political decisions, it can enlighten the decision makers.

With these considerations in mind, the Office of Technology Assessment (OTA) examined the controversies involved in using CEA, commissioned a number of case studies, and examined its application in governmental and nongovernmental programs. The study was confined to the application of CEA/CBA to medical technology, that is, the drugs, devices, medical, and surgical procedures used in medical care. The analysis of many reports revealed a wide spectrum of studies ranging from simple identification of costs of different programs assuming equal efficacy to a detailed analysis of different outcomes using different technologies and comparing the costs of each in relation to the specific outcomes. A number of methodological problems were identified, including paucity of data on efficacy, differing outcome measures for diagnostic technologies, inability to deal with issues of equity, lack of agreement on discount rate to be used, and different impacts on different population groups. Nevertheless ten basic principles were identified that should be followed in carrying out CEA analysis.

1. *Define problem*: The problem should be clearly and explicitly defined and the relationship to health outcome or status should be stated.
2. *State objectives*: The objectives of the technology being assessed should be explicitly stated and the analysis should address the degree to which the objectives are, or are expected to be, met.
3. *Identify alternatives*: Alternative technological means to accomplish the objectives should be identified and subjected to analysis. When slightly different outcomes are involved, the effect this difference will have on the analysis should be examined.
4. *Analyze benefits/effects*: All foreseeable benefits/effects (positive and negative outcomes) should be identified, and when possible, should be measured. Whenever possible, and if agreement on the terms can be reached, all benefits should be valued in common terms to make comparisons easier.
5. *Analyze costs*: All expected costs should be identified and whenever possible should be measured and valued in dollars.
6. *Differentiate perspective of analysis*: When private or program benefits and costs differ from social benefits and costs, and when a private rather than a program perspective is appropriate for the analysis (and vice versa), the nature of the perspective should be identified.
7. *Perform discounting*: All future costs and benefits should be discounted to their present value.
8. *Analyze uncertainties*: Sensitivity analysis should be con-

ducted. Key variables should be analyzed to determine the importance of their uncertainty to the results of the analysis. A range of possible values for each variable should be examined for effects on results.

9. *Address ethical issues*: Ethical issues should be identified, discussed, and placed in appropriate perspective relative to the rest of the analysis and the objectives of the technology.

10. *Discuss results*: The results of the analysis should be discussed in terms of validity, sensitivity to changes in assumptions, and implications for policy or decision making.

We found that many methodological weaknesses are obscured by deriving a numerical bottom line. Consideration should therefore be given to identifying clearly all elements included in or affected by the decision, rather than aggregating a complex set of calculations and possibly ignoring those factors that do not lend themselves to quantification.

Further research into the methodology for performing CEA can be expected to resolve some problems, but others, such as valuing intangibles, appear to be inherent in the analysis itself. Researchers must recognize variations in the use of CEA that are dependent on the particular stage in the life cycle of a technology at the time of application. The possibility of affecting the course of a technology's diffusion and use might be greater in the early stages of its development, but uncertainties as to efficacy and costs will make evaluation difficult. CEA may be more valid later in a technology's life cycle, but the usefulness of the information gained in affecting diffusion and policy will be more limited.

The current and potential use of CEA in a number of health programs was examined. CEA as a cost-control aid would appear to be applicable to reimbursement programs such as Medicare and Blue Cross/Blue Shield. Historically, however, costs have not been an explicit element in the reimbursement decision-making process. Rather, reimbursement decisions have been based on data relating to safety, efficacy, stage of development, and acceptance by the health care community.

Applying cost-effectiveness criteria to reimbursement decisions will therefore necessitate examining how, when, and by whom the technology is used. The technology under consideration usually cannot be viewed in isolation but can be examined under specific indications for use. One example might be the problem of skull x-rays. Historically, skull x-ray examinations have been used as a standard radiological procedure in the evaluation of patients with head injury. The yield of abnormal findings, however, relates to the presence of a number of signs, symptoms, or risk factors. Thus the cost effectiveness of skull x-rays will depend on the stringency of the indications used for performing the examination. CEA also will be strongly affected by the outcome measure used in the analysis. Is diagnostic accuracy a sufficient outcome measure, or should impact on treatment decision and clinical result be the endpoint of the analysis? These issues would need to be addressed before CEA could be used in any decision affecting reimbursement for skull x-rays.

Thus before CEA can be used to determine reimbursement policy under a fee-for-service system, the development of specific procedure and treatment protocols is necessary. For example, in 1978 when the Health Care Financing Administration (HCFA) approved payment for body CAT scans, the use of this technology was limited to certain organ systems. This decision constituted a break with previous precedent and has led to concern over Medicare's involvement in defining the practice of medicine.

Partly in response to these problems, alternative reimbursement systems have been recommended, two of which are prospective budgeting and capital cost controls. Under these systems hospitals would be aware of the funding available for the coming year and could, in cooperation with their medical staffs, determine which services should be provided. Cost-effectiveness analysis could then be used locally to determine which competing technologies would receive priority or what limits would be placed on the utilization of resources. By these means the decision-making process would be placed in the hands of the users at the local level.

Reimbursement by capitation method through the Health Maintenance Organization (HMO) mechanism also has been gaining favor for the same reason. Clearly HMOs have a direct economic incentive to provide cost-effective care and therefore would be expected to find CEA a useful tool. However, the Office of Technology Assessment's review of HMOs decision-making criteria revealed that because of competition with providers operating in a fee-for-service environment, they do not commonly weigh benefits against costs in determining what medical services to offer. Rather, they seek ways to provide benefits comparable to fee-for-service medicine at the lowest cost feasible.

Another potential program area for the use of CEA is the Professional Standards Review Organization (PSRO) program. Such analyses could be useful in three areas: (1) the development of standards of care; (2) the internal management of individual PSROs; and (3) an evaluation of the national effort. Individual PSROs should not be expected to carry out their own CEAs. Rather, the national PSRO program could select key technologies for analyses, commission the appropriate studies, and make the information available to the local programs. In evaluating the national PSRO program, efforts to date have been directed toward determining whether the program spends more money than it saves. These analyses, however, have not addressed cost savings in relation to the impact of PSROs on the quality of care or health outcomes. Although extremely difficult, this issue must be addressed in any evaluation of a program designed to improve the quality of medical care.

Although the National Health Planning and Resources Development Act explicitly states that resources are to be allocated in a more efficient manner, and costs and benefits should be weighed in the decision-making process, OTA found that the Health Systems Agency rarely looked beyond capital costs in carrying out their economic analyses. No example of balancing costs against health benefits in the certificate of need process was identified. Thus the planning process is oriented toward health needs, which will continue to be the case in the absence of a budget ceiling.

It has been suggested that the Food and Drug Administration (FDA) undertake cost-effectiveness analyses prior to licensing new drugs or devices. Such an undertaking is clearly beyond the Agency's legislative authority, and the methodological problems involved in such an undertaking would be enormous. In using CEA for new drugs, the FDA would be required to look at all the costs associated with a new drug or device, as well as to predict what the future costs would be. All indications and conditions of use would require such an examination, which would appear to be impractical and probably counterproductive. However, the Agency does attempt to identify those therapeutic breakthrough drugs, which clearly are more efficacious than currently available drugs, and to give them priority in the review process.

In summary, the current use of formal CEA/CBA in decision making is the exception, not the rule. However, some subset of the issues relating to costs do enter into many health care decisions. Despite the methodological problems, the quality and validity of decisions appear to be increased by analyses that force a structuring of the decision-making process, and thereby provide a framework for identifying and considering as many relevant costs and benefits as is feasible. Clearly CEA is more valuable and more likely to be used under a constrained budget where trade-offs have to be made, rather than where constraints are nonexistent or indirect. It is difficult to state in an absolute sense that y effect is worth x cost. However, at some point, stating the most effective way to spend x dollars for health care may be essential.

Part Six

CASE STUDIES

Intensive Care Unit Workshop
Burn Care Unit Workshop
End-Stage Renal Disease Workshop
Workshop on Prenatal Diagnosis
Workshop on the Artificial Heart

INTENSIVE CARE UNITS TODAY

by William A. Knaus and George E. Thibault

Hospitals in past eras were last resorts; families accepted admission only after attempts at home care had failed. It was recognized that hospital treatment had a limited ability to cure disease, and dying patients remained at home and were treated by family or friends. As Lewis Thomas wrote recently, "In my own clinical years and in the wards at the Massachusetts General . . . students were taught by Harvard's most expert clinicians, but all of the teaching was directed at the recognition and identification of disease. Therapy was an afterthought, if it was mentioned at all."[1]

Today patients and physicians turn to hospitals more frequently and with greater expectations. Although many hospitalized patients still die, it is only after aggressive attempts have been made to save them. When these particular therapeutic attempts are successful, hospitals gain a general reputation of being capable of performing miracles. When heroic efforts fail, or when the emphasis is on the disease instead of the patient, modern hospital care can still be viewed as valueless and even cruel.

As suggested by Louise Russell,[2] intensive care units are the hallmark of a modern hospital. Although almost unknown in 1960, more than 5,600 ICUs can be found in 6,300 acute care hospitals in the United States. These 55,000 ICU beds account for five percent of all U.S. hospital beds. According to the American Hospital Association, ICU beds increased by three percent in 1979 while total nonfederal, short-term beds declined.[3]

In today's intensive care unit, a fine line separates benefit from

This report was made possible by a grant from the Health Care Financing Administration, #18-P-97079/3-02.

The authors would like to acknowledge the contributions of Marc R. Chassin, Professional Standards Review Organization, Department of Health and Human Services; Elizabeth Draper, Douglas P. Wagner, and Jack E. Zimmerman of the ICU Research Unit, The George Washington University Medical Center; Albert G. Mulley, Health Practices and Evaluation Unit, Massachusetts General Hospital; and John H. Siegel, Department of Surgery, State University of New York at Buffalo.

harm and enthusiasm from criticism. Many physicians point to the development of ICUs as one of the most important medical advances within the last 20 years.[4] Conversely, critics claim that ICUs concentrate many elaborate and expensive treatments in one place but do so without any documented benefit for patient care.[5]

What prompted this sudden popularity in ICUs? Why are they still in demand? What kinds of services do they deliver and what types of patients do these units treat? What is the evidence indicating their value? Why do they have so many supporters and so many critics?

This paper will examine these questions by reviewing the history of intensive care and summarizing the relevant literature. Our perceptions of the current status of ICUs will be presented and our feelings that evaluations of ICUs should receive priority in the future will be discussed.

Definitions

Intensive care refers to two distinct clinical activities: coronary care units (CCUs) and general medical-surgical or multidisciplinary intensive care units, commonly referred to as medical intensive care units (MICUs), surgical intensive care units (SICUs), or simply intensive care units (ICUs).

CCUs treat a small range of patients, mainly heart attack victims and those with related cardiac problems. Overall, CCU patients are not as critically ill as ICU patients, although individual admissions can be similar. The therapeutic services provided within a CCU are fewer than those available in a multidisciplinary ICU, and unlike ICUs, emphasis is placed on the diagnosis of acute myocardial infarctions.

Most multidisciplinary ICUs treat patients for whom the diagnosis is obvious. Patients suffering from acute drug overdoses, respiratory failure, gastrointestinal bleeding, and diabetic ketoacidosis are treated in a medical ICU. A surgical ICU treats patients recovering from open heart surgery, neurosurgery, and other major or complicated operations.

Other more specialized units exist in addition to CCUs and ICUs. Some treat only burn patients while others specialize in trauma or neurosurgical cases. A few beds within a general medical-surgical ICU occasionally are designed as trauma, burn, or even CCU beds. These variations among different hospitals make a precise division of the 55,000 intensive care beds difficult. On the basis of available information, we estimate that there are approximately 16,000 CCU beds and 39,000 multidisciplinary or specialized ICU beds nation-wide.[3]

While there are many combinations and organizations of intensive care units, there is one unifying concept: Intensive care is concentrated care. ICU patients may differ in respect to diagnoses or severity of illness, but they all share one common characteristic: They all need, or are perceived to need, more attention than is available on a general hospital ward. The demand for this concentrated attention prompted the move toward intensive and coronary care. This concentrated attention is provided in part by high technology machines (such as mechanical ventilators, or various monitoring devices) but to a large degree is the result of close attention from nurses and doctors.

Development of ICUs

One of the earliest references to the concept of general intensive care is by Florence Nightingale. In 1852 she wrote that, "it was valuable to have one place in the hospital where postoperative and other patients needing close attention could be watched." Under Nightingale's leadership, a few such special wards were established in English hospitals during the late eighteenth and early nineteenth century.[6]

In 1930 a surgeon named Kirschner extended the concept to Germany. In 1942, stimulated by the shortage of hospital personnel created by World War II, the first postanesthesia recovery room was created in the United States at St. Mary's Hospital in Rochester, Minnesota. In that same year the Coconut Grove fire in Boston prompted the development of a burn unit at the Massachusetts General Hospital.[6] Following World War II and the enthusiasm generated by battle triage experience, a study maintained demand for special care by suggesting that one third of the deaths occurring within 24 hours of surgery could be prevented if the patients received better nursing care.[7] This study provided an obvious incentive for hospitals to concentrate nurses and patients in postoperative recovery rooms. The number of recovery rooms increased during the 1950s but, as shown in Figure 1, the turn toward ICUs occurred somewhat later.

In a review of ICU development Russell[2] found that state and local hospitals, as well as those hospitals operating for profit, tended to adopt intensive care later and commit relatively fewer beds to it than nonprofit hospitals. Hospitals with medical school affiliations generally opened intensive care units before community hospitals.

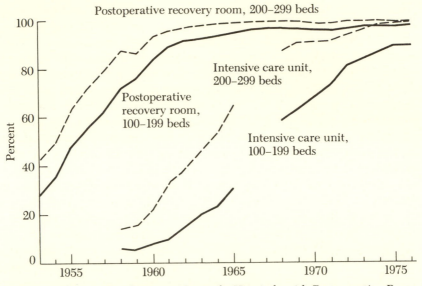

Figure 1 Percentage of Private Nonprofit Hospitals with Postoperative Recovery Rooms (1953–1976) and Intensive Care Units (1958–1976), by Size of Hospital. *Note:* Gaps in the lines reflect years in which the facility was omitted from the survey. *(SOURCE: Louise B. Russell,* Technology in Hospitals: Medical Advances and Their Diffusion, *Figure 3-1, p. 42. Washington, D.C.: Brookings, 1979. Copyright © 1979 by the Brookings Institution.)*

The latter finding is consistent with one of the main driving forces behind ICUs: the knowledge of and desire for new technologies.

The mechanical respirator was one of the most important of these new technologies. Mechanical ventilators, which created a negative pressure around the patient, had been available since the 1930s to treat polio victims whose respiratory muscles had been paralyzed. These iron lungs required that the patient be placed entirely inside the respirator. Placing critically ill or postoperative patients in iron lungs, however, was incompatible with the other specialized care they also needed.[6]

An important breakthrough in mechanical ventilation came during a 1952 polio epidemic in Copenhagen. Among the patients stricken were 31 cases where extensive respiratory paralysis developed; twenty-seven of these patients died. The twenty-eighth patient, a 12-year-old girl, was also near death when the hospital's chief epidemiologist asked Dr. Bjorn Ibsen, an anesthesiologist, to recommend therapy. Employing an artificial breathing technique then commonly used for surgical anesthesia, Dr. Ibsen pushed oxygen into the young girl's paralyzed chest; she survived. Application of positive pressure ventilation quickly became the standard in cases of severe paralytic polio. As a result, the mortality rate declined from 87 percent to less than 40 percent.[8]

To replace the number of persons who used their hands to breathe for paralyzed patients (at one time 250 nurses and 250 medical students were used to provide continuous care to 75 patients in Copenhagen), a machine was developed. At first the technology was quite simple. A pump mechanism weighing less than 20 pounds delivered a breath by the stroke of a piston and was monitored by watching the patient's chest rise and fall. As our understanding and treatment of respiratory failure improved, however, so did the complexity of the ventilator. Alarms were added, humidification of gas was necessary for prolonged ventilation, and precise control of the volume of gas delivered became essential.[6]

As respirators increased in complexity and changed in design, indications for their use also expanded. Patients suffering from respiratory failure following surgery or persons in shock from a variety of causes became candidates for ventilatory support.

At first, patients requiring mechanical ventilation were treated on medical or surgical wards. One or two such patients, however, could demand much of a nurse's time, making it impossible for her to care for other patients. As a result, hospitals and physicians began concentrating ventilator patients in special areas. The demand for intensive care grew.

An interest in aggressive coronary care paralleled the development of assisted ventilation and the resultant growth of intensive care units. The concept of coronary care was first reported in the medical literature in 1963 by Dr. Hughes Day from Kansas City.[9] The coronary care unit (CCU) was based on the principle that continuous electrocardiographic monitoring of patients with acute myocardial infarction would enable physicians to identify and treat life-threatening arrhythmias, thereby reducing hospital mortality. Initially patients were kept in the CCU for their entire hospital stay; subsequently the length of CCU stay for most patients was reduced to three to five days when the highest risk for arrhythmias was found to occur in the first few days following the myocardial infarction. The underlying concept of coronary care, like that of mechanical ventilation for respiratory failure, was so attractive that it was rapidly accepted without direct proof of benefit. In less than a decade coronary care units became the accepted standard of care for all patients with known or suspected myocardial infarctions. During this period of time much more was learned about the natural history of acute myocardial infarctions, and concepts of therapy changed from a passive approach (wait for complications and then treat) to an anticipatory approach (identify high risk patients and intervene prophylactically).

Following this initial enthusiasm for multidisciplinary intensive care and coronary care units, other technological advances occurred which insured the continued demand for these units. These advances included new surgical procedures such as coronary artery bypass, as well as nonsurgical life-support mechanisms such as dialysis for renal failure and intra-aortic balloon pumping for cardiogenic shock.

Other technological advances facilitated diagnosis and management of critically ill patients and justified intensive care. An example of such a technology is the Swan-Ganz pulmonary artery catheter. Introduced in 1970, the catheter enables right heart catheterization to be done at the bedside, thereby providing continuous intracardiac and intrapulmonary pressure monitoring.[10] This information assists the physician in determining how much fluid or other therapy the patient may need. The catheter has achieved widespread use with an estimated two million catheters placed in the past decade.

The Swan-Ganz catheter is interesting in that it not only helped to increase demand for ICUs but represented a technology which, although relatively inexpensive (approximately $100 per catheter plus a variable professional fee for insertion), at the same time may be associated with high indirect or induced costs which can occur in several ways:

1. Additional nursing and technical personnel may be required for the care of patients once the catheter is inserted.
2. Additional laboratory tests may be ordered to avoid complications, confirm findings, or follow therapy initiated because of the presence of the catheter.
3. Complications may occur which necessitate other treatments and prolong ICU or hospital stay.[11]
4. The monitoring itself may prolong ICU stay (or in some cases may be the sole justification for ICU care).
5. The placement of the catheter may be the initial step in a commitment to a more aggressive therapy and may lead to other more costly interventions.

Thus in a period of less than 40 years, stimulated by technical advances like the Swan-Ganz catheter and mechanical respirator, ICUs rapidly evolved from a few beds set aside for postoperative care to multipurpose, multispecialty units comprising as many as 10 percent of the beds in some tertiary hospitals. Today they are the accepted standard of care for acutely ill patients.

Studies of Efficacy of ICUs

The turn toward ICU and CCU occurred without clear evidence that the patient received any direct benefit, except for polio victims and persons developing ventricular fibrillation. The few studies available

Table 1   Comparison of Mortality for Patients Treated with Mechanical Ventilation: June 1965–August 1968

	General Hospital Areas			On RICU		
	No. of Patients	No. of Deaths	Percentage of Deaths	No. of Patients	No. of Deaths	Percentage of Deaths
COPD	11	6	55	27	5	19
Neurologic disorders	46	35	76	26	5	19
Pneumonia	7	5	71	10	5	50
Drug ingestion	6	2	33	2	1	50
Miscellaneous	32	28	88	14	8	57
Total	102	76	75	79	24	30

SOURCE: R.M. Rogers, C. Weiler, and B. Ruppenthal (1972). Impact of the respiratory intensive care unit in survival of patients with acute respiratory failure. *Chest* 62:96.

represent well-intentioned inquiries but are limited by design, methodology, and the rapid changes in ICU treatments.

A study by Rogers and colleagues[12] at the Hospital of the University of Pennsylvania evaluated the efficacy of respiratory intensive care (RICU). They examined age, diagnosis, outcome, and type of ventilator used among 212 respiratory failure patients between 1965 and 1968. These patients were all treated on a general medical or surgical ward without the skills of specialized respiratory physicians or nurses. In the second part of this study, initiated after the RICU opened in 1968, the first 200 patients were followed and their outcomes determined. When the study was completed, the findings showed that 75 percent of the patients in the study group had died compared to 30 percent of the patients treated in the RICU (see Table 1).

In the discussion accompanying this study, Dr. Rogers and his co-authors were careful not to ascribe the difference in survival directly to the geographic isolation of the respiratory care unit. They pointed to the "increased awareness and skill of the professional and non-professional staff of the hospital in respiratory care" that resulted from concentrating staff and patients in an ICU.[12]

This study failed to control for severity of illness and for other potentially important differences in the two patient populations. Questions regarding why and when patients were placed on respirators, their underlying physiologic abnormalities, and duration of treatment were not addressed. These different factors could have introduced important biases into the study design. For example, the authors admitted that patients were diagnosed as having respiratory insufficiency earlier and were treated more aggressively after the respiratory team was established than before. This might have introduced an important selection bias by bringing to the unit more patients who were destined to survive, regardless of therapy. Nevertheless, the marked improvement in survival strongly suggests that an intensive care unit is indicated for patients being treated for respiratory failure. What then is the specified treatment for those patients who may be acutely ill but for whom a respirator is not needed?

A study performed by Dr. Paul Griner at the University of Rochester, similar in design and timing to that of Rogers, addresses this question.[13] Griner compared patients with acute pulmonary edema before and after an ICU was opened at Strong Memorial Hospital. His research indicated that mortality did not change after the unit was opened but remained stable at 8 percent.

As can be seen in Table 2, Griner did find that ICU patients were treated more aggressively. ICU admissions were placed more frequently on mechanical ventilators, had more arterial blood studies drawn, remained in the hospital longer than ward patients, and incurred substantially higher charges. In this study, as in the Rogers report, the issue of severity of illness was not addressed. In his discussion Griner states that patients should not be any sicker in 1971 (when the ICU was available) than they were in 1969 (when it was not).[13]

Although limited in both design and scope, these two studies realistically summarize early attempts at evaluation of ICUs. They suggest that intensive care may be essential for some patients but may be of marginal demonstrable value for others. These same conclusions appear in other studies that sought to demonstrate the value of coronary care units.

In the decade following Day's 1963 publication,[9] many nonrandomized studies were done comparing the outcome of CCU and non-CCU care in patients with myocardial infarction.[14-20] In most of these studies, patients were assigned to CCU or non-CCU care based on physician preference or bed availability (there were fewer CCU beds then). Within the approximately 2,500 patients included in these surveys, the cumulative hospital mortality for non-CCU cases was 34 percent and for CCU cases, 21 percent. This difference is certainly significant and could be explained by the effectiveness of the CCU in eliminating sudden arrhythmic death. The difference could be explained, however, by variations in patient population or by other factors unrelated to CCU care. A study by Hill,[21] in which no difference in outcome was found once the data were analyzed according to patient age, supported this possibility. It is noteworthy that the total mortality in this study (both CCU and non-CCU populations) was 16 percent—considerably lower than the hospital mor-

Table 2 Experience of Patients Admitted with Diagnosis of Acute Pulmonary Edema Before and After Opening of Intensive Care Unit

	Before Opening of Unit. 1969–1970	After Opening of Unit		Total Admissions 1970–1971
		Unit Admissions	Nonunit Admissions	
Episodes (number)	38	18	20	38
Deaths (number)	3	2	1	3
Intubations (number)	8	14	1	15
Blood gas analyses (number)	21	98	13	111
Complications (number)	2	3	1	4
Survivors 6 months after discharge (percent)	73	—	—	67
Mean time (and range) in emergency room hours)	11.8 (1–29)	4.3 (1–12)	11.7 (4–24)	7.7 (1–24)
Mean time (and range) on unit* (days)	—	4.8 (1–12)	—	—
Mean day (and range) permitted up in chair*	5.3 (2–13)	6.1 (3–12)	4.4 (2–10)	5.2 (2–12)
Mean duration (and range) of hospitalization* (days)	13.7 (2–39)	18.3 (9–35)	13.9 (5–28)	16† (5–35)
Mean (and range) of total hospital bill* (dollars)	1844.00 (295–4379)	3448.00 (1909–5811)	1893.00 (603–3363)	2694.00† (603–5811)

Source: P.F. Griner (1972). Treatment of acute pulmonary edema: Conventional or intensive care. *Ann. Intern. Med.* 77:504.

* Excludes patients who died during hospitalization.
† $P > 0.10$ for the difference between 1969 to 1970 and 1970 to 1971.
‡ $P < 0.01$ for the difference between 1969 to 1970 and 1970 to 1971.

tality reported in the studies a decade earlier. This finding suggests a general decline in hospital mortality from myocardial infarction independent of CCU care per se and may be attributable to better diagnosis and therapies. Such advances (if they are real) may not have occurred without the benefit of knowledge gained from the early years of coronary care units; we may now be seeing the application of the principles of CCU care outside of the CCU.

The efficacy of the CCU has been called into question by two randomized studies[22,23] from Britain comparing home and hospital care for patients with myocardial infarction. Each of these studies failed to show a difference in outcome between the two treatment groups. In each study, patients were screened for an absence of complications prior to randomization; thus each study started with relatively low-risk patients, a group for whom a difference in outcome might be difficult to show. Both studies involved a small number of patients, and the health care system in Britain was geared to provision of nurse and physician surveillance for home care of patients with myocardial infarction. Despite these qualifications, however, the results of these studies prompt us to question whether CCU treatment is necessary for all patients with myocardial infarction.

ICU Utilization

The rapid growth of ICUs continued as more specialties became convinced that the benefits found in coronary care units could be extended to their patients. In the early 1970s, however, concern was raised over the costs generated by this large and growing investment. Investigators began to recognize that a small number of patients—those with the poorest outcomes—consumed a disproportionate amount of ICU resources. In 1972 Cullen and his colleagues,[24] in the postoperative acute care unit at the Massachusetts General Hospital, identified 226 critically ill patients with a one-year survival rate of 25 percent and a full recovery rate of 12 percent; their mean hospital bills were in excess of $14,000.[24] This study was followed by others citing the inverse relationship between cost and survival.[25,26]

These observations prompted attempts to improve the utilization of ICUs by identifying patients who could and could not benefit from intensive care. Rather than concentrating on patients who might not require intensive care, such as the low-risk, British heart attack victims, American researchers chose the other end of the spectrum—patients who had the greatest risk of dying.

Investigators performed sophisticated physiologic studies and constructed elaborate severity of illness indexes that would enable them

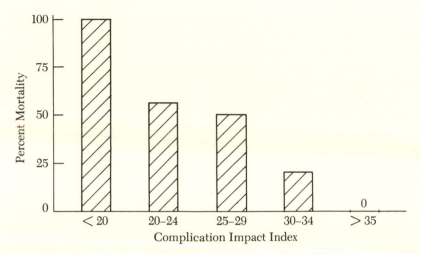

Figure 2 Mortality Percentage Plotted as a Function of Various Ranges of the Complication Impact Index (CII). Patients can be separated into groups in which all or none survive. Linear relationship between mortality and subgroupings further supports the validity of this approach. *(SOURCE: J. M. Civetta, Selection of patients for intensive care. In* Recent Advances in Intensive Therapy, *ed. I. M. Ledingham. New York: Churchill Livingstone, 1977, p. 16.)*

to predict the probability of survival in critically ill patients.[27–31] By a variety of statistical techniques, they were able to identify ICU patients with a 94 percent, 95 percent, and even 100 percent chance of dying. Such projections were possible shortly after ICU admission. Figure 2 illustrates one example of these efforts, the Complication Impact Index developed by Joseph Civetta.[27]

None of these severity of index scales, however, gained acceptance, and to our knowledge none has ever been used extensively outside the developing institution. One reason for the failure of most of these prognostic scales was that their utilization of complicated statistical techniques was hard for physicians to adopt. Another reason was that the formal use of individual prognosis, like that of acceptable risk, was not yet part of American medical practice.[32,33] American physicians have always felt obligated to provide maximal effort for their patients. Thus a severity of illness index indicating 100 percent mortality implies survival is unprecedented, not impossible. Finally, ethical issues are involved in the decision to withhold therapy in acutely ill patients.[34]

If physicians are unwilling to withhold therapy in acute situations, however, because of their inability to predict the outcome with certainty, they do appear willing to make similar judgments in patients with chronic, terminal illness. Physicians on the staff of the Sloan-Kettering Memorial Cancer Center in New York developed a classification system based on the prognosis of the patient's cancer.[35] Under their system all admissions were placed into one of four classes: Class A—good long-term prognosis; Class B—limited long-term prognosis; Class C—short-term prognosis poor; and Class D—no further therapy.

A patient whose underlying prognosis was poor (Class D) was not a candidate for transfer to the ICU, even if complication developed in the hospital. A subsequent survey[26] at Memorial suggested that implementation of this more selective admission policy reduced the in-unit mortality from 22 percent to 18 percent and increased by 10 percent the number of ICU patients discharged from the hospital alive.

Although the situation at Memorial was unique, the concept of limiting ICU admission based on poor prognosis is becoming increasingly accepted. Acute problems developing in institutionalized patients may go recognized but untreated. Studies[36] evaluating the opportunity for survival of young adults suffering from progressive cystic fibrosis have led to informal policies limiting their treatment on ventilators. Patients with long-term and presumably end-stage chronic pulmonary disease are, at times, denied ICU admission, not only because the ICUs have little to offer such patients, but because more nurses, physicians, families, and patients have experienced despair and increased suffering when heroic treatment fails in the face of chronic disease.[37]

Because of these experiences, which were brought to national attention by the Karen Ann Quinlan case, we believe that progress will continue in limiting the use of ICUs for treating hopelessly ill patients.[38] Experience to date, however, also suggests that physician practices will be slow to change, and the legal and ethical issues will need to be resolved more clearly.

A more immediate approach to improved ICU utilization lies in identifying low-risk patients who are placed in the ICU for monitoring purposes alone. These patients are stable at ICU admission but are at risk for developing a complication or disease that would require intensive care.

In a recent survey[39] of 624 patients in the multidisciplinary ICU of the George Washington Hospital, approximately 46 percent were found not to be critically ill. These patients were admitted for a variety of reasons, including recovery from major elective surgery and self-inflicted drug overdoses. As a group they were at low risk for requiring subsequent active treatment. According to the definitions employed in the study, only 24 of 289 initially monitored patients, or 17 percent, needed intensive treatment prior to discharge. A survey[40] of medical ICUs/CCUs in the Massachusetts General Hospital, with 2,693 consecutive admissions, found that 77 percent of the patients were admitted for monitoring; only 10 percent of these monitored patients subsequently received major interventions.

These studies suggest that even though critically ill patients remain the largest individual consumers of intensive services, low-risk patients now may be the largest aggregate consumers of care, representing a change in the indications for intensive care that has occurred over the past two decades. This change results from ICU beds becoming more plentiful and from a perception, increasingly held by physicians, that conventional nursing services are not able to treat many of these noncritically ill patients. This trend is documented by comparing the above data with findings from a surgical intensive care unit from 1968 to 1971 in which only 21 percent of patients were admitted for monitoring.[41] An increase in monitoring also may be seen in the review of the coronary care unit literature,[14-20] which shows that the incidence of documented myocardial infarction (MI) in patients admitted with suspected MI has fallen from greater than 50 percent to levels between 20 and 30 percent. This decline in diagnosis of MI has occurred despite the fact that the tests now available for the diagnosis are more sensitive than they were a decade ago.

These findings indicate that the threshold for admitting patients to an ICU is lower today than during the time the units were first developed. From the perspective of the individual physician, the reasoning behind increased admission is clear: (1) there are some complications for which effective therapy is now available in an intensive care setting; (2) my patient has some unknown risk of such a complication; (3) since I cannot predict what is going to happen to my patient, I should not take any chances and should place him/her in the ICU. Given this sequence of thought, a more efficient utilization of the ICU for these patients lies in identifying which complications are, indeed, better treated in an ICU and, more importantly, improving the physicians' ability to predict which low-risk patients will or will not have such complications.

Early studies at improving ICU utilization by aiding physicians' ability to predict low risk have employed two kinds of approaches. The first approach has attempted to better define diagnostic criteria for admission to a CCU. To assist in selecting patients for CCU admission, Dr. Pozen and his colleagues[42] developed a simple arithmetic algorithm increasing the precision with which physicians either choose patients for the CCU or send them home. Pozen tested his instrument in the Boston City Hospital emergency room. During one month, residents evaluating patients were free to make use of computer prognosis when deciding whom to admit; during the next month they made their decisions unaided. After a year the results were compared. During months when the predictive instrument was available, overall diagnostic accuracy increased from 83 percent to 91 percent and the admission rate fell from 26 percent to 14 percent, without sending any more patients home with infarctions in control months versus experimental ones.

The second approach reviewed information generated soon after ICU admission and has identified low-risk patients who can be rapidly transferred elsewhere. This approach has been used by Mulley and his colleagues[43] at the Massachusetts General Hospital to identify low-risk, suspected MI patients who could be transferred from the CCU within 24 hours rather than after the standard three days. A similar approach is being pursued by two multidisciplinary ICUs in Washington and Boston to identify low-risk neurosurgery patients who might receive more rapid triage out of intensive care or might not be admitted in the first place. These later two studies,[44,45] although performed independently, both identified identical subgroups of neurosurgical admissions whose potential for developing serious problems was extremely small.

To date, no randomized studies have been done to test the hypothesis that low-risk patients can be identified and safely cared for outside the ICU. In the face of physician reluctance to formally take such risks, the medical profession will have to perceive real advantages to such studies. The impetus may be provided by the following three observations.

1. Some patients may actually be harmed by a stay in the ICU. This harm may be the consequence of unnecessary tests or therapy that would not have been administered in another setting or of psychological stress from a stay in the ICU.[46]
2. With more patients being admitted to ICUs and with a more stringent examination of capital expenditures and the decreased supply of trained nurses, the supply of ICU beds or the personnel to staff them may be decreased. Physicians will then need to make triage decisions and will want better information for these decisions.[47]
3. With the rising cost of medical care, physicians increasingly will be asked to take a more active role in cost containment. Since the ICU is one of the most rapidly rising segments of the total health care bill, it is a logical place to start such scrutiny.

The Cost of Intensive Care

Table 3 compares room charges to ICU charges for 18 hospitals in the Washington, D.C., metropolitan area. ICU charges are, on the average, approximately three times larger. Of the total days spent

Table 3   Comparison of Room Versus Intensive Care Unit Charges for Eighteen Hospitals in the Washington, D.C., Metropolitan Area

Hospital	Semiprivate Room	Intensive Care Unit
Alexandria	125	335
Arlington	126	350
Capitol Hill	180	350
Childrens	220	750
Circle Terrace	122	352
Commonwealth Doctors	118	350
Fairfax	118	350
Georgetown University	198	460
George Washington University	234	690
Greater Southeast	180	388
Howard University	215	500
Montgomery General	139	208
Mount Vernon	118	350
Prince George's General	120	490
Sibley Memorial	120	372
Southern Maryland	135	405
Suburban	112	283
Washington Hospital Center	202	574

Source: Based on information supplied to Group Hospitalization, Inc., Washington, D.C.

in a short-term hospital, five percent were spent in ICUs. Therefore, projecting the threefold increase, we estimate that the national bill in 1979 for ICU expenditures was $12.8 billion. This represents approximately 15 percent of the total nationwide hospital charges.

Due to these large national costs, we feel that the growth in the number of the ICUs merits close examination. The cost of each new or converted ICU bed is estimated to be from $44,000 to $75,000.[48] If the three percent per year growth in ICUs recorded in 1979 continues, 1,650 new beds a year, or a capital expenditure of $123,750,000, will result. These large investments in technology will need careful review regardless of specific future national health policy decisions.

An examination of ICU costs by the Arthur D. Little Co.[48] found that the largest single component of an average daily cost of $664 was ancillary services at $410. Facilities and equipment were estimated at $28, overhead at $46, and expenditures for personnel at $180. Although this limited survey is not applicable on a nationwide basis, the concentration and high use of ancillary services within most ICUs, highlighted by the study, has important implications in terms of cost savings. If we assume that not admitting a patient to an ICU or CCU will reduce by only 10 percent the total amount of ancillary services they receive, the savings could conservatively be estimated at $468 million. Furthermore, the possibility exists that many patients (particularly the low-risk patients) could receive more efficient care outside the ICU setting, even if the care itself were not different.

Future Evaluations of ICUs

The task of ICU evaluation has been overwhelming because of the multiplicity of variables and the impossibility of conducting "experiments" on acutely ill patients. Our research and that of others, however, suggests that the task, though difficult, is not impossible, and certain principles can help to guide future efforts.

1. *Intensive care units should be viewed as an aggregate technology aimed at physiologic monitoring and treatment.* ICUs exist to monitor and treat one or more of the body's major physiologic systems; CCUs concentrate on the cardiovascular system. ICU treatment encompasses many different organ systems. Because of the concentration on physiologic monitoring and treatment, the diagnosis that brings a patient to the ICU is not as important as it is in the regular hospital setting. Because they are selected by their need for discrete physiologic monitoring or treatments, a smaller range of patients can be found in ICUs than in a general hospital.

Whether treating medical or surgical cases, ICUs employ similar physiologic principles of life support and draw on a limited number of interventions. If this fact is recognized, many of the seemingly endless varieties of equipment, personnel, and institutional arrangements for intensive care become less confusing. In this regard a major contribution was the development of the Therapeutic Intervention Scoring System, a method of quantifying resource utilization.[24]

This does not mean that future research efforts can ignore the important clinical distinctions between units that treat heart attack victims and units that treat burn victims. The need to carefully describe the characteristics of units being compared will always remain.

2. *Detailed evaluation and assessment of intensive care is possible using current methodology.* Because ICUs concentrate on physiologic abnormalities, improvements in patient classification is possible. The ICU research unit at George Washington University has recently developed the APACHE (Acute Physiology and Chronic Health Evaluation) classification system for critically ill patients.[49] APACHE combines an acute physiology score with a letter classification indicating the preadmission health status of the patients. In validation studies, the acute physiology score of APACHE has been strongly related to both patient outcome and therapeutic requirements and should be an important tool in measuring severity of illness among

various critically ill patient groups. Use of the APACHE system has already provided new insight into the relationship between intense medical care and outcome.

In a recent study of 613 ICU patients,[50] the APACHE classification was combined with Cullen's Therapeutic Intervention Scoring System, indicating that the association between intense medical treatment and survival is not as dismal as was once thought. In contrast to previous findings suggesting that more treatment decreased a patient's chance of survival, this study found that the relationship between increasing treatment and the probability of death was non-linear, with some patient groups having a higher probability of survival with increased therapy. Future studies measuring severity of illness will help to better elucidate the efficacy of intensive care for specific patient groups.

Since the majority of CCU patients suffer from chest pain suspicious of acute myocardial ischemia, their major differences are also in respect to their severity of illness. Reliable and well-validated systems are available to measure the severity of illness of CCU admissions.[51,52] Using these systems, comparisons could be made among units in regard to outcome. The influence of new therapies could also be determined, and various studies suggest patients who might be early subjects for these investigations.[42,43] Finally, the increased use of severity of illness indexes can reduce dependence on randomized clinical trials. By controlling for severity of illness in similar patients who receive different therapies because of naturally occurring variations in medical care, the incremental value of medical care can be estimated.[50]

3. *Evaluation of intensive care units can be justified on the basis of cost savings, but this is not the only reason.* The continuing demand for ICU beds is another reason. Currently the number of ICU beds is increasing at a rate faster than regular hospital beds.[3] Unless indications for ICU admission are better defined, this demand will continue to grow. In the future, hospitals may designate 50 percent of their beds to intensive care; once built, these units will be used by an increasing number of patients. ICU evaluative research may help to reduce this demand for more beds and to make triage decisions if and when there are fewer beds.*

The decreased ability of nurses on regular hospital floors to treat

---

* In some instances ICU evaluative research will result in the conclusion that *more* intensive care is needed rather than less. A study underway at Massachusetts General Hospital is attempting to identify the risk factors in patients who have serious complications after discharge from a coronary care unit. If these patients can be identified prospectively, they may be candidates for longer ICU stays and more intensive therapy.

critically ill patients is a noneconomic cost of the emphasis on and increased demand for intensive care. As was feared in the late 1960s, the concentration of labor-intensive patients in special care units reduces the nursing skills available on the general hospital ward,[53] and the increasing shortage of hospital nurses creates concern for this trend.[54] The interaction of an increasing demand for ICU nurses, decreased skills on the part of floor nurses, and a diminishing total supply of nurses could help precipitate a future crisis in hospital care. Many ICUs already have shortages of nurses and, as a result, an increased risk of complications. A recent report from an intensive care unit at the University of Pittsburgh found a high rate of human error and equipment malfunction;[55] 65 percent of the adverse incidents occurred during times when availability of nurses was not sufficient.

Another reason for scrutinizing ICUs is that many aspects of ICU care are dangerous. Although most procedures are performed at low risk, we are now beginning to realize that expertise in invasive procedures varies widely, requiring constant supervision. Reports from a number of centers have recently emphasized the wide range of patient hazards encountered within an ICU. Although, as expected, those patients at highest risk for complications are also those who are most critically ill, the increased application of diagnostic and treatment technologies increases the risk for all patients.[56] Likewise, the acute confusion and disorientation, "ICU psychosis," that occurs as a result of acute illness and patient isolation in the ICU has been reported for all classes of ICU admissions.[46]

Finally, some physicians complain that they also become isolated once their patient is admitted to an ICU.[37] It is difficult and sometimes impossible for physicians who have had long associations with patients to maintain control over situations where protocols and technology appear to dominate. Although we feel this latter problem is avoidable in a properly run unit, isolation of the physician, as well as the patient, should be considered as additional costs of intensive care.

4. *Evaluation of intensive care units will require more research into decision analysis and acceptable risk.* We predict that the greatest advances in the utilization of intensive care units will occur through a better description of the low-risk patient combined with more selective admission policies for chronically ill patients. We have seen formal and informal moves in this direction by the medical community;[35,36] both involve areas where probability estimates and decision analysis could prove helpful.[57] Is there an acceptable risk for not placing patients in an ICU? How should we integrate probability estimates into clinical decision making? New ways to provide information to physicians are needed so they can approach some clinical decisions in a probabilistic way. Technology is now available to develop computer databanks to store, retrieve, and analyze infor-

mation so that a physician can use a computer terminal to establish the probability of a given diagnosis or outcome for the patient in question. More experience is needed, however, to incorporate the "loss function" of the physician or patient in a given situation. What is the relative negative value assigned to making an error in prediction? A standard which is highly specific but less sensitive will run the risk of missing a diagnosis or complication but will not inappropriately diagnose (or treat) many patients. Conversely, a highly sensitive but less specific standard will miss a few people at risk but will also include many who are not.

Therefore, in different situations, different values may have to be assigned these two outcomes.[58] The use of different diagnostic tests to predict the probability of myocardial infarction and its complications may serve as an example. On admission we would want to administer a very sensitive test so as to miss as few patients as possible during the critical first 24 hours. We realize that more patients than will actually have the diagnosis of complication will be admitted. With the decline in acute risk that occurs after 24 hours in the CCU, we want the diagnostic test to be as specific as possible so that this diagnosis is not unnecessarily assigned to a patient. As research is unlikely to enable us to be 100 percent sensitive and 100 percent specific in our predictions, such choices will have to be made. At that time data must be generated to improve both the sensitivity and specificity of our predictions.

Conclusions

Drawing on a number of important technologic developments, ICUs have brought new capabilities to modern medical care. In the near future the efficacy of ICUs should become increasingly obvious. Preliminary evidence of an increase in survival for trauma and cardiovascular surgery patients over the past five years is available from a study by John Siegel and associates.[59] Recent analysis of survival of burn patients from 1965 to 1979 demonstrated that overall survival rates are increasing and hospitalization times are decreasing at all levels of burn severity.[60] The increased use of CCUs may have played a role in the recent decline in mortality from heart disease.[61]

By reversing acute physiologic derangements, and aggressively detecting problems before they progress, ICUs do save lives. While their services should be available to patients who can benefit from them, ICUs should not be used for patients who are either too healthy or too ill. To approach this goal, physicians will require more precise information regarding the need for ICU services by specific patient groups. A framework for integrating this information into clinical decision making will be essential.

In the future it may not be enough to say, "Build more ICUs" or "Don't build more ICUs." It will not be sufficient to state, as Massachusetts recently did, that ICUs be limited to 8 percent of total hospital beds or that the number of beds be related to overall hospital utilization.[62] We will need better information as the basis of our national decisions.[63] The goal of ICU research is to enable both physicians and others to make informed decisions that will ultimately lead to better and more cost-effective medical care.

Finally, as we have seen, ICUs are influenced more by events than time, and medical technology in the United States is constantly expanding its capabilities. Although many advantages will be possible because of support from ICUs, to expand ICUs in terms of numbers and services without a systematic attempt at evaluation is to ignore both scientific and social obligations. We have the tools to begin that evaluation today, and appropriate research can lead to the development of new and better ICUs tomorrow.

Endnotes

1. Thomas, L. (1980). The right track. *Wilson Quarterly* 4:87–98.
2. Russell, L.B. (1979). *Technology in Hospitals*. Washington, D.C.: The Brookings Institution.
3. Hospital Statistics (1979). Chicago: American Hospital Association.
4. Del Guercio, L.R.M. (1977). Triage in cold blood. *Crit. Care Med.* 5: 167–169.
5. Illich, I. (1976). *Medical Nemesis: The Expropriation of Health*. New York: Pantheon Books.
6. Hilberman, M. (1975). The evolution of intensive care units. *Crit. Care Med.* 3:159–165.
7. Ruth, H.S., Hawgen, F.P., and Grave, D.D. (1947). Anesthesia study commission: Findings of eleven years activity. *JAMA* 135:881–884.
8. Lassen, H.C.A. (1953). A preliminary report on the 1952 epidemic of poliomyelitis in Copenhagen with special reference to the treatment of acute respiratory insufficiency. *Lancet* 37–41.
9. Day, H.W. (1963). An intensive coronary care area. *Dis. of the Chest* 44:423.
10. Swan, H.J.C., Ganz, W., Forrester, J., Marcus, H., Diamond, G., and Chonette, D. (1970). Catheterization of the heart in man with use of flow-directed balloon-tipped catheter. *N. Engl. J. Med.* 283:447.
11. Swan, H.J.C. and Ganz, W. (1979). Complications with flow-directed balloon-tipped catheters. *Ann. Intern. Med.* 91:494.
12. Rogers, R.M., Weiler, C., and Ruppenthal, B. (1972). Impact of the respiratory intensive care unit in survival of patients with acute respiratory failure. *Chest* 77:501–504.

13. Griner, P.F. (1972). Treatment of acute pulmonary edema: Conventional or intensive care. *Ann. Intern. Med.* 77:501–504.
14. Kimball, J.T., Klein, S.W., Stingfellow, C.A., and Killip, T. (1966). Comparison of coronary unit and regular hospital care in acute myocardial infarction (P). *Circ. Abstracts of the 39th Scientific Sessions.* Vols. XXXIII and XXXIV, pg. III, 143.
15. Meltzer, L.E. (1968). *Acute Myocardial Infarction,* eds. D.G. Julian and M.F. Oliver. Edinburgh: Churchill Livingstone Ltd.
16. Sloman, G. and Brown, R. (1970). Hospital registration in patients with acute myocardial infarction. *Am. Heart J.* 79:756–761.
17. Marshall, R.M., Blount, S.G., and Genton, E. (1968). Acute myocardial infarction: Influence of a coronary care unit. *Arch. Int. Med.* 122:432–435.
18. Christiansen, I., Iversen, K., and Skouby, A.P. (1971). Benefits obtained by the introduction of a coronary care unit. *Acta Med. Scand.* 189:285.
19. Hofvendah, S. (1971). Influence of treatment in a coronary care unit on prognosis in acute myocardial infarction: A controlled study in 271 cases. Dept. of Medicine, Karolinska Institute at Serafimerlasaretter, Stockholm.
20. Fagin, I.D. and Anandiah, K.M. (1971). The coronary care unit and mortality from myocardial infarction: A continued evaluation. *J. Am. Geriatr. Soc.* 19:675.
21. Hill, J.D., Holdstock, G., and Hampton, J.R. (1971). Comparison of mortality of patients with heart attacks admitted to a coronary care unit and an ordinary medical ward. *Br. Med. J.* 2:81.
22. Mather, H.G., Morgan, D.C., Pearson, N.G., Read, K.L.Q., Shaw, D.B., Steed, G.R., Thorne, M.G., Lawrence, C.J., and Riley, I.S. (1976). Myocardial infarction: A comparison between home and hospital care for patients. *Br. Med. J.* 1:925.
23. Hill, J.D., Hampton, J.R., and Mitchell, J.R.A. (1978). A randomized trial of home-versus-hospital management for patients with suspected myocardial infarction. *Lancet* 22, April.
24. Cullen, D.J., Ferrara, L.C., Briggs, B.A., Walker, P.F., and Gilbert, J. (1976). Survival, hospitalization charges and follow-up results in critically ill patients. *N. Engl. J. Med.* 294:982.
25. Civetta, J.M. (1973). The inverse relationship between cost and survival. *J. Surg. Res.* 14:265.
26. Turnbull, A.D., Carlon, G., Baron, R., Sichel, W., Young, C., and Howland, W. (1979). The inverse relationship between cost and survival in the critically ill cancer patient. *Crit. Care Med.* 7:20.
27. Civetta, J. (1977). Selection of patients for intensive care. In *Recent Advances in Intensive Therapy,* ed. I.M. Ledingham. New York: Churchill Livingstone.
28. Afifi, A.A., Sacks, S.T., Liu, V.Y., Weil, M.H., and Shubin, H. (1971). Accumulative prognostic index for patients with barbiturate glutethimide and meprobamate intoxication. *N. Engl. J. Med.* 285:1497.
29. Cullen, D.J., Ferrara, L.C., Gilbert, J., Briggs, B.A., and Walker, P.F. (1977). Indicators of intensive care in critically ill patients. *Crit. Care Med.* 5:173.
30. Bartlett, R.J., Gazzaniga, A.B., Wilson, A.F., Medley, T., and Wetmore, N.

(1975). Mortality prediction in adult respiratory insufficiency. *Chest* 67: 680.

31. Shoemaker, W.C., Chang, P., Czer, L., Bland, R., Shabot, M., and State, D. (1979). Cardiorespiratory monitoring in post-operative patients. I. Prediction of outcome and severity of illness. *Crit. Care Med.* 7:237.

32. Imbus, S.H. and Zawack, B.E. (1977). Autonomy for burned patients when survival is unprecedented. *N. Engl. J. Med.* 297:308.

33. Cohen, H. (1977). Response to autonomy for severely burned patients. *N. Engl. J. Med.* 297:1182.

34. Critical Care Committee of the Massachusetts General Hospital (1976). Optimum care for hopelessly ill patients. *N. Engl. J. Med.* 295:362–369.

35. Turnbull, A.D., Goldiner, P., Silverman, D., and Howland, W. (1976). The role of an intensive care unit in a cancer center. *Cancer* 37:82–86.

36. Davis, P.B. and diSant'Angnese, P.A. (1978). Assisted ventilation for patients with cystic fibrosis. *JAMA* 239:1851–1854.

37. Schoenfeld, M.R. (1978). Terror in the ICU. *Forum* 1:14–17.

38. Reiser, S.J. (1977). Therapeutic choice and moral doubt in doing better and feeling worse. *Daedalus* 106:47–56.

39. Knaus, W.A., Wagner, D.P., Draper, E.A., Lawrence, D.E., and Zimmerman, J.E. The range of intensive care services today. *JAMA*, in press.

40. Thibault, G.E., Mulley, A.G., Barnett, C.O., Goldstein, R.L., Reder, V.A., Sherman, E.L., and Skinner, E.R. (1980). Medical intensive care: Indications, interventions, and outcomes. *N. Engl. J. Med.* 302:938–942.

41. Pessi, T.T. (1973). Experiences gained in intensive care of surgical patients: A prospective clinical study of 1,001 consecutively treated patients in a surgical intensive care unit. *Ann. Chir. Gynaecol.* (Suppl.) 62:1–72.

42. Pozen, M.W., D'Agostine, R.B., Mitchell, J.B., Rosenfeld, D.M., Guglielmino, J.T., Schwartz, M.L., Teebagy, N., Valentine, J.M., and Hood, W.B. (1980). The usefulness of a predictive instrument to reduce inappropriate admissions to the coronary care unit. *Ann. Intern. Med.* 92:238–242.

43. Mulley, A.G., Thibault, G.E., Hughes, R.A., Barnett, G.O., Reder, V.A., and Sherman, E.L. (1980). The course of patients with suspected myocardial infarction: The identification of low-risk patients for early transfer from intensive care. *N. Engl. J. Med.* 302:943–948.

44. Knaus, W.A., Draper, E., Wagner, D.P., and Zimmerman, J.E. (1981). Neurosurgical admissions to the ICU: Intensive monitoring versus intensive therapy. *Neurosurg.* 8:438–442.

45. Teplick, R., Caldera, D.R., Gilbert, J.P., and Cullen, D.J. (1979). Benefit of elective intensive care admission after certain operations. (Abstract), *Anesthesiology*, S150.

46. Hackett, T.P. (1976). The psychiatrist's view of the ICU: Vital signs stable but outlook guarded. *Psychiatric Annals.* 6:10.

47. McPeek, B., Gilbert, J.P., and Mosteller, F. (1980). The clinician's responsibility for helping to improve the treatment of tomorrow's patients. *N. Engl. J. Med.* 302:630.

48. Arthur D. Little, Inc. (1979). Planning for intensive care units: A technical assistance document for planning agencies. DHEW Publication No. (HRS) 79–14020. Washington, D.C.: U.S. Government Printing Office.

49. Knaus, W.A., Zimmerman, J.E., Wagner, D.P., Draper, E.A., and Lawrence, D.E. APACHE—Acute physiology and chronic health evaluation: A physiologically based classification system. *Crit. Care Med.*, in press.

50. Scheffler, R.M., Knaus, W.A., Wagner, D.P., and Zimmerman, J.E. Severity of illness and the relationship between intense treatment and survival, forthcoming.

51. Peel, A.A.F., Semple, T., Wang, I., Lancaster, W.M., and Dall, J.L.G. (1962). A coronary prognostic index for grading the severity of infarction. *Br. Heart J.* 24:745.

52. Norris, R.M., Brandt, P.W.T., Caughey, D.E., Lee, A.J., and Scott, P.J. (1969). A new coronary prognostic index. *Lancet* 1:274.

53. Skidmore, F.D. (1973). A review of 460 patients admitted to the intensive therapy unit of a general hospital between 1965 and 1969. *Br. J. Surg.* 60:1–16.

54. Knaus, W.A. (1980). Nurses can no longer be taken for granted. *The New York Times*, 24 November.

55. Abramson, H.S., Wald, K.S., Grenvik, A., Robinson, D., and Snyder, J.V. (1980). Adverse occurrences in intensive care units. *JAMA* 244:1582–1584.

56. Puri, V.K., Carlson, R.W., Barder, J.J., and Weil, M.H. (1980). Complications of vascular catheterization in the critically ill. A prospective study. *Crit. Care Med.* 8:495–499.

57. Weinstein, M.C., and Fineberg, H.V. (1980). *Clinical Decision Analysis*. Philadelphia: W.B. Saunders.

58. Bendixen, H.H. (1977). The cost of intensive care. In *Costs, Risks, and Benefits of Surgery*, eds. J.P. Bunker, B.A. Barnes, and F. Mosteller, pp. 372–384. New York: Oxford University Press.

59. Siegel, J.K., Cerra, F.B., Moody, E.A., Shetge, N., Coleman, B., Garr, L., Shubert, M., and Keane, J.S. (1980). The effect on survival of critically ill and injured patients of an ICU teaching service organized about a computer-based physiologic care system. *J. Trauma* 20:7:558–579.

60. Feller, I., Tholen, D., and Cornell, R.G. (1980). Improvements in burn care, 1965 to 1979. *JAMA* 244:2074–2078.

61. Stern, M.P. (1979). The recent decline in ischemic heart disease mortality. *Ann. Int. Med.* 91:630–640.

62. Determination of need guidelines for intensive and coronary care units. Department of Public Health, State of Massachusetts, 1978.

63. Relman, A.S. (1980). Intensive care units: Who needs them? *N. Engl. J. Med.* 302:965–966.

CAN RANDOMIZED TRIALS OF INTENSIVE CARE MEET ETHICAL STANDARDS?

by Sankey V. Williams

The innovative clinicians who first use new procedures often become convinced of their usefulness early on, and are eager to apply them to control groups because of their reputed effectiveness. Randomized trials of these new diagnostic and therapeutic procedures are difficult to conduct. Procedures which have become widely accepted by practicing clinicians are even more difficult to study by means of randomized trials. Widespread acceptance implies that the procedures are useful and that a denial of these procedures to control groups is therefore unethical. This situation obtains when intensive care units are considered for the treatment of patients with acute myocardial infarction. Despite suggestive evidence that standard hospital care is as cost effective as intensive care for uncomplicated acute myocardial infarction, no randomized trial has been reported and none is currently in progress in this country. Such a trial would have to be large, complex, and expensive, and these factors alone are strong enough to prevent such trials from being conducted. The initial and most important barrier, however, is that satisfactory measures have not been found to answer honest concerns about the ethics of conducting such a study.

Over the past twenty years, medical intensive care units have become standard treatment for patients with acute myocardial infarc-

This work is supported by the National Health Care Management Center, Grant #HS02577 from the National Center for Health Services Research, Office of the Assistant Secretary for Health, DHHS. The opinions, conclusions and proposals in this article do not necessarily represent those of the organizations that funded this work.

The following people deserve recognition for the contributions they have made to the effort which led to this article: Ellen Smith, Lisa Kauffman, Osler Peterson, and Peter Plantes.

tion.[1] As with many technological innovations, however, rapid diffusion of intensive care units did not wait for rigorous evidence of effectiveness. Several reports in the late 1960s described reductions of 50 percent in early hospital mortality, largely because of successful therapy of life-threatening arrhythmias.[2,3] These results led to an enthusiastic acceptance of intensive care units, although most studies used weak research designs with nonequivalent control groups. Comparisons were made between either institutions with or without intensive care units or before and after the establishment of a unit within a single institution. Two studies were made in which patients were assigned to intensive care units or standard wards by bed availability; these studies also demonstrated improved survival for patients who received care in the special units.[4,5]

More recently several studies suggest intensive care may not be necessary for all patients with myocardial infarctions. Hill and others[6] performed a study in which patients with myocardial infarctions who were admitted to the intensive care unit were compared with patients admitted to the general medical ward due to the shortage of intensive care beds. When patients were stratified by age, no significant differences between the two groups were seen in early morbidity or mortality. In Britain two randomized, controlled studies of home versus hospital care for patients with uncomplicated myocardial infarctions demonstrated no significant difference in mortality between two groups of patients carefully selected for the absence of complications.[7,8] These results suggest that hospital and thus intensive care may not improve survival for this subgroup of patients.[9]

While the benefits of intensive care have not been adequately documented, their high costs have been. In 1979 estimated costs for construction of intensive care units were as high as $75,000 per bed, with an additional $25,000 for equipment.[10] Daily charges for intensive care at the Hospital of the University of Pennsylvania in Philadelphia are $371 compared with $215 for a semiprivate room (not including the cost of ancillary services). Others have found the daily cost of intensive care to be approximately three times the cost of care in a standard hospital bed.[1,11,12]

Conflicting evidence exists in regard to the ability of intensive care units to improve the outcome of patients with uncomplicated acute myocardial infarction. At the same time the high cost of intensive care diverts resources from other, potentially more useful, patient care services. Under these circumstances it seems appropriate to define the role for intensive care using a controlled clinical trial in which some patients would be randomly allocated either to the intensive care unit or to a standard hospital bed. Elsewhere in this volume, Sackett argues for the use of randomized clinical trials in situations where weaker research designs have not been able to provide definitive answers. Most clinicians, however, remain convinced of the probable benefit of intensive care and cite their own experience and the widespread acceptance of intensive care units by their colleagues as reasons why such a trial would not meet ethical standards.

This paper describes a strategy for determining whether ethical standards can be met by a randomized clinical trial of intensive care for acute myocardial infarction. The strategy consists of three stages: (1) selection of appropriate hospitals for study; (2) development of a measure to predict which patients will have uncomplicated infarctions and thus will qualify for the study; and (3) use of an observational study to confirm that intensive care does not provide a benefit large enough to be measured easily.

Selecting Units for Study

Perhaps the most important stage involves selecting appropriate hospitals for study. At least five criteria must be considered.

1. Large numbers of patients will be needed to reach convincing conclusions.
2. The institutions chosen should represent the mix of hospitals that now care for these patients.
3. Special monitoring equipment should be available.
4. Each hospital should have a policy that allows patients to be discharged prematurely when the intensive care unit is full and a new patient requires admission.
5. Hospital administrators, physicians, and patients must be willing to participate in such a study.

To calculate the number of patients needed for such a study, the following assumptions have been made:

1. During the hospital stay the mortality rate in the control group is 4 percent and the rate of other adverse outcomes is 30 percent.[6,7,13,14]
2. The acceptable error for concluding that a difference exists when, in fact, there is no difference (the alpha error) is 5 percent.
3. The acceptable error for concluding that no difference exists when, in fact, there is a difference (the beta error) is 10 percent.

Each group would require 476 patients to detect a change in adverse outcomes from 30 percent in the control group to 40 percent in the study group. Although these are large groups, the caseloads of some hospitals are sufficient to make recruitment realistic. In one medical intensive care unit at Massachusetts General Hospital, a mean of 524 patients are admitted each year with precordial pain and suspected acute myocardial infarction.[15] Assuming that one half of such patients would qualify for the study and that, of these, one half would consent to participate in the study, the question about possible differences in complication rates between the two types of care could be answered in two years if data were collected in four such hospitals.

Not all hospitals included in the study, however, should be university medical centers. If the conclusions are to be widely used for changing patient management, the results must reflect the treatment provided in other types of hospitals, particularly community hospitals, where the majority of such patients receive their care.

There are additional restrictions on hospital selection. Patients randomized to standard wards for their care may require therapy for life-threatening cardiac arrhythmias. To protect them, special monitoring equipment should be available. Selection may have to be restricted to hospitals which are able to offer this special monitoring to patients who are randomized to the control group.

In recent years many hospitals have begun to use computer-assisted telemetry equipment that allows patients to be monitored for cardiac arrhythmias while they are in standard hospital beds. This equipment combines conventional monitoring leads with a small battery-powered radiotransmitter that can be worn by the patient, and a continuous signal is transmitted to a central receiver in the intensive care unit. The frequency and configuration of the heart's electrical signals can be examined continuously by a small computer with the capability of identifying potentially dangerous arrhythmias, including, for example, frequent premature ventricular contractions and ventricular tachycardia. The computer can be set to sound an alarm as soon as a potentially dangerous arrhythmia occurs, thus allowing nurses and physicians to respond promptly with appropriate therapy. In most hospitals, this equipment is used to supplement the monitoring equipment available in the intensive care unit. Generally patients with an acute myocardial infarction are admitted directly to the unit where they are monitored for several days. They are then discharged from the unit to a standard hospital bed where the telemetry equipment is used to continue cardiac monitoring.

When used for patients located near the unit and supported by well-trained personnel, this telemetry equipment permits care which approximates the care delivered in the intensive care unit, if the number of arrhythmias detected and the response time for therapy are considered.* Telemetry equipment allows a relatively large

---

* Stephen Corday, M.D., personal communication.

number of patients to be cared for by a relatively small number of intensive care nurses and physicians, thus reducing personnel costs. Because of the dramatic decrease in the cost of information processing, this equipment is less expensive to use than the full range of facilities provided in the unit.

Before considering a hospital for inclusion in the study, one additional special feature should be present. To conduct an observational study to confirm the need for a randomized controlled trial, the unit must be busy enough to occasionally discharge patients prematurely from the unit to standard beds. Premature discharge would occur when the unit is full and new patients who are more critically ill require admission. This will be explained in more detail when the third stage of the strategy is discussed.

Finally, to recruit hospitals to the study, physicians and hospital administrators must be willing to endorse the study and to work with the study group. Detailed planning will be necessary to adapt the overall protocol to the individual hospital, and continuous efforts will be necessary to maintain satisfactory data collection during the study.

Developing Measures to Predict Complications

Only patients who are considered low risks for developing complications should be included in the study. This group is least likely to benefit from intensive care, and there is the least amount of concern that they will receive care on a standard ward.

Numerous measures have been reported to predict the outcome of patients with acute myocardial infarction. Killip[16] introduced a simple classification system based largely on clinical evidence of the extent of left ventricular dysfunction at the time of admission to the hospital. Norris[17] developed a similar but more complicated index that includes both clinical signs and laboratory studies. Pozen and others[18] have described an instrument that improves diagnostic accuracy when used in the emergency room to evaluate patients with suspected acute myocardial infarction. Mulley and his colleagues[19] have recently described a simple way to identify low-risk patients within 24 hours of admission to the intensive care unit.

These measures are not sufficient, however; to answer questions about the selection of appropriate patients for study, these procedures should be tested on patients who are being admitted to the study hospitals. After a review of the patients' records, these procedures can establish the accuracy of measures used in the populations considered for study, thus minimizing any chance of entering patients who might suffer adverse outcomes as a result of the study.

Consideration should also be given to developing a new predictive index that includes a measure of time from onset of symptoms until the patient presents for medical care. Half of the deaths from acute myocardial infarction occur within the first two hours of the onset of identifiable symptoms, and many of these are the result of ventricular arrhythmias which could be prevented by monitoring and treatment.[20] After two hours, the death rate drops sharply, and left ventricular failure replaces arrhythmia as the most important cause of death.

Conducting an Observational Study

The final stage of the strategy includes a prospective, observational study to confirm that differences in outcome between intensive and standard care are not large. If the hospitals have been selected properly, when the intensive care unit is full and a new patient requires admission, the patient discharged prematurely from the unit will be the one who is best able to tolerate the discharge.

Two groups will be selected from all patients in the unit with suspected myocardial infarction: a study group of patients who are discharged prematurely and a control group of patients who could have been discharged prematurely but were not because their beds were not needed for new admissions. Patients to be included in the study group are identified by the physician who discharges them prematurely. Patients are assigned to the control group on the first day the physician states they could have been discharged prematurely.

Information about each patient's clinical status can be collected prospectively on the day of admission and each day the patient remains in the unit. Information concerning clinical outcomes during the hospital stay can be collected retrospectively from the patient's medical record (Table 1). These outcomes will include traditional measures such as death, cardiac arrest, and ventricular arrhythmias, as well as measures designed to identify more sensitive indicators of morbidity. These consist of diagnostic studies used to evaluate new or continuing symptoms, such as diagnostic enzyme studies and electrocardiograms for recurrent chest pain. They also include therapeutic maneuvers used to treat symptoms—for example, nitrates or narcotics for chest pain and diuretics for congestive heart failure.

If this observational study finds that patients discharged prematurely from the unit have more complications or worse outcomes than those who could have been discharged prematurely but were not, then proceeding with a randomized trial will not be ethical. If the observational study fails to find a difference between the two groups, all reasonable efforts will have been exhausted and plans for the randomized trial can proceed.

Table 1 Outcome Measures

Severe Adverse Clinical Outcomes	Other Adverse Clinical Outcomes	
	Direct measures	Indirect measures
Cardiac arrest	Atrial arrhythmias requiring therapy	Diagnostic cardiac enzymes
Ventricular tachycardia	Angina pectoris	Electrocardiograms
Premature ventricular contractions requiring therapy	Pericarditis	Arterial blood gases
	Pneumonia	Nitroglycerin
Second degree and third degree heart block	Psychiatric problems requiring new drug therapy (other than benzodiazepines) or a psychiatric consult	Initial antiarrhythmic dose
Pulmonary edema		Narcotics
Extension of infarct or new infarct		Initial Digitalis dose
Cardiogenic shock		Initial diuretic dose
Papillary muscle rupture		
Rupture of ventricular septum		
External heart rupture		
Arterial embolus		
Return to intensive care unit		
Death		
Pulmonary embolus		

Conclusion

The strategy just described may answer ethical concerns that have been raised about conducting randomized clinical trials for patients with uncomplicated acute myocardial infarction. Careful selection of hospitals for participation in the study will ensure that the results will be scientifically sound and that valid conclusions can be drawn, whatever the outcome. No study that risks the lives of the patients is ethical if this criterion is not met. Careful selection of patients will ensure that only those patients who are likely to do well with standard hospital care will be included in the study. The observational study is a final check against the possibility that the experience in the hospitals being studied is somehow different from the experience reported in the literature.

Endnotes

1. Russell, L.B. (1979). *Technology in Hospitals: Medical Advances and Their Diffusion.* Washington, D.C.: The Brookings Institution.
2. Yu, P.N., Bielski, M.T., Edwards, A., Friedberg, C.K., Grace, W.J., January, L.E., Likoff, W., Scherlis, L., and Weissler, A.M. (1971). Resources for the optimal care of patients with acute myocardial infarction. *Circulation* 43:A171–A183.
3. Gordis, L., Naggan, L., and Tonascia, J. (1977). Pitfalls in evaluating the impact of coronary care units on mortality from myocardial infarctions. *Johns Hopkins Med. J.* 141:287–295.
4. Hafvendahl, S. (1971). Influence of treatment in a CCU on prognosis in acute myocardial infarction. *Acta Med. Scand.* 519(suppl.):1–78.
5. Christensen, I., Iverson, K., and Skouby, A.P. (1971). Benefits obtained by the introduction of a coronary care unit. *Acta Med. Scand.* 189:285–291.
6. Hill, J.D., Holdstock, G., and Hampton, J.R. (1977). Comparison of mortality of patients with heart attacks admitted to a coronary care unit and an ordinary medical ward. *Br. Med. J.* 3:81–83.
7. Hill, J.D., Hampton, J.R., and Mitchell, J.R.A. (1978). A randomized trial of home-versus-hospital management for patients with suspected myocardial infarction. *Lancet* 1:837–841.
8. Mather, H.G., Morgan, D.C., Pearson, N.G., Read, K.L.Q., Shaw, D.B., Steed, G.R., Thorne, M.G., Lawrence, C.J., and Riley, I.S. (1976). Myocardial infarction: A comparison between home and hospital care for patients. *Br. Med. J.* 1:925–929.
9. Peterson, O.L. (1978). Myocardial infarction: Unit care or home care? *Ann. Intern. Med.* 88:259–261.
10. Bureau of Health Planning. (1979). Planning for coronary care units: A technical assistance document for planning agencies. Health Planning Methods and Technology Series, HRP–0101001, DHEW Publication No.

(HRA) 79–14019, Washington, D.C.: U.S. Government Printing Office, January.

11. Griner, P.F. (1975). Medical intensive care in the teaching hospital: Cost versus benefits. *Ann. Intern. Med.* 78:581–585.

12. Bloom, B.S., and Peterson, O.L. (1973). End results, cost and productivity. *N. Engl. J. Med.* 288:72–78.

13. Bigger, J.T., Heller, C.A., Wenger, T.L., and Weld, F.M. (1978). Risk stratification after acute myocardial infarction. *Am. J. Cardiology* 42:202–210.

14. Bloch, A., Maler, J., Haissley, J., Felix, J., and Blackburn, H. (1974). Early mobilization after myocardial infarction. *Am. J. Cardiology* 34:152–157.

15. Thibault, G.E., Mulley, A.G., Barnett, G.O., Goldstein, R.I., Reder, V.A., Sherman, E.L., and Skinner, E.R. (1980). Medical intensive care: Indications, interventions, and outcomes. *N. Engl. J. Med.* 302:938–942.

16. Killip, T., Kimball, J.T. (1967). Treatment of myocardial infarction in a coronary care unit. *Am. J. Cardiology* 21:457–464.

17. Norris, R.M., Brandt, P.W.T., Caughey, D.E., and Lee, A.J. (1969). A new coronary prognostic index. *Lancet* 1:274–281.

18. Pozen, M.W., D'Agnostino, R.B., Mitchell, J.B., Rosenfeld, D.M., Guglielmino, J.T., Schwartz, M.L., Teebagy, N., Valentine, J.M., and Hood, W.B., Jr. (1980). The usefulness of a predictive instrument to reduce inappropriate admissions to the coronary care unit. *Ann. Intern. Med.* 92:238–242.

19. Mulley, A.G., Thibault, G.E., Hughes, R.A., Barnett, G.O., Reder, V.A., and Sherman, E.L. (1980). The course of patients with suspected myocardial infarction. The identification of low-risk patients for early transfer from intensive care. *N. Engl. J. Med.* 302:943–948.

20. Armstrong, A., Duncan, B., Oliver, M.F., Julian, D.G., Donald, K.W., Fulton, M., Lutz, W., and Morrison, S.L. (1972). Natural history of acute coronary heart attacks: A community study. *Br. Heart J.* 34:67–80.

OVERVIEW: THE BURN CARE WORKSHOP

by Louise B. Russell

A thorough assessment of any technology requires information about the incidence of the condition the technology is designed to treat or diagnose, the results of using the technology, and its costs. The papers presented at the Burn Care Workshop by John Locke, Andrew Munster and Bernard Linn focus on each of these aspects of burn care in turn. Together they provide a preliminary description of burn care from these perspectives and illustrate for a particular case some of the difficulties of technology assessment.

Each author was asked to present information cross-classified as much as possible by age and percent of body surface area burned (%BSA), two characteristics of burned patients that are important for the course and outcome of treatment. Information about any medical technology needs to be broken down in some such fashion in order to permit more detailed judgments about the appropriate use of the technology.

The preliminary description provided by these papers will be tested, sharpened, and expanded in detail as data become available from a series of burn care demonstration projects financed by the Division of Emergency Medical Services in the Department of Health and Human Services. Contracts were awarded to six demonstration sites in 1977. A major purpose of these demonstrations is to collect extensive and consistent data on burns and burn care. Initial findings from all six sites were scheduled to have been presented to Congress in February of 1981. Reports and studies from the staffs of the demonstration projects will continue to add to our knowledge of burn care in future years.

John Locke's paper on incidence draws on preliminary data from the New England demonstration site, which covers the six New England states. These data are believed to represent 88 to 90 percent of the burns that occurred in the region between June 1, 1978, and May 31, 1979. They show that 20 patients were hospitalized with burns for every 100,000 population in the region, and that more than 60 percent of the hospitalized patients had burns over 10 percent or less of their body surface area.

In the discussion that followed the presentation, Dr. Munster observed that the incidence rates were lower than had been expected when the demonstration projects began. He also pointed out that the textbook criteria state that, except in special cases, burns of 10 percent or less in children and 15 percent or less in adults do not require hospitalization.[1] The preliminary data for New England thus suggest that too many burn patients are hospitalized. Dr. Linn added that similar results had appeared in his study of Florida, where more than 50 percent of the hospitalized patients had burns of 10 percent or less.[2]

Dr. Linn's paper documents some of the costs of burn care using data from the 1973 Florida study. This study focused on the costs of initial hospitalization and did not collect information about physicians' fees, the costs of repeat hospital stays, or care after discharge from the hospital. The data show that a burn patient spent twice as long in the hospital as the average patient, and the charges for care were twice as high. Charges were higher for patients with more serious burns, for those who had more operations during their stay, and for those who were treated in special burn units. Data from the demonstration projects may shed new light on the independent effects of these and other factors discussed in the paper and may provide additional information on the other costs of caring for burn patients. Dr. Linn also notes the important finding, reported earlier,[2] that mortality rates did not differ between patients who received care in special burn units and those who did not.

The earlier reports of this finding prompted Dr. Munster to investigate further the differences that care in special burn units might make. Mortality is clearly important and is easy to measure, but for those patients who survive their burns, outcomes must be measured in other ways. For these patients the important outcomes are morbidity from the burn, scarring, contractures that restrict their ability to use their hands, arms and legs, and other consequences of the injury that alter their appearance and ability to function. In his paper Dr. Munster describes the difficulties of trying to measure these outcomes, a necessary first step in linking them to the kind of care received. His preliminary work in this area produced two interesting results; because only 32 patients were involved, these results must be considered tentative until they can be confirmed by more extensive studies. The first result was that significant improvement continued for at least one year after the burn injury occurred so that evaluations made upon discharge from the hospital underestimated the degree of recovery eventually achieved. The second finding was that the severity of the burn was not related to the patient's degree of recovery.

If confirmed by further studies from the demonstration projects, the results reported in these papers will have important implications for the provision of burn care. The papers also offer considerable insight into the care and effort that must be invested in a study to produce useful results. Indeed, a single study is seldom definitive on any issue. Rather, it is usually necessary for a series of studies with similar findings to accumulate before a conclusion can be accepted with confidence, and even then the conclusion may have to be amended if the technology continues to improve. Because of the resources required by good assessments, the costs and benefits of technology assessment, as well as those of the technologies being assessed, should be reconsidered from time to time.

Endnotes

1. Artz, C.P., Moncrief, J.A., and Pruitt, B.A. (1979). *Burns: A Team Approach*. Philadelphia: W.B. Saunders Co.
2. Linn, B.S., Stephenson, S.E., Jr., and Smith, J. (1977). Evaluation of burn care in Florida. *N. Engl. J. Med.* 296:311–315.

THE EPIDEMIOLOGY OF BURNS
IN NEW ENGLAND

by John A. Locke

Physicians, hospital administrators, health planners, public health professionals, educators, and others who deal with the burn injury have long recognized the need for adequate data delineating the nature and extent of this injury problem. With present costs of facilities, staff, and resources required to assure the delivery of contemporary health care, the availability of data for the planning, delivery, and evaluation of burn care assumes an importance which combines the demands of cost effectiveness and optimum patient care. Those who deal with prevention of the burn problem also need adequate burn data to guide and evaluate their efforts.

While important clinical advances in burn care have consistently been reported in recent years, data describing epidemiologic characteristics of the burn problem have been limited in frequency, scope, and duration. Frequently studies of burn occurrences have been based on a group of patients, a series of patients seen at one or more hospitals, or on limited population groups such as those in multicounty areas. Rarely has burn incidence been determined for the population of an entire state for an entire year. This situation began to change in 1977 with the inauguration of the National Burn Demonstration Project.

Description and Data Collection

The National Burn Demonstration Project is a major burn data collection effort which seeks to establish broad baseline information regarding the epidemiological characteristics of the burn problem and to describe its impact on the existing system for the delivery of burn care. Funded by the Division of Emergency Medical Services of the Department of Health, Education, and Welfare, the Project is being implemented at six sites across the country, with a combined population of approximately 28 million. The New England site is the largest of the six demonstration sites, covering the six New England states with an aggregate population of approximately 12 million and comprising nearly 45 percent of the entire national study.

In brief, the Project involves the identification of burn cases presenting for treatment at acute care hospitals; the review of relevant medical and financial records; and the completion of abstracts describing the patient, the circumstances of the accident, the extent of the resulting burn injury, the nature and duration of treatment required, and a summary of the financial charges and reimbursements associated with the burn injury.

The data system used at all six demonstration sites was developed, and is maintained, by an independent contractor, Arthur Young & Co. of Washington, D.C. The forms, data collection procedures, data management and documentation procedures, and data support resources are described in a comprehensive procedures manual published by the data system contractor. Important among the study's provisions are procedures and assurances regarding the confidentiality of patients and facilities. The use of a unique patient identification number and special tracking codes makes it possible to follow the movement of a specific patient as he or she appears at any data trap at facilities within the study. These data traps include the emergency department, inpatient, rehabilitation/outpatient, and postmortem.

Preparing for project operations, the New England Regional Burn Program secured the voluntary participation of 95 percent of the approximately 250 acute care hospitals in New England. In unison with the other five demonstration sites, data collection efforts were scheduled for May 1, 1978 through June of 1980.

Data collection operations in New England are conducted by approximately two dozen data accessors located throughout the six-state region. The accessors routinely visit participating hospitals to carry out tasks related to case finding, record review, and data collection. All completed abstracts are subjected to review and computer system edits to assure completeness and correctness. Quality control measures have been carried out to ascertain the completeness of case finding and the level of accuracy and uniformity in data coding practices.

Study Findings

While the Burn Demonstration Project collects data from four data traps, this review of preliminary data deals only with emergency department and inpatient cases. Due to systematic delays in reporting cases at the rehabilitation/outpatient and the postmortem data traps, we will have to defer the examination of those data. For the sake of convenience and to facilitate interpretability, the present review is based on burns occurring during a one-year period, June 1, 1978 through May 31, 1979.

As further qualification of the present data, it is estimated that approximately 5 percent of eligible cases are in temporary suspense, pending correction of errors or completion of missing data elements. A case-matching edit which links data providing information on the patient's state, county, and community of residence has been completed for approximately 93 percent of the cases on the master file. For these reasons, and the fact that approximately 5 percent of the acute care hospitals in New England are not participating in the Burn Demonstration Project, we believe that 10 percent to 15 percent of the burns occurring in the study period are not presently included in our data base.

Under a feature unique to the New England site, data pertaining to burns treated in emergency departments (ED) were collected under a sampling plan whereby ED burns were reported on one day out of every ten. In estimating the total number of ED burns occurring in the period under review, projections made from data collected under the ED sampling plan are subject to a standard error of approximately 1.5 percent. Emergency department data presented here have *not* been corrected for sampling error. In the tables included in this report, ED data are reported as the actual number of ED cases encountered under the sampling plan. Rather than multiplying the number of ED cases by ten to reflect the sampling fraction, the sampling data is expanded by a factor of ten by reducing the denominator in rate calculations by a factor of ten, that is, 10^5 to 10^4.

Number of Burn Occurrences

During the one-year period, a total of 4,192 New England residents were treated and released (T&R) for burn injuries at the emergency departments of participating New England hospitals; 2,359 New England residents were hospitalized for burn care as inpatients. Since the emergency department burns were reported under a 10 percent sampling plan, the 4,192 ED cases represent an estimated 42,000 such injuries for the 12 months studied.

Using the 1970 U.S. Census data as the basis for estimating the population exposed to the risk of burn injury, the reported cases represent incidence rates of:

35.4 $ED_{T\&R}$ burns/10^4 person-years, and
19.9 inpatient burns/10^5 person-years

Age

Table 1 gives the number and rate of ED and inpatient burns by age groups for the one-year period, while Figure 1 shows these rates graphically. The group of infants and toddlers (age 0 to 2) is disproportionately numerous in both categories of cases, while the elderly group (age 65 and over) shows a high incidence only for inpatient burns. For both categories of injury, adolescents (age 13 to 19) and young adults (age 20 to 44) are the only other age groups showing higher than average rates of incidence. The increased susceptibility and vulnerability of the very young and the elderly to the occurrence and consequences of burn injury contribute to the high incidence in these age groups, while it seems likely that increased exposure to the

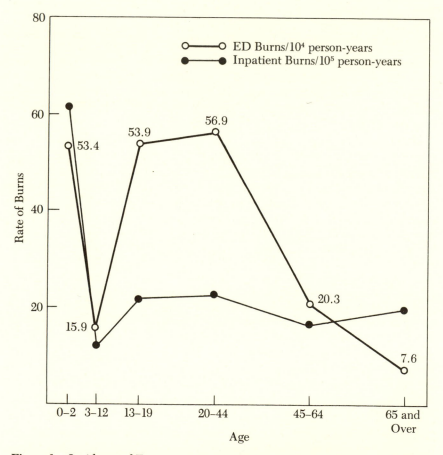

Figure 1   Incidence of Emergency Department and Inpatient Burns, New England, June 1, 1978 to May 31, 1979, by Age Groups. (*SOURCE: New England Regional Burn Program, National Burn Demonstration Project, March 1980.*)

Table 1   Number and Rate* of Emergency Department and Inpatient Burns, New England, June 1, 1978–May 31, 1979, by Age Groups

Age Group	Population†	ED Burns		Inpatient Burns	
		Number	Rate	Number	Rate
0– 2	577,090	308	53.4	358	62.0
3–12	2,268,294	361	15.9	279	12.3
13–19	1,526,859	823	53.9	311	20.4
20–44	3,676,058	2,092	56.9	807	22.0
45–64	2,523,845	512	20.3	369	14.6
65 and over	1,269,517	96	7.6	235	18.5
Total	11,841,663	4,192	35.4	2,359	19.9

SOURCE: New England Regional Burn Program, National Burn Demonstration Project, March 1980.

* Rate for ED Burns is #ED Burns/10^4 person-years; rate for Inpatient Burns is #Inpatient Burns/10^5 person-years.
† Population data from 1970 U.S. Census.

Table 2   Emergency Department and Inpatient Burns, New England, June 1, 1978 to May 31, 1979, Male:Female Ratio

Age Group	ED Burns			Inpatient Burns		
	Male	Female	M:F Ratio	Male	Female	M:F Ratio
0–2	175	133	1.32	219	139	1.58
3–12	211	150	1.41	183	96	1.91
13–19	489	334	1.46	255	56	4.55
20–44	1,330	762	1.75	635	172	3.69
45–64	290	222	1.31	262	107	2.45
65 and over	37	59	0.63	100	135	0.75
Total	2,532	1,660	1.53	1,654	705	2.35

SOURCE: New England Regional Burn Program, National Burn Demonstration Project, March 1980.

risk of burn injury through a greater range of activity contributes to the level of burn incidence observed among adolescents and young adults.

Sex

In the present data, burn injury appears to be a condition which largely affects the male population in New England. Table 2 shows a preponderance of males for all age groups under 65 years and for both ED and inpatient burns. The ratio of males to females ranges from 1.3 to 4.5 among the various age groups. More females than males were observed only among the elderly.

The disproportionate involvement of males is especially evident among the category of inpatient burns, and most noticeably among those of working age. As will be noted subsequently, burn etiologies commonly associated with work related injuries (for example, chemical, electrical, flash, scald/steam) account for a smaller proportion of inpatient burns than ED burns, but the present data do not yet permit full examination of the role of occupational hazards in the occurrence of inpatient burns.

Etiology

The data system established for the Burn Demonstration Project provides ten categories for the coding of burn etiology, plus an additional category where burn etiology is other or unknown. Table 3 gives the number and rate of emergency department burns and inpatient burns for each etiological category. Liquid scalds, hot surface, and chemical burns are the most numerous types of ED burns, while flame and liquid scalds are the dominant types of inpatient burns. In each instance the above-named etiological types account for more than half of the burn injuries in their respective groups.

Tables 4 and 5 examine etiology in terms of age by giving the

Table 3   Number and Rate° of Emergency Department and Inpatient Burns, New England, June 1, 1978 to May 31, 1979, by Etiology

	ED Burns		Inpatient Burns	
Etiology	Number	Rate	Number	Rate
Chemical	706	59.6	151	12.8
Electrical	75	6.3	91	7.7
Cold	24	2.0	26	3.0
Flame	369	31.2	703	59.4
Hot Surface	736	62.2	214	18.1
Flash	165	13.9	173	14.6
Scald/Steam	163	13.8	66	5.6
Scald/Liquid	909	76.8	668	56.4
Scald/Grease	356	30.1	148	12.5
Radiation	394	33.3	32	2.7
Other/Unknown	295	24.9	77	6.5
Total	4,192	354.0	2,359	199.2

Source: New England Regional Burn Program, National Burn Demonstration Project, March 1980.

° Rate for ED Burns is #ED Burns/10^5 person-years; rate for Inpatient Burns is #Inpatient Burns/10^6 person years.

Table 4 Number and Rate* of Emergency Department Burns, New England, June 1, 1978 to May 31, 1979, by Etiology and Age

Etiology	0 to 2 Years		3 to 12 Years		13 to 19 Years		20 to 44 Years		45 to 64 Years		65 Years and Over		Total	
	No.	Rate	No.	Rate	No.	Rate	No.	Rate	No.	Rate	No.	Rate	No.	Rate
Chemical	28	48.5	23	10.1	161	105.2	410	111.4	78	31.0	6	4.7	706	59.6
Electrical	7	12.1	14	6.2	5	3.3	38	10.3	11	4.4	0	—	75	6.3
Cold	0	—	5	2.2	5	3.3	13	3.5	1	0.4	0	—	24	2.0
Flame	10	17.3	37	16.3	78	51.0	181	49.2	52	20.6	11	8.7	369	31.2
Hot Surface	129	223.6	97	42.7	117	76.5	291	79.1	87	34.5	15	11.8	736	62.2
Flash	1	1.7	11	4.8	26	17.0	107	29.1	16	6.3	4	3.1	165	13.9
Scald/Steam	3	5.2	4	1.8	36	23.5	91	24.7	25	9.9	4	3.1	163	13.8
Scald/Liquid	95	164.6	105	46.3	151	98.7	407	110.6	117	46.4	34	26.8	909	76.8
Scald/Grease	11	19.1	21	9.3	84	54.9	181	49.2	49	19.4	10	7.9	356	30.1
Radiation	7	12.1	17	7.5	106	69.3	234	63.6	27	10.7	3	2.4	394	33.3
Other/Unknown	17	29.5	27	11.9	54	35.3	138	37.5	49	19.4	9	7.1	294	24.8
Total	308	533.8	361	159.0	823	537.9	2,091	568.2	512	203.2	96	75.6	4,191	354.0

SOURCE: New England Regional Burn Program, National Burn Demonstration Project, March 1980.

* Rate for Emergency Department burns is #ED Burns/10^5 person-years.

Table 5   Number and Rate* of Inpatient Burns, New England, June 1, 1978 to May 31, 1979, by Etiology and Age

Etiology	0 to 2 Years		3 to 12 Years		13 to 19 Years		20 to 44 Years		45 to 64 Years		65 Years and Over		Total	
	No.	Rate	No.	Rate	No.	Rate	No.	Rate	No.	Rate	No.	Rate	No.	Rate
Chemical	7	12.1	6	2.6	26	17.0	74	20.1	34	13.5	4	3.1	151	12.8
Electrical	17	29.5	24	10.6	12	7.8	29	7.9	8	3.2	1	0.8	91	7.7
Cold	0	—	0	—	4	2.6	21	5.7	6	2.4	5	3.9	36	3.0
Flame	21	36.4	86	37.9	108	70.6	272	73.9	131	52.0	85	66.9	703	59.4
Hot Surface	50	86.7	19	8.4	17	11.1	63	17.1	31	12.3	34	26.8	214	18.1
Flash	2	3.5	23	10.1	39	25.5	72	19.6	29	11.5	8	6.3	173	14.6
Scald/Steam	5	8.7	0	—	7	4.6	36	9.8	12	4.8	6	4.7	66	5.6
Scald/Liquid	227	393.4	92	40.5	49	32.0	136	37.0	87	34.5	77	60.6	668	56.4
Scald/Grease	17	29.5	19	8.4	31	20.3	56	15.2	21	8.3	4	3.1	148	12.5
Radiation	1	1.7	3	1.3	9	5.9	14	3.8	3	1.2	2	1.6	32	2.7
Other/Unknown	11	19.1	7	3.1	9	5.9	34	9.2	7	2.8	9	7.1	77	6.5
Total	358	620.5	279	122.9	311	203.3	807	219.3	369	146.4	235	185.0	2359	199.2

Source: New England Regional Burn Program, National Burn Demonstration Project, March 1980.

* Rate for Inpatient Burns is #Inpatient Burns /10⁶ person-years.

Table 6   Leading Categories of Emergency Department and Inpatient Burns
by Age of Patient and Etiology

Age Group	ED Burns	Inpatient Burns
0 to 2 years	Hot Surface Scald/Liquid	Scald/Liquid Hot Surface
3 to 12 years	Scald/Liquid Hot Surface	Scald/Liquid Flame
13 to 19 years	Chemical Scald/Liquid	Flame Scald/Liquid
20 to 44 years	Chemical Scald/Liquid	Flame Scald/Liquid
45 to 64 years	Scald/Liquid Hot Surface	Flame Scald/Liquid
65 years and Over	Scald/Liquid Hot Surface	Flame Scald/Liquid

SOURCE: New England Regional Burn Program, National Burn Demonstration Project,
March 1980.

number and rate of burns for each etiological category and age group.
Table 6 summarizes this information by listing the two types of burns
that account for 40 to 70 percent of the injuries in each age category.
It is striking to note that liquid scalds are the first or second leading
type of burn in all age groups for both emergency department and
inpatient burns. Chemical burns are the leading type of ED burn for
ages 13 to 44, and appear to reflect, in part, the occurrence of work-
related injuries. Among inpatients, flame burns are the leading type
of burn for all ages beyond 12 years.

Month of Injury

Figure 2 depicts the pattern of ED and inpatient burn occurrence
throughout the year. Both classes of burns show a distinct seasonal
increase during the summer months, with approximately one third of
all cases occurring during the months of June, July, and August. For
both ED and inpatient burns, the maximum number of cases was
found in July.

Within these data, we were not surprised to see that the most frequently occurring type of burn (liquid scald and hot surface for ED
burns; and flame and liquid scalds for inpatient burns) displayed a
similar pattern, that is, peak occurrence during the summer and
greatest frequency generally during the month of July. Contrary to
the general trend we noted that among inpatient burn cases, the
preponderance of flame burns over liquid scald burns existed entirely

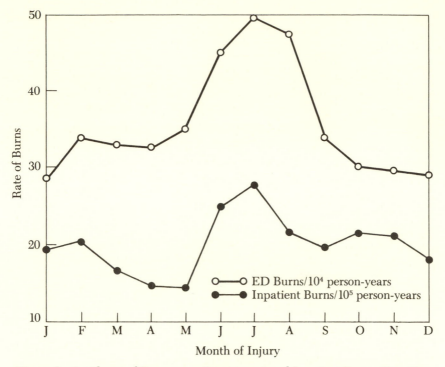

Figure 2   Incidence of Emergency Department and Inpatient Burns, New England, June 1, 1978 to May 31, 1979, Showing Month of Injury. *(SOURCE: New England Regional Burn Program, National Burn Demonstration Project, March 1980.)*

within the three northern New England states where, for the peak month of July, flame burns were more than double the number of liquid scalds.

Work Status

Work-related burn injuries were recorded when that information was available on the medical records under review. Work status was determined for 75 percent of ED cases and for nearly 95 percent of inpatient cases. The data indicate that more than 40 percent of ED burns were work-related while fewer than 25 percent of inpatient burns occurred at work.

Table 7 lists the number of ED and inpatient cases for which work status was determined and the percentages of those cases which were work related. For ED burns more than 50 percent of cases were work related for four types of burns (chemical, electrical, flash, and steam scald); for inpatient cases, only chemical and steam scald burns were reported as work related in more than 50 percent of cases. Reference

Table 7 Emergency Department and Inpatient Burns, New England, June 1, 1978 to May 31, 1979, by Etiology and Work Status

	ED Burns		Inpatient Burns	
Etiology	*Number*	*Percent Work Related*	*Number*	*Percent Work Related*
Chemical	548	64.4	135	59.3
Electrical	65	50.8	89	36.0
Cold	19	15.8	35	28.6
Flame	272	30.9	643	16.2
Hot Surface	607	31.1	204	21.6
Flash	124	55.6	155	35.5
Scald/Steam	125	58.4	60	50.0
Scald/Liquid	693	39.5	639	15.8
Scald/Grease	273	49.1	139	34.5
Radiation	298	15.4	28	0
Other/Unknown	181	46.4	66	18.2
Total*	3,205	41.9	2,193	24.9

SOURCE: New England Regional Burn Program, National Burn Demonstration Project, March 1980.

* Total based on cases for which work status was determined.

to Tables 4 and 5 reveals that these types of burns occurred mostly among adults of working age.

Burn Severity

The severity of burn injury is a difficult variable to identify, based on data from the Burn Demonstration Project. Work is underway in New England and in other demonstration sites to develop more adequate methods for indexing the severity of burn injuries. With the data presently available, it is possible to comment on severity of injury only in terms of patient disposition and, for inpatient cases, in terms of the extent of burn injury expressed as percent of body surface area burned.

With reference to patient disposition, the New England data indicate that approximately 4 to 5 percent of burn patients seen in hospital emergency departments were admitted for inpatient care. This would suggest that one out of every twenty burn patients required hospitalization.

While the extent of body surface area burned is not reported for patients treated at hospital emergency departments, the data on the distribution of inpatient cases according to percent of body surface area burned indicate that approximately one third of hospitalized patients had burns in excess of 10 percent BSA: 63 percent of the patients had involvement of 10 percent or less of their body surface; 22 percent had involvement of between 11 and 20 percent; 8 percent had involvement of between 21 and 40 percent, and the remainder were not specified.

Population Characteristics

The occurrence of burns is currently being analyzed in association with population characteristics available through U.S. Census publications. Figure 3 shows the incidence rate for ED and inpatient burns

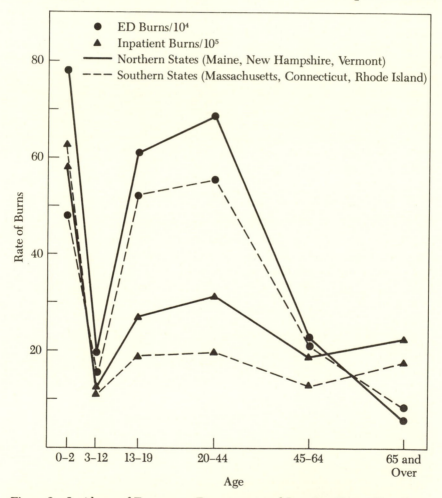

Figure 3   Incidence of Emergency Department and Inpatient Burns, New England, June 1, 1978 to May 31, 1979: Comparison of Northern Three States and Southern Three States. (*SOURCE: New England Regional Burn Program, National Burn Demonstration Project, March 1980.*)

Table 8   Incidence* of Emergency Department and Inpatient Burns: New
England, June 1, 1978 to May 31, 1979, Comparison of Northern
Three States and Southern Three States, by Etiology

Etiology	ED Burns			Inpatient Burns		
	Northern States	Southern States	N/S Ratio	Northern States	Southern States	N/S Ratio
Chemical	77.7	55.5	1.40	19.8	11.2	1.77
Electrical	6.0	6.4	0.94	12.0	6.7	1.79
Cold	4.6	1.4	3.29	5.5	2.5	2.20
Flame	38.2	29.6	1.29	75.4	55.8	1.35
Hot Surface	77.3	58.8	1.31	20.7	17.5	1.18
Flash	16.1	13.4	1.20	22.1	12.9	1.71
Scald/Steam	17.9	12.8	1.40	9.2	4.8	1.92
Scald/Liquid	80.5	75.9	1.06	59.3	55.8	1.06
Scald/Grease	27.6	30.6	0.90	13.3	12.3	1.08
Radiation	41.9	31.3	1.34	6.4	1.9	3.37
Other/Unknown	29.0	24.0	1.21	8.3	6.1	1.36
Total	416.7	339.7	1.23	252.1	187.3	1.35

SOURCE: New England Regional Burn Program, National Burn Demonstration Project,
March 1980.

* Rate for Emergency Department burns is #ED Burns/10^5 person-years; rate for
Inpatient burns is #Inpatient Burns/10^6 person-years.

for the three northern New England states (Maine, New Hampshire,
and Vermont) and the three southern New England states (Massachusetts, Connecticut, and Rhode Island).

The data on which Figure 3 is based show that the incidence of
ED burns in the northern states is nearly 23 percent higher than that
in the southern states. Among inpatient burns the incidence rate is
more than 34 percent higher in the northern part of the region. A
striking and largely consistent pattern of higher incidence in the
northern states is evident when rates of ED and inpatient burns are
examined by etiologic categories. The data presented in Table 8
shows the preponderance of burns in northern New England, especially among inpatient burns.

The two groupings of New England states can also be character-
ized in terms of: more densely populated versus less densely populated; urban versus rural; more affluent versus less affluent; differing
housing characteristics; and other environmental and socioeconomic
factors. The possibilities for comparisons are numerous and fascinat-
ing. Multiple linear regression is now being conducted to identify
associations of possible importance in understanding the burn problem. If this effort proves fruitful, this type of analysis could extend
to the level of counties, cities, and towns, and for the Boston Standard Metropolitan Statistical Area, to the level of census tracts.

Discussion

With New England's participation in the Burn Demonstration Project, we have moved rapidly from a paucity to an abundance of descriptive burn data. Indeed, another full year of burn data will soon be available. Additional data will not only allow for a more thorough examination of patterns cited on the basis of a greatly enlarged volume of data, but will provide important insight into clinical and treatment issues. Information regarding transfer patterns and other characteristics of the system for the delivery of burn care will also be available.

While it is premature to attempt definitive interpretation of the preliminary data, a few general patterns are evident. From the first year of data collection we have seen more emergency department burns and fewer inpatient burns than we had expected. Based on guidelines provided for developing proposals for participation in the Burn Demonstration Project, we had expected to encounter approximately 30 inpatients per 100,000 person-years of exposure. Even allowing for unreported cases, the rate of inpatient cases appears to be as much as 20 percent below expected levels. To date, a more striking trend in the burn data is the apparent excess in the number of ED cases relative to the number of inpatients. A more careful, in-depth examination of the burn problem in New England should offer a clearer explanation for the observed distribution of cases.

In a discussion of newly acquired data it is important to recognize that the data must be obtained at public expense and cost should be considered. Project operations involved in producing the 12 months of data reported here cost an estimated $400,000. The resulting cost of approximately $8 for each burn case studied is not a true indication of the project's cost and worth. In a real and important sense, a comprehensive study of burns, or of any health condition, provides relevant information about the entire population within which the study is conducted. When viewed from this perspective, the data reported here cost slightly more than 3 cents per capita.

Even when seen in this more modest cost perspective, however, it appears unlikely that a descriptive burn study of this scope and duration will be repeated in the near future. Continuing burn data must be sought through alternative approaches, including mandatory or voluntary burn reporting.

Massachusetts, Connecticut, and several other states now require the reporting of burn injuries. Experience with burn reporting in Massachusetts has been spotty, in my view. Due to an absence of effective coordination of the reporting program, burns have been under-reported, and some hospitals do not routinely report the burns they do treat. Our experience in the Burn Demonstration Project suggests that hospitals will sustain a remarkable level of effort in providing data if they receive some positive encouragement and feedback of information. Most importantly, they must perceive that the data are valuable and will be useful in the larger context of improved health care.

The major advantage of mandatory reporting is its relatively low cost. If properly managed, a burn-reporting program, both for the hospitals and for appropriate state government agencies, in the long run will cost far less than a return to the days when burn data simply were not available.

DEVELOPMENT OF AN OUTCOME SCALE FOR BURNS

by Andrew M. Munster, Betsy C. Blades, Joyce Mamon, and Anne Kaszuba

Measuring the outcome of care is important in assessing the efficacy of the delivery of tertiary care. When tertiary care is directed at a disease or an injury with a high mortality rate, a reduction in the mortality rate can be considered a helpful index of measurement. When the mortality rate is low, the measurement of outcome must be directed at surviving patients. Our preliminary attempts to measure the outcome of therapy for a highly specialized area of tertiary care, the treatment of thermal injury, will be described.

Burn injuries severe enough to cause hospitalization affect about 200,000 persons in the United States each year.[1] Approximately 4,000 of these patients are treated in one of the 165 specialized centers or burn units available throughout the country, which are concerned with highly specialized treatment and geared toward high technology, monitoring, and support services; the remainder are treated in community and general hospitals.

Burns are usually measured by percentage, which indicates the percent of the body surface area burned, and by depth, the partial or full thickness of the injury. These measurements can be used as indications for admission to a specialized center. However, many other variables affect the outcome of the injury.[2] The mortality rate for burn injuries reached its current level after a considerable drop in the 1960s, following the introduction of effective topical chemotherapy and continued to decline steadily, following the introduction of improved resuscitation and nutritional support.

Measurements of the efficacy of burn treatment have relied principally on mortality statistics, taking into consideration the size and depth of the burn and the age of the patient. The cost efficiency of burn centers in reducing mortality rates has been questioned by a recent study.[3] However, demonstrating the effectiveness of burn centers may be difficult using the mortality rate as a measurement.

We know that the survival rate of hospitalized burn patients is 84 percent and the mortality rate is 16 percent. In addition, the survival rate of patients with burns under 30 percent of the total body surface is 100 percent and the mortality rate of patients with burns over 70 percent of the body surface is 100 percent. Therefore substantial numbers of patients with burns between 30 percent and 70 percent of the total body surface must be available for analysis to evaluate results.

A major burn is a highly crippling injury. Even a minor burn can be crippling if a patient's career is affected. All patients suffer a degree of physical, psychosocial, and emotional upset, and the injury can result in lifelong disability, minimal disability, or no disability. In a specialized burn center, psychosocial and emotional support, occupational and physical therapy, vocational guidance, and rehabilitation are aimed at minimizing disability and strengthening support as survivors begin to readjust to the outside world. This aspect of care has never been measured; it is far more difficult and complicated than simply measuring mortality rates.

Three years ago we attempted to determine the feasibility of measuring the quality of life in surviving burn patients. If the quality of life of survivors could be measured, then a valuable tool could be developed for comparing the performance of specialized centers with that of general and community hospitals, for documenting changes in the therapeutic efficacy of one unit over a period of time, for comparing various specialized centers, and for helping to pinpoint areas of therapeutic weakness within one hospital.

Pilot Study

Our initial efforts were based on 32 adult patients discharged from the Baltimore Regional Burn Center with burns ranging from 9 percent to 63 percent of the total body surface.[4] A physician director, a social worker, and an occupational therapist collaborated to devise an interview schedule that included five equally weighted performance areas derived from the literature and assumed to define comprehensively the quality of life. These areas were: work, dependency, physical/joint function, psychosocial function, and self-image.

Patients were interviewed according to the schedule to determine their preinjury performance level or baseline. They were then asked to describe current functioning in these areas, and scores were assigned according to predetermined criteria. Post-burn performance equal to performance at the preinjury level was given a score of 100 points. However, when patients showed improvement, they were able to score above baseline up to a maximum of 148 points. The pilot instrument is illustrated in Table 1.

This small initial study yielded two rather surprising results. First, a statistically significant improvement was observed in patients interviewed at 12 months postdischarge compared with those interviewed at 3 months postdischarge. This result suggests that continuous improvement may be possible for at least 12 months after discharge. Second, no statistically significant difference in performance was seen between patient groups when the analysis was based on total percent of the body surface affected by injury. In other words, patients with small burns did not necessarily do better than patients with large burns. Finally, as expected, some patients actually outscored their pre-burn performance following the injury, suggesting that the injury acted as a motivating factor for improved performance. We were unable, however, to determine whether these patients benefited from any of the specialized services available. The study demonstrated the feasibility of measuring outcome and set the stage for a major attempt to develop a valid and reliable scale to measure the outcome of the injury in surviving burn patients.

Table 1   Questionnaire for Evaluation of Outcome of Burn Injury

Questionnaire Item	Evaluation Categories	Value in Points
Section One: Work		
1. 25% higher status or job compared with preinjury, or 25% increase in hours worked		30
2. 6 to 25% higher		25
3. −6% to +6%		⊘20 °
4. 25% to 6% loss in hours or wages		15
5. Greater than 25% loss but some work		10
6. Unemployed		0
Section Two: Dependence		
7. Improvement compared with preinjury		25
8. Same as preinjury		⊘20
9. Needs minor help, such as with household tasks, shopping, care of children but otherwise at preinjury level		15
10. Needs major help—another person has moved in, or patient has had to move		10
11. Totally dependent on others' help		5
12. Institutionalized		0
Section Three: Function		
13. Normal in all limbs or preinjury		⊘20
14. 25% limitation in one major joint		15
15. 25% limitation in no more than 2 major joints or 25–40% in one joint		10

Table 1 (continued)

Questionnaire Item	Evaluation Categories	Value in Points
16. 25–40% limitation in 2 joints or 40–60% limitation in one joint		7
17. 60% limitation in 1 to 2 joints		5
18. 60% limitation in 2 joints		0
Section Four: Personal and Social		
19. Marital status		
Married now, not prior or engaged or living with or otherwise connected, and not before		10
20. No change from preinjury or widowed		⑤
21. Separated or divorced since injury		0
22. Involvement in family activities:		
a. Meals with family		
b. Child care	i. Definite increase	10
c. Maintenance chores	ii. Some increase	7
d. Planning activity	iii. No change	⑤
e. Activity at home	iv. Some decrease	3
f. Activity outside	v. Definite decrease	0
g. Extended family events		
23. Involvement in social activities:	i. Definite increase	10
a. Frequency of visits with friends	ii. Some increase	7
b. Visits by phone	iii. No change	⑤
c. Activities by group (church, clubs, etc.)	iv. Some decrease	3
d. Hobbies and education	v. Definite decrease	0
24. Sexual Activity	i. Definite increase	6
	ii. Some increase	4
	iii. No change	③
	iv. Some decrease	2
	v. Definite decrease	0
25. Drug and Alcohol Use	i. Increase	0
	ii. No change	②
	iii. Some decrease	4
	iv. Marked decrease	5
Section Five: Subjective		
26. Satisfaction with socioeconomic status compared with preinjury	i. Better, excellent	6
	ii. Better, good	4
	iii. As before	③
	iv. Worse, but fair	2
	v. Much worse	0
27. Satisfaction with family and social life compared with preinjury	i. Excellent, better	6
	ii. Good, better	4
	iii. As before	③
	iv. Worse	2
	v. Much worse	0

Table 1   (continued)

Questionnaire Item	Evaluation Categories	Value in Points
28. Somatic—sleep, crying spells, nightmares	i. Less than before ii. Same as before iii. More than before	5 ③ 0
29. Mood and outlook for the future	i. Bright, better than ever ii. At preinjury level iii. Poor, worse than before	5 ③ 0
30. Self image—patients' own opinion of appearance, attractiveness, ability to cope, compared with preinjury	Score subjectively on a scale of 10 for best and 0 for worst, with 8 being level of preinjury 0–10	

Note: At the present time, the mean score for patients discharged from the Baltimore Regional Burn Center approximately 5.4 months after injury, with a mean burn size of 29 percent is 74 percent rehabilitation according to this scoring system. We do not yet have enough patients to break this data down by percent burn sustained.

100 points = Preinjury level
148 points = Maximum possible

° Circled scores indicate preinjury level.

Development of a Burn-Specific Scale

As the scale is still being developed, we will address only our objectives, methodology, and some of the difficulties encountered in the endeavor.

Central Questions

Initially we were interested in learning whether any of the existing measurements of health outcome or health status were sufficiently sensitive to be applied to a group of patients with such a specific injury, or whether the development of a new burn-specific scale was necessary. Other questions also needed answers. Given the varied injuries and personalities of the patients who survived, what factors really mattered to patients in regard to the quality of their lives following injury? Were there substantive differences of opinion in terms of relevancy between professionals taking care of these patients and the patients themselves? Did the distribution of the burn or the time postdischarge make a difference?

Procedure

A panel of judges consisting of 15 professionals and 20 expatients was selected to answer these questions. The 15 professionals were experienced burn physicians and surgeons and allied health pro-

fessionals with expertise in burn care. The 20 patients were separated into the following four groups of five each with the following
characteristics: Group 1—one year from discharge with predominantly facial or hand burns; Group 2—under one year from discharge with predominantly facial and hand burns; Group 3—over
one year from discharge with predominantly trunk burns; and
Group 4—under one year from discharge with predominantly trunk
burns. The purpose of this grouping was to encompass the greatest
possible variation of cosmetic and functional impairment and to
eliminate the time factor, which was shown by the pilot study to
play a possible role in post-burn performance.

Three existing validated scales of disability were used for comparison of relevance with a group of burn-specific items developed
by our staff. A questionnaire was drawn up consisting of 136 items:
the abbreviated Sickness Impact Profile,[5] the Katz Activities of Daily
Living Scale (abbreviated to 6 items),[6] and the General Index of
Well Being Scale constituting 67 items.

We conducted a series of interviews with several of our own patients and staff, asking them to free associate about matters which,
in their opinions, were relevant to the health and happiness of burn
patients. After examining replies and eliminating duplications, we
developed 175 burn-specific items which were combined with items
from the above-mentioned scales for a questionnaire consisting of
386 items. Care was taken not to disturb the validated order of the
three existing questionnaires, and items were coded so that the
judges had no knowledge of the sources of the questions. In addition, five "ringer" items were added to identify judges who had
not paid attention to the questionnaire or had misunderstood its purpose. Judges were instructed to rate the items on an 11-point scale
with equal intervals, 11 being most relevant and 1 being least relevant in evaluating outcome: The instructions requested the judges
to read these items for relevance to a general burn patient population, not necessarily to themselves. The five "ringer" items were:

1. I am bothered by the increasing cost of living.
2. I prefer color to black and white television.
3. I need a bigger refrigerator.
4. I would like to see the Grand Canyon.
5. I am worried my child is not going to pass his/her first-grade
   exams.

Results

Of the 35 questionnaires sent out, 27 were returned fully completed
—16 from patients and 11 from professionals. An examination of the

Table 2   Median Score and Q Value for Each Instrument

Instrument	Number of Items	Median Score	Median Q Value
Sickness Impact Profile	136	6.5	3.4
Activities of Daily Living	6	10.6	1.0
General Well Being Index	67	7.8	2.7
Burn-Specific Items*	175	7.5	3.0

Note: N = 24 respondents.

* "Ringer" items are excluded.

response distribution for the five "ringer" items revealed that two of these items were not as irrelevant as had been anticipated: "I am bothered by the increase in cost of living" was given an 11 rating by six respondents and "I prefer color to white television" was rated 11 by four respondents. These two items were excluded from all analysis. Each of the remaining three items in this group was given an 11 by one respondent. The remaining items were used to identify respondents whose relevance scores were of questionable validity. Thus the final analysis was based on data from the remaining 24 respondents.

Because the exploratory data analysis indicated skewed response distributions from many questionnaire items and because the sample size was small, the median was selected as the measure of central tendency. For measuring dispersion we chose the semi-interquartile range (Q) $Q = \dfrac{Q3 = Q1}{2}$ where $Q3 = $ 75th centile and $Q1 = $ 25th centile. The Q value was used to select items of the highest agreement between judges (low Q) from items with high median scores.

Table 2 shows the median scores and Q values for each instrument. The highest median score, 10.6, was noted for the six items of Activities of Daily Living (ADL). Similarly the median Q value was extremely low, 1.0. The median score of the items in the index of General Well Being was 7.8, with a Q value of 2.7. The Burn-Specific items had a median score of 7.5 and a Q value of 3.0. The items included in the Sickness Impact Profile (SIP) had a median score of 6.5 and a 3.4 Q value.

Rank-order listings of items with a median score of 9.0 or greater revealed 6 items from the SIP, all 6 from the ADL, 13 from the Index of General Well Being Scale, and 24 from our Burn-Specific items.

Central tendency and dispersion were measured for certain item categories which could be separated clearly, that is, emotions, feelings and social interaction. Generally the median scale value of the burn-specific items was somewhat higher than the items in the burn Sickness Impact Profile in the same category, but the dispersion was fairly high for both sets.

When items with significant differences in median scores for patients with trunk versus face or hand burns were analyzed ($N = 7$ in each group), only six items from the entire questionnaire were rated higher, with a significance level of 5 percent or less, and all six were made by those patients in the face and hand group. Similarly, when items with significant differences in median scores were analyzed in patients with burns sustained more than one year ago as contrasted with patients with burns suffered less than one year ago, only six items showed statistically significant differences. When differences between patients and professionals were analyzed, 28 of the 136 items from the Sickness Impact Profile were rated higher by patients; two of the six items from the Activities of Daily Living were rated significantly higher by patients. In the scale of General Well Being, 17 of 67 items showed statistically significant differences, with higher ratings given by patients. With regard to Burn-Specific Items, 38 of 175 showed statistically significant differences in responses, overwhelmingly rated higher by professionals than patients. Considering that the entire questionnaire consisted of 386 items, the differences in opinions between patients and professionals cannot be considered significant.

Table 3   Factors in Comparison of Sickness Impact Profile and Clinician Assessment

	SIP	*Clinician Assessment*
Focus:	Behavior	Behavior and feelings
Orientation:	Assessment based on current state only	Assessment based on change from preinjury
Definition:	136 items 12 categories (permits greater definition in comparable areas)	17 response ratings 5 categories
Direction:	Negative state	Positive and negative change
Value:	Weighted by category and item	Equally weighted categories Equally weighted items within categories
Refinement:	Validity and reliability testing	Inclusion based on review of literature and arbitrary inclusion
Rating type:	Objective	Permits objective and subjective ratings

**Table 4 Comparison of Scores for Seven Patients
by Sickness Impact Profile (SIP) and Clinician Assessment**

Age	Sex	Percent of Body Surface Area Burned	Time Post-burn	SIP (total percent of illness)*	Clinician Assessment (percent of preinjury)†
45	Male	15%	2.5 months	22.8	63.1
19	Female	13%	12 months	11.5	94.5
25	Male	12%	12 months	5.9	84.0
28	Male	30%	6 months	8.9	97.7
49	Male	12%	6 months	4.2	85.7
22	Female	20%	3 months	27.5	81.0
36	Male	25%	2 months	53.8	54.0

* Optimal score = 0.   † Optimal score = 100.

Interviews with Recently Discharged Patients

We chose the Sickness Impact Profile to further evaluate the validity of an existing scale in the measurement of outcome in burn patients. Paired interviews with a small group of seven patients was conducted. The Sickness Impact Profile was administered and an evaluation was made by a clinician, using the pilot study assessment instrument previously mentioned.

Table 3 shows several substantive differences between the Sickness Impact Profile and the clinician assessment; gross scores are shown in Table 4. In general patients seemed cooperative, were not fatigued unduly by the interviews, and coped with interviewing techniques well.

A close examination of the items and categories in the two scales yielded the following information. Both methods appeared to identify areas of depression, anxiety, nervousness, sleeplessness, lack of work, some gross physical limitations, and decreased social interaction. The clinician's assessment was more sensitive in identifying scarring, concern about appearance, alcohol abuse following the injury, and problems of self image. The Sickness Impact Profile was more sensitive in identifying some physical limitations and certain psychosocial interactions.

Discussion

The study just described is not conclusive in determining the value of an established outcome scale, such as the Sickness Impact Profile, for burn patients. No longitudinal survey compared a burn-specific scale with the Sickness Impact Profile; merely a survey of judgment with regard to relevance was made. Some preliminary conclusions, however, can be drawn from this study:

1. Items specific to the burn injury are thought to be relevant by a substantial number of judges.
2. The highest scoring items on relevance constitute a composite of questions which include physical items (the ADL), psychological items (the GWB scale), general questions of health (the SIP), and items which are specifically aimed at components of burn injury.
3. Patients and professionals generally agree that these questions are relevant to the outcome of burn injury.
4. No substantial difference of opinion on relevance was seen between patients, based on either the distribution of their burn or the amount of time that had elapsed since injury.
5. Many similarities exist between the SIP and the clinician's assessment methods in the overall evaluation of patients, but the sensitivity of the two instruments with regard to specific domains varies.

These various scales, including a scale constructed of burn-specific items, appear to measure different components of the injury. Ideally the most sensitive scale would incorporate elements from each of these scales.

Several issues and problems should be addressed in this continuing investigation. A major question is whether any one scale currently exists—most of these scales have taken years of work to assemble and validate—which is applicable to the survivors of burn injury, as well as other injuries and illnesses necessitating tertiary care. At the present time we can only state our belief that scales which measure the quality of life in survivors play a necessary role in the assessment of high technology medical care and should be included in designs which evaluate the effectiveness of such technologies.

Endnotes

1. Baxter, C.R., Marvin, J.A., and Curreri, P.W. (1974). Early management of thermal burns. *Postgrad. Med.* 55:131.
2. Zawacki, B.E., Azen, S.P., Lubus, S.H., and Chang, V.T. (1979). Multifactorial profile analysis of mortality in burned patients. *Ann. Surgery* 189:1.
3. Linn, B.S., Stephenson, S.S., Bergstresser, P., and Smith, J. (1979). Dollars spent relate to outcomes in burn care? *Med. Care* 17:835.

4. Blades, B.C., Jones, C., and Muster, A.M. (1979). Quality of life after major burns. *J. Trauma* 19:556.
5. Department of Health Services, University of Washington (1975). Sickness Impact Profile, *AJPH*, 65:1304.
6. Katz, S., Ford, A.B., Moskowitz, R.W., Jackson, B.A., and Jaffee, M.W. (1963). Studies of illness in the aged: The index of ADL. *JAMA* 185:914.
7. Fazio, F. (1977). A concurrent validational study of the NCHS, General Well Being Schedule, DHEW Pub. No. (HRA) 78–1347, U.S. Dept. of HEW.

COSTS OF INPATIENT BURN CARE

by Bernard S. Linn

Current studies of the health care field focus on cost containment or quality assurance, but too often address these problems independently. Analyses of the health care system can be misleading if the impact of cost on the quality of services provided is not considered. Proponents of higher quality should consider the probability of resulting increases in cost; and conversely, those interested in reducing costs should assess the impact this might have on the quality of services.[1]

The relationship between cost and quality of care can be studied in the treatment of burned patients. Since age of the patient and percent of body surface area burned are indicators of severity, and both are quantitative variables, differences in severity of the burn injury can be adjusted for in comparisons of quality or outcomes of care. Furthermore, outcome (survival) can be readily observed with burned patients.

The cost of burn care is high[2,3] and varies with such factors as the admission status of the patient, specifically, age and severity of injury; with intensity and length of inpatient care; and with the eventual outcome of burn care. In reporting our study we attempt to relate some of the costs in treating burns with selected indicators of severity of injury, health care processes, and patient outcomes.

Method

The data presented here are part of a larger study of burn care undertaken in Florida.[4] Information was gathered on all patients admitted over a one-year period to 73 selected hospitals throughout

---

This project was supported by a grant HS-01801 from US Public Health Service, Department of HEW, Health Services Research.

the state. Nurses employed on a full-time basis in different regions of the state obtained information from medical records following a standard research protocol. Each nurse was supervised by a surgeon who helped design the study and also served as a member of an executive committee that met regularly to monitor its progress. Data from medical records included 35 patient variables, some of which were: demographic items, burn history, admission status, inpatient services, and outcome. Data collected on hospitals included 20 variables describing such items as size, staff, special intensive care facilities, patient turnover, and charges.

Eight patient variables were selected for further analyses in relation to the costs of burn care. These variables were considered indicators in the four areas to which they pertained: (1) *admission status* referred to severity of burns by percent body surface area (BSA) burned and age of patient; (2) *intensity and duration of treatment* was determined by number of operations and length of stay; (3) *outcome* referred to survival rate; and (4) *charges* were measured by the amount charged for care (excluding the physician charges). Although other facts were available in each of these areas, the above factors were considered adequate indicators. Using only a few variables in our analysis proved to be more manageable than working with a larger number.

As hospital charges have increased by nearly 13 percent per year since the time of this study, the dollar figures based on this data, collected in 1973, could conservatively be doubled to reflect 1980 charges.[5] It should be emphasized that in this study charges represent the bill for hospital care only and are expressed as follows: total charges, which are correlated with length of stay and thereby indicate the magnitude of charges for burn care; or per diem charges, which provide a more standard unit for comparison among patients or among hospitals. Physician charges, loss of income, or less tangible elements, such as psychological trauma, were not included.

Our first step in analyzing the data was to determine the relationship of the total charges and per diem charges to the patient's length of stay, number of operations, percent BSA burned, survival, and age; Pearson product moment correlations were used.

We also analyzed the data from three other perspectives. The first analysis dealt with differences in charges, length of stay, number of operations, and survival according to age and percent BSA burned. The four age categories—under 16, 16 to 44, 45 to 64, and over 64— corresponded to those commonly used by the National Center for Health Statistics in their surveys. Although it might have been desirable to report the very young and very old as separate groups, we felt that the sample would have been too small for meaningful analyses. In the subsequent analyses, however, age was used as a continuous variable, allowing for more complete measurement in terms of variance.

Percent BSA burned was divided into six groups: less than 10 percent; 10 to 19 percent; 20 to 29 percent; 30 to 39 percent; 40 to 49 percent; and 50 percent and over. Data were then analyzed using a 4×6 factorial design for multivariate analysis of variance, which tested for differences between the four age groups and between the six BSA groups, and for any significant interaction between BSA and age. In the analysis we attempted to learn whether charges, length of stay, number of operations, and survival differed in accordance with age and severity of injury, and whether the differences were consistent within each cell of the analysis.

In our second analysis of the data, we divided the patients according to place of treatment (a burn unit or a hospital) and survival of the burn injury. In this case we used a 2×2 factorial design for multivariate analysis of variance to determine differences among groups in terms of total charges, per diem charges, percent BSA burned, number of operations, age, and length of stay. In this analysis, we attempted to answer the following questions: In terms of the variables, were there any differences between those patients who did or did not survive and those who were treated in hospitals as opposed to burn units? Were the findings related to survival consistent for both unit and hospital groups?

In the third analysis, the data were divided by region (all seven regions of the state) and by survival, using a 2×7 factorial design for multivariate analysis of variance. The same variables were then examined as described above. The question addressed here was whether there were differences in the variables according to region and survival and whether findings relating to survival could be replicated in each region.

Results

The Hospitals. Hospitals ranged from large, public institutions with round-the-clock emergency services, medical school affiliates, and special burn units to small, private hospitals with virtually no emergency services. Admissions averaged about 12,000 per year, and approximately 23 (2 percent) were burned patients. More specifically, hospitals averaged 51,138 emergency room visits per year. Of these, 208 (4 percent) were for burn injuries. Length of stay averaged 7.5 days for all patients, with mean inpatient revenue amounting to about $9 million a year and revenue write-offs to about $1 million.

The Patients.   Of the 1,656 patients studied, the average age was 31 years, two thirds were male, and about one third had had emergency treatment for burns prior to admission. Over 50 percent of the burns occurred at home, primarily due to flame or hot liquids; approximately 25 percent of the patients were black. Burns of more than 30 percent BSA were sustained by one of every eight patients, and 25 percent of the patients were admitted with less than 5 percent BSA; 6 percent of the patients died as a result of their burns, and about 50 percent had complications.

Comparison of Burn Patients with other Hospital Patients.   Not only did burn patients stay twice as long as patients admitted for other reasons (15.7 days as against 7.5 days), but their average hospital bill was also twice that of the other patients ($1,817 as against $954). Furthermore, revenue lost per burned patient was $766 (36 percent of the billed amount), three times greater than the average percent of billed, lost revenue from other categories of patients and six times greater in actual dollar amounts.

Patients were excluded from the analyses if we were unable to determine the date of occurrence for the burns or the burns were considered "old" (had occurred more than two days prior to admission). Beause patients with "old" burns (N = 194) were being readmitted for care of the same injury, we wanted to avoid two pitfalls: (1) counting the same person twice in terms of outcome if he or she had been treated in another study hospital; or (2) dealing with patients whose initial burn data and charges were missing due to treatment outside our study group. In trying to recreate a homogeneous group for analysis, only "new" burns, which represented a first admission for treatment, were included.

Correlates of Total and Per Diem Charges.   Table 1 shows the correlation between total and per diem charges and the other variables studied. All of the eight items correlated significantly with total charges. As might be expected, length of stay had a higher correlation with total charges than any other variables; the higher the charges, the longer the stay. The number of operations and percent BSA burned also correlated significantly with higher costs, as did burn unit treatment, death, and older age.

Fewer significant correlates of per diem charges were seen and these were found to correlate significantly with mortality, increased severity of injury, burn unit care, and older age. Number of operations and length of stay understandably were not associated to any great extent with per diem charges.

Age and Severity.   Table 2 compares charges, length of stay, number of operations, and percentage of survivors cross-classified by age and percent BSA burned. Table 3 presents significance levels for differences among the four age groups and among the six BSA

Table 1   Correlates of Total and Per Diem Charges*

Variable	Total Charges	Per Diem Charges
Percent of BSA Burned	.35	.33
Number of Operations	.44	.03
Died	.17	.46
Length of Stay	.79	−.04
Burn Unit Care	.23	.20
Age	.12	.08

Note: Pearson r correlations: $r = .06$ is significant at $P < .05$, $r = .08$, $P < .01$.

* Total and per diem charges were correlated at $r = .27$.

burn groups, indicating any statistically significant interactional effects between age and severity.

Table 2 shows that charges increased with age and severity of burns until the death rate increased substantially at about 40 to 49 percent BSA burned in the oldest group, and at about 50 percent BSA burned in other groups. Length of stay similarly increased: shorter length of stay resulting from a high mortality rate was directly associated with the charges for the young, middle-aged, and elderly groups. However, charges continually increased with BSA burned for patients 16 to 44 years of age. The most severely injured showed a sharp drop in survival and shorter length of stay but had more operations.

The number of operations by age was not significantly different but severity was a factor. In children and young adults, the number of operations increased in severity to nearly 50 percent BSA burned. In the middle and older age groups, the number of operations rose to the point of 40 percent BSA burned, and then dropped, probably due to the decrease in survival. In those groups of 45 to 64 years old with 40 to 49 percent BSA burned, the death rate remained constant, number of operations decreased and charges increased.

Table 3 summarizes these observations and shows significant differences in survival by age and percent BSA burned, with survival differing more than any other variable among the six BSA groups. Charges and length of stay differed both by age groups and percent BSA groups but did not always change consistently within groups. In fact, the large number of interaction effects found indicates that the relationship between age and severity varies among the groups. In general, charges, mortality, and length of stay tended to increase with age and severity in certain groups, but not necessarily in other groups.

Specialized Treatment for Burns.   Table 4 presents charges, percent BSA, number of operations, age, and length of stay cross-tabulated by burn unit or hospital treatment and survival. Overall,

Table 2   Comparison of Charges, Length of Stay, Number of Operations and
Survival by Age and Severity of Burn

Comparison Items, by % of BSA Burned	Age Groups			
	Under 16 $N = 22\%$	*16 to 44* $N = 49\%$	*45 to 64* $N = 18\%$	*Over 65* $N = 11\%$
Less than 10 (N = 41%)				
Charges	786	753	1492	1941
Length of Stay	9.2	8.4	13.9	18.9
# Operations	.6	.6	.8	.8
Survival	100%	100%	99%	92%
10 to 19 (N = 30%)				
Charges	1560	1633	2017	1890
Length of Stay	16.4	13.6	19.1	26.3
# Operations	1.0	.8	1.3	1.2
Survival	100%	99%	96%	86%
20 to 29 (N = 13%)				
Charges	1874	2195	2214	2865
Length of Stay	28.1	20.0	25.1	21.0
# Operations	1.3	1.1	1.1	1.2
Survival	97%	100%	93%	92%
30 to 39 (N = 7%)				
Charges	5017	3695	4250	6931
Length of Stay	38.5	29.2	30.6	41.3
# Operations	1.9	1.8	3.3	3.4
Survival	92%	95%	81%	44%
40 to 49 (N = 3%)				
Charges	5439	3677	6496	5035
Length of Stay	47.4	28.4	41.0	24.5
# Operations	3.8	1.2	1.8	.9
Survival	100%	94%	80%	37%
Over 50 (N = 5%)				
Charges	2726	4937	3689	934
Length of Stay	27.2	22.4	31.0	2.8
# Operations	1.6	2.0	.8	.5
Survival	80%	43%	50%	09%

Table 3   Statistical Significance (F-ratios) of Main Effects of Age and Severity of Burn, and Interaction of These Factors for Selected Variables

Variable	Main Effects		Interaction of Age and % BSA Burned
	Age	% BSA Burned	
Charges	4.54**	15.04***	3.85**
Length of Stay	2.67*	9.82***	2.38*
Operations	.73	10.90***	2.43*
Survival	32.98***	106.16***	4.19**

* P < .05
** P < .01
*** P < .001

Table 4   Comparison of Patients in Burn Unit or Hospital by Charges, Percent
BSA Burned, Number of Operations, Age, and Length of Stay

Variable	Groups		F-Ratios		
	Unit (21%)	Hospital (79%)	Main Effect		
			Unit/Hospital	Survival	Interaction
Charges (total)			42.24***	24.69***	.46
Lived (93.4%)	3039	1463			
Died (6.6%)	4222	3172			
Charges (per diem)			23.39***	12.28***	.69
Lived	136	96			
Died	347	241			
Percent BSA Burned			.64	97.07***	13.28***
Lived	13	13			
Died	60	42			
Number of Operations			.70	.10	.02
Lived	1.56	.92			
Died	1.50	.83			
Age			2.07	51.38***	2.70
Lived	30.31	31.82			
Died	45.17	55.91			
Length of Stay			20.14***	2.51	2.81*
Lived	22.33	15.25			
Died	12.18	13.17			

* P < .05
** P < .01
*** P < .001

survival was 93.4 percent—86 percent for burn units and 95 percent for hospitals. About one fifth of the patients received treatment in specialized facilities located in three of the hospitals in large metropolitan areas. The best predictor of survival was percent BSA burned, followed by age, and charges. Specifically, a higher percent BSA burned, older age, and higher charges related to mortality, while number of operations and length of stay did not relate significantly to survival.

Charges, length of stay, and number of operations discriminated significantly between burn unit and hospital care, with higher charges, more operations, and longer stays associated significantly with the burn units. In comparing those patients cared for by burn units or those treated in hospital, age was not a distinguishing factor.

In terms of percent BSA burned, an interesting finding emerged: no main effect difference was found between unit and hospital care, although a highly significant interactional effect occurred, indicating that for those patients who *lived*, no meaningful difference existed between the burn unit and hospital in terms of percent BSA burned. The percent BSA burned averaged 13 percent for each group, with a standard deviation of 12 for each. However, for patients who died, the percent BSA burned was significantly higher in the burn units (60 percent with a standard deviation of 28) compared with the hospitals (42 percent BSA, with a standard deviation of 29). The number of operations was higher for these patients in the burn units, and they lived the same length of time as those patients with close to 20 percent less BSA burned who were treated in hospitals.

Our findings demonstrated that total charges differed: Patients in the burn units were charged twice as much as those in the hospitals. Although age was not a distinguishing factor, the number of operations was higher in the burn units and the length of stay was longer (22 days compared to 5 days). This raises questions about the locus of care for patients with less severe burns or smaller percent BSA burned.

Regional Variations. In Table 5 the same variables are cross-classified by survival and by region. The number of operations and charges differed by region, but age and length of stay did not. Previous examination of the data indicated that charges were associated with age and length of stay; factors other than these might then account for charge differences by region. Overall, findings related to survival were consistent across region: number of operations was not associated with survival in any particular region; for those patients who died, age as well as charges were higher. Percent BSA burned, however, showed a significant interaction between region and survival. In region 4, the percent BSA burned was essentially

Table 5 Comparison of 7 State Regions by Percent BSA Burned, Number of Operations, Length of Stay, Age, and Charges

Variable	State Regions							F-Ratios	
								Main Effect by Region	*Interaction*
	1	2	3	4	5	6	7		
Percent BSA Burned									
Lived	11.9	14.4	12.1	18.0	12.2	17.2	9.6	2.96**	11.74***
Died	84.3	50.1	38.4	18.6	55.7	54.0	45.5		
Number of Operations									
Lived	1.2	.8	.7	.7	1.3	2.1	.6	8.27***	.53
Died	2.3	.6	.0	.4	1.4	1.4	1.0		
Length of Stay									
Lived	14.5	17.8	15.7	16.7	20.3	15.7	11.9	1.62	.72
Died	27.0	10.3	10.2	17.0	13.2	6.6	20.0		
Age									
Lived	31.4	29.9	28.9	33.3	35.1	32.2	33.5	1.45	1.83
Died	29.3	62.7	43.4	62.8	47.1	53.4	63.5		
Charges (total)									
Lived	1204	1820	1498	1516	2644	1649	1048	4.71***	1.95
Died	7819	3375	2526	2317	4634	1914	2588		
Charges (per diem)									
Lived	80.3	82.3	93.6	89.2	132.2	103.1	87.3	5.09***	2.10
Died	269.6	337.5	252.6	136.3	356.5	273.4	129.4		

* $P < .05$
** $P < .01$
*** $P < .001$

the same among those patients who lived and those who died, as opposed to all other regions in which the percent BSA burned was significantly lower for patients who survived. Charges also tended to be lower, suggesting that the quality of care might need further examination in this region.

Discussion

These data show that burn treatment is costly in both absolute and relative terms. Charges tend to increase with severity of burn and age of the patient until about 40 percent BSA burned is reached in adults, and total charges decrease as length of stay goes down because of increased mortality. However, when looking at those patients who died from their burns, total charges as well as charges per day are greater in this group. The particular advantage of looking at per diem together with total charges may be seen when examining those patients with the shorter survival times, for example, very old patients with burns of over 50 percent BSA.

Survival rate is best for children and worst for older patients. Severe burns can be successfully treated in children while minor injuries can be fatal in the elderly. In another study[6] I have discussed some of the reasons why burns are more severe in the elderly, but in general their decreased ability to respond quickly to environmental stimuli and the degenerative changes in the skin due to aging are factors.

Our study shows that burn units are more expensive than hospitals and are well equipped to treat patients with severe burns. The patients who died in these specialized centers averaged 60 percent BSA burned, compared to those who died in hospitals with about 42 percent BSA burned. Patients with the more severe burns were kept alive for the same length of time as those treated in the hospital; however, for surviving patients (86 percent), both the average age and percent BSA burned were identical to those of surviving patients in hospitals with no special burn facilities and where cost of care was less. Furthermore, number of operations and length of stay were greater for patients treated in burn units.

We do not want to suggest that all the patients should have been treated in hospitals without burn units. Consideration should be given to the fact that percent BSA burned and age are not the only criteria of severity. The depth of the burn may have been greater in this group, even though our earlier analyses[7] did not suggest this. Also, in prior analyses,[4] pulmonary injury and disease (10 percent in this sample) were not significantly correlated with outcome, although Zawacki, et al.[8] found bronchopulmonary diseases, abnormal

P_aO_2, and airway edema important discriminators of mortality in burned patients. Also, patients with injuries affecting a small percent of body surface area may have been treated in a burn unit because of the location (hands or lungs) or the source (electrical) of the burn injury. Although these factors cannot be excluded without further analyses, it seems likely that a large proportion of burn unit patients could have been treated just as well and at less expense in hospitals.

Along these same lines, many of the patients admitted to hospitals with less than 5 percent BSA burned (25 percent of admissions) could very possibly have been treated on an outpatient basis, further reducing the charges of hospitalization.

In another paper[4] I discussed the importance of continuing education for emergency room physicians in assessing the need for the initial management of burns. In this two-year study,[9] completed in 1979, emergency room physicians of ten hospitals were offered an educational program, and ten hospitals without such a program were used as the control group. Our findings showed definite improvements; the survival rate of burned patients treated by the physicians who had received training increased. Thus it may be possible to influence charge reduction by offering educational programs at the emergency room level.

Other issues, such as the high cost of gathering good data, might also be discussed. Because nurses understand hospital procedures and medical treatment, they make ideal data collectors; hiring these professionals on a full-time basis is therefore a costly but necessary component in acquiring accurate information.

Another issue which might be considered is the limitations in the data collected for this study. Personal interviews, although costly, are more complete and reliable than medical records for gathering information. Our assessment of the quality of care based on survival alone is another limiting factor. Ideally, such variables as disability, satisfaction with care, and social adjustment should be included in defining outcome and evaluating quality.

Finally, a crucial question seems to be whether increased charges are associated with better survival rates. Our data seem to indicate that this may not be the case for very severely burned patients, but it does not mean that the quality of care offered should be any less.

Endnotes

1. Flood, A.B., Ewy, W., and Scott, W.R., Forrest, W.H., and Brown, B.W. (1979). The relationship between intensity and duration of medical services and outcomes of hospitalized patients. *Med. Care* 17:1088.

2. Oilstein, R.N., Crikelair, G.F., and Symonds, F.C. (1971). The burn center concept. *Hosp. Management* 21 (Feb.–Mar.).

3. Stone, N.H. and Boswick, J.R. (1970). Specialized burn care. *Surg. Clin. N. Am.* 50:1437.

4. Linn, B.S., Stephenson, S.E., and Smith, J. (1977). The assessment of needs for burn care. *N. Engl. J. Med.* 296:311.

5. U.S. Department of Health, Education and Welfare (1978). *Health, United States*. PHS Pub. No. 78–1232.

6. Linn, B.S. (1980). Age differences in the severity and outcome of burns. *J. Am. Geriatr. Soc.* 28:118.

7. Linn, B.S., Stephenson, S.E., Bergstresser, P.R., and Smith, J. (1979). Do dollars spent relate to outcomes in burn care? *Med. Care* 17:835.

8. Zawacki, B.E., Azen, S.P., Inbus, S.H., and Chang, Y.C. (1979). Multifactorial probit analysis of mortality in burned patients. *Ann. Surgery* 189:1–5.

9. Linn, B.S. (1980). The impact of continuing education on the process and outcome of emergency room burn care. *JAMA* 244:565–570.

THE END-STAGE RENAL DISEASE WORKSHOP: INTRODUCTION

by Richard A. Rettig

The workshop attempted to establish a basis for discussing selective issues associated with the treatment of end-stage renal disease. Successive presentations by myself; Dr. Eli H. Friedman,* Department of Medicine, SUNY Downstate Medical Center; and Dr. Robert J. Wineman provided general overviews on federal government policy, clinical practice, and patient population, all in historical perspective. Dr. Edmund G. Lowrie focused on the difficulties of prescribing dialysis treatment on an individualized basis and on the biocompatibility of dialyzer materials.

The purpose of pursuing the above approach was to stimulate workshop discussion of a number of the issues involved in assessing the treatment of end-stage renal disease. But no attempt was made to assess either dialysis or transplantation in a comprehensive way. The paper following the introduction, prepared after the workshop, sets forth the many factors that complicate the conceptualization, designs, and implementation of any comprehensive assessment.

The central policy issue for any new medical technology or procedure is whether or not, and how, to finance patient treatment. That issue was resolved for end-stage renal disease when Congress, in 1972, authorized Medicare coverage for those under 65 years of age having permanent kidney failure. The high cost of that program, stemming from high annual costs per patient and a steadily growing patient population, has kept the Medicare End-Stage Renal Disease (ESRD) program on the public agenda.

One major dynamic underlying the treatment of end-stage renal disease patients, as Friedman noted in his conference presentation,

---

* Dr. Friedman's presentation is not included as a chapter in this volume.

has been the continuing improvement of existing means of treatment and the emergence of new treatment modalities or of old modalities in new forms. Constant technical and clinical change, often of an incremental nature, typifies the evolution of most medical technologies. Another dynamic, as Wineman notes, is the extent of the changes that occur over time in the patient population. In the case of end-stage renal disease, changes have been observed in the race, sex, age, and medical status of patients. Other procedures or technologies can be expected to witness similar changes in the patients who benefit from them. Finally, as Lowrie's discussion of biocompatibility points out, no sooner are old problems laid to rest than new ones come into view, especially where treatment involves extended interaction between the patient and technology.

The workshop and the papers in this volume sought to capture some of the complexities associated with assessing medical technology in the instance of end-stage renal disease.

CRITICAL ISSUES IN THE ASSESSMENT OF END-STAGE RENAL DISEASE

by Richard A. Rettig

End-stage renal disease, or permanent kidney failure, has been treated for two decades by hemodialysis, peritoneal dialysis, and kidney transplantation. Regarding assessment of the pertinent medical technology, both this disease and its treatment should have ceased to attract attention long ago. In actuality, this is not the case.

Several factors complicate the assessment of end-stage renal disease. First, though a single disease is involved, several different treatment procedures are needed, in contrast to an assessment involving a single medical technology. Both hemodialysis and peritoneal dialysis are provided on an inpatient and outpatient basis in both institutional and home settings. Transplantation, similarly, has two sources of donated organs—a living relative or a cadaver.

These treatments interact in ways that further complicate assessment. Potential transplant recipients, especially those awaiting a cadaver kidney, are often maintained on hemodialysis until a suitable kidney becomes available. If a transplant fails, moreover, the patient returns to dialysis—or dies.

Another complicating factor is that the determination of appropriate treatment is dependent on the patient's condition. In general, transplant recipients are selected from those patients with end-stage renal disease who have relatively few other medical complications and who are seldom over 60 years of age. Similarly, home dialysis patients usually are clinically and psychologically stronger than patients treated in an institutional setting; data in the United States and Europe show that home dialysis patients consistently experience higher survival and lower morbidity than patients treated in institutional settings.[1] Peritoneal dialysis, and particularly continuous ambulatory peritoneal dialysis (CAPD), is indicated for patients who are older, have poor cardiovascular systems, and are diabetic.[2] The relative efficacy of these treatments is therefore difficult to measure and compare.

The attitudes of physicians toward the various treatment modalities is an important determinant. Transplant surgeons and nephrologists place different values on transplantation and dialysis. Moreover, as noted by Simmons and colleagues,[3] physicians tend to see the patients who represent the failures of the others, thus finding powerful reinforcement for their respective views. Nephrologists differ in their philosophies regarding the relative merits of home versus center dialysis.[4] Differential financial incentives to physicians to treat patients in these two settings make the assessment problem all the more complicated.

When successful, transplantation is clearly the most attractive treatment for end-stage renal disease and is the only procedure that can genuinely be deemed a cure. For a successful transplant recipient, the quality of life is substantially better than it is for a patient undergoing dialysis. The highest probability for a successful transplant occurs when the organ is donated by a living relative; however, these living-relative donor transplants have declined proportionately to fewer than 30 percent of all transplants performed in the United States.[4]

Cadaver transplantation can be successful, but for much of the past decade, the one-year survival of the transplanted kidney (the *graft* survival rate) has been below 50 percent.[5,6] The one-year *patient* survival rate has increased to over 85 percent and 90 percent in the major centers, reflecting a widespread shift in the philosophy of treating the rejecting kidney.[7] Previously, substantial efforts were made to prevent rejection of the transplanted kidney by the administration of heavy doses of immunosuppressive drugs and steroids. Quite often, however, the patient died in the process. Today physicians are more willing to accept the rejection of the kidney and save the patient.

This combination of low graft survival and high patient survival means that the effectiveness of cadaver transplantation is uncertain. The possible outcomes include success or failure of the transplanted kidney, resulting in a return to dialysis, or death. The second outcome is most likely in that an expensive surgical operation often will be preceded by dialysis treatment and followed by the same treatment after the transplanted kidney fails. Only within the past two or three years has there been evidence of one-year cadaver kidney survival reaching above 50 percent, a development associated mainly with a yet unexplained correlation between the number of pretransplant blood transfusions and graft survival.[8]

The efficacy of dialysis also is a matter of continuing discussion. In rudimentary terms, dialysis prolongs life, a fact that has been true for nearly two decades. On the other hand, no satisfactory measure of the adequacy of dialysis exists, either for frequency or duration.[9] Moreover, the quality of life for dialysis patients is questionable, especially for those who are older and have other medical complications. Finally, limiting the measurement of efficacy to the effects of medical treatment is conceptually deficient. Rehabilitation of patients to a state approximating their activity level and occupational status before treatment requires more and different resources —including a desire to be rehabilitated by the patient—than simple medical treatment provides.[10]

A further complication in the assessment of the treatment of end-stage renal disease stems from the limited ability of the federal government to generate data on these issues.[11,12] The problems faced by Medicare's End-Stage Renal Disease (ESRD) program in creating and maintaining a satisfactory data system should be examined, as other technologies, procedures, or diseases might have similar problems of assessment.

Despite the difficulties, the attempts to assess treatment has continued because end-stage renal disease has become the classic example of a life-saving medical technology that is extremely expensive to finance. The Health Care Financing Administration estimates expenditures of nearly $1.5 billion in 1980 for an estimated 64,000 beneficiaries of the Medicare ESRD program. If the dollars involved were one third of this amount, or the beneficiary population were twice the size, interest most likely would be much less. However, the fact that relatively few individuals are being kept alive at a relatively high cost stimulates assessment efforts.

In my opinion society requires two types of assessments: (1) detailed analyses on selected aspects of treating the end-stage renal disease patient, and; (2) more interpretative analyses of the affect of the end-stage renal disease experience on medicine and society, relative to the allocation—or rationing—of scarce medical resources.

Four critical issues have been identified as being central to the first type of analysis: innovation, diffusion, utilization, and cost.

Innovation

Both transplantation and dialysis represent major innovations in clinical medicine, each emerging as "established" therapy during the early 1960s, after a decade or more of "experimental" research.[13,14] Transplantation required the development of a surgical technique, the recognition of the central role of the immune system in the acceptance or rejection of the transplanted kidney, and the discovery of the suppression of the immune-response rejection process by drugs. The earliest, successful transplants were performed on identical twins at the Peter Bent Brigham Hospital in Boston, and the procedure pointed out the importance of immunological compatibility between recipient and donor. The discovery that the immunological rejection response could be controlled by drugs eliminated the use of whole-body irradiation for immunosuppression and ushered in the possibility of using cadaver kidneys as a donor source.

Hemodialysis required the development of a technology to cleanse the blood of toxic waste outside the body. The first artificial kidney machine was built by Willem Kolff in Holland during World War II.[15] Treatment of permanent kidney failure became possible when, in 1960, Belding Scribner invented the simple access device that made it possible to connect a patient to the machine on a recurring basis without repeated surgery.[16]

Both treatments have experienced continuous incremental technical and clinical change. The rapid scientific advances in immunology have created the possibility for more effective matching between donor and recipient, resulting in better success for transplantation. To date, more questions than solutions have been raised by scientific advances in immunology. One important clinical development, mentioned earlier, has been the changed philosophy of dealing with rejection of the transplanted kidney.[7] Improved survival of cadaver transplant recipients has resulted from this change in philosophy. Substantial refinement in dosage rates for immunosuppressive drugs has occurred also.

For many patients receiving transplants, the survival of cadaver transplants is directly related to the number of blood transfusions occurring before surgery.[8] This recent discovery is notable for several reasons: it reversed the view prevalent in the early 1970s that pretransplant blood transfusions were to be avoided; it was empirical, deriving initially from a retrospective regression analysis of data by Terasaki[6] (since confirmed prospectively in both animals and humans in the United States and Europe); and it still lacks a satisfactory scientific explanation.

A number of changes have occurred in hemodialysis. In 1966, for instance, physicians at the Bronx Veterans Administration Hospital developed a subcutaneous fistula as a blood access means.[17] This fistula involves a surgical knotting of an artery and vein, usually in the forearm, into which a two-way flow needle is inserted. Patients are connected to the machine in this manner. This procedure was a significant improvement over the mechanical arteriovenous cannulae-and-shunt apparatus originally introduced by Scribner. It has become the access means of choice whenever possible, as it has few of the drawbacks of infection and clotting, and vulnerability to bumping associated with the shunt.

Important design changes have occurred in dialyzers, which are disposable membranes through which the blood is passed and cleansed of toxic elements. A hollow fiber dialyzer, which resembles a bundle of clear, cellulosic straws packed together in a compact cylinder, was introduced in the early 1970s. This new dialyzer combines greater membrane surface area in a smaller volume and, with certain other innovations, facilitates faster dialysis times.

In reponse to the incentives of Medicare reimbursement, many clinicians have reused dialyzers, which involves: (1) cleaning a once-used dialyzer; (2) storing it in a sterile solution; (3) removing it from storage; and (4) using it again on the same patient and repeating the procedure several times. Despite economic considerations, questions have been raised about the risk of reuse to the patient. Although reuse has occurred for some years and is now probably more extensive than ever, existing data show minimal patient mortality or morbidity due to the practice. Nevertheless, both a patient organization—the National Association of Patients on Hemodialysis and Transplantation (NAPHT)—and an industry trade association—the Health Industries Manufacturers Association —have opposed reuse, and Congress has requested a study of the matter.[18,19] The National Institutes of Health are supporting such a study at present. In this volume Lowrie presents data suggesting that reuse may be less damaging to patients than initial use. Friedman has supported the position of the NAPHT "that dialyzer reuse be viewed as an experimental undertaking requiring informed patient consent."[18]

Medical technology experiences continuous incremental changes over time;[20,21] some of these changes enhance quality, others decrease costs. This pattern of technical change, though not highly visible, can have substantial economic effects. Indeed, some evidence suggests that prices for dialysis supplies have steadily declined in actual (deflated) dollars since the inception of the Medicare ESRD program in 1973.

One issue raised by these experiences and changes is whether scientific progress in biomedical research is likely to generate major, new innovations in the prevention or treatment of end-stage renal disease. The National Institute of Arthritis, Metabolism and Digestive Diseases (NIAMDD) published a report that addresses this issue and comments on the gap between medical technology and an understanding of the disease process:[22]

Nowhere is the gap in understanding more apparent than in the case of end-stage kidney disease (permanent kidney failure). The tech-

nique of dialysis prolongs the lives of patients with this fatal condition, but it provides little insight into the nature of the underlying disorder. Dialysis is very expensive but does not cure the disease or return patients to a normal state of health, and the treatments must be continued indefinitely for the patient to survive. Transplantation has the potential to cure the patient with end-stage renal disease, but thus far the supply of donor kidneys is inadequate, and there are too many treatment failures (most often caused by rejection of the transplanted organ) to call the technique a means of general cure. Nevertheless, dialysis and transplantation are at present our only effective methods of prolonging and improving the quality of life of patients with these otherwise fatal diseases.

The report also summarizes the limits of dialysis and transplantation: "For most problems fundamental knowledge is limited, and currently no rational approach to treatment or cure of the diseases that lead to chronic renal failure can be devised. Significant practical advances must await the acquisition of more fundamental knowledge." The overall report identified research needs but was cautious about predicting major advances in the near term.

Diffusion

Rogers and his colleagues[23,24] conducted studies on the diffusion of innovation based on four essential elements: the innovation itself; communication about the innovation; the time involved; and the social system into which the innovation was introduced. Another study by Coleman, Katz, and Menzel[25] traced the diffusion of a new drug, over time, among physicians in four midwestern communities. The differences in behavior characterizing early and late adopters of the drug were analyzed, as were the patterns of influence among physicians affecting the adoption decision.

One major problem with this research, however, is that it assumes a fairly elementary pattern of social innovation. Innovations studied in the literature, for example, are rather simple in character.[26] In the treatment of end-stage renal disease, however, no single innovation exists. There are two basic forms of treatment; two donor sources for transplantation; and two types of dialysis. Hemodialysis uses three types of dialyzers—coil, parallel plate, and hollow fiber; and both hemodialysis and peritoneal dialysis can be performed in the institutional or home setting. The innovation is the substitution of one of several medical procedures for permanent loss of kidney function, hardly simple in nature.

Communication patterns among key individuals influencing the diffusion of dialysis and transplantation have never been fully analyzed. Several different communications processes, however, operate simultaneously.[27] At the time the two procedures emerged, much attention within the medical community was devoted to arguing about the status of the procedures: Were they experimental or established treatments?[14] Those who quickly concluded that the treatments were efficacious then instructed their colleagues at other institutions in their use and established one pattern of communication. Often, at least for dialysis, the early adopters were young clinicians who preferred to set up clinical practices rather than pursue careers in academic medicine. In contrast, those who regarded the procedures as experimental tended not to adopt them (or did so very late), continued their strong research interests, and sometimes sought to dissuade younger physicians from following the clinical pathway. These two patterns of communication differed widely, and it was not until the passage of the Medicare kidney amendment in 1972 that academics began to make their peace with clinical practitioners, especially in the acceptance of dialysis. These different communications patterns, however, can be better understood by turning briefly to a description of the social system from which they emerge.

The nature of the social system for treating end-stage renal disease is complex, and during the 1960s it evolved in several ways.[27] One avenue of diffusion was within the Veterans' Administration. From 1963, when plans for 30 or more dialysis treatment centers were first announced, until the early 1970s, the VA created a number of hospital-based dialysis treatment centers, supported several key kidney transplantation centers, and initiated a major effort in home dialysis. Within the Public Health Service, the social system evolved as research grants and contracts supported both dialysis and transplantation investigators at certain medical schools, and demonstration grants permitted the development of both hospital-based and home dialysis treatment capabilities.

Developments within academic medicine influenced the diffusion of treatment for end-stage renal disease. In the 1950s and 1960s, departments of medicine were undergoing substantial internal differentiation into subspecialties.[28] Nephrology was one of these subspecialties, created mainly in response to the emergence of dialysis. If the clinical roots of nephrology were in the technology of dialysis, the intellectual roots were in renal physiology, one of the major scientific fields of medicine. Indeed, many renal physiologists displayed great ambivalence, and often sharp hostility, toward dialysis. Differentiation within academic medicine, therefore, facilitated the diffusion of dialysis by providing an institutional base for nephrology. That diffusion was constrained by the reaction to dialysis from within departments of medicine, especially in the internal competition for space and faculty positions and the external quest for research support.

Since the Medicare ESRD program became operative in mid-1973, the social system of clinical nephrology interacted with the reimbursement system and resulted in the proportionate decline of home dialysis and the flourishing of proprietary outpatient dialysis.[4,29] According to the data presented by Wineman in this volume, patients dialyzed at home account for approximately 13 percent of the total patient population, while those receiving treatment on an outpatient basis make up approximately 35 percent of the total patient population.

No adequate study of the diffusion of treatment for end-stage renal disease has been done, and several reasons have been suggested: (1) the innovation is not a single, simple procedure, but a set of different procedures applied in various ways; (2) the communication processes among key figures are complex and difficult to trace; (3) the social system has been outlined in only rudimentary ways; and (4) the external factors influencing diffusion, especially government financing of care, have been analyzed extensively but not systematically within a diffusion framework.[11]

In a careful empirical study, Russell[30] analyzed the diffusion of dialysis units in the hospital sector in the United States. However, her study did not include the diffusion of home dialysis, nonhospital outpatient dialysis, nor transplantation. Consequently, the study revealed both the utility and limits of the diffusion approach.

It is questionable whether these diffusion studies have any value. For descriptive purposes, those conducted to date in medical technology are not highly instructive about the dynamics of change in medical practice. Several hypotheses about the diffusion patterns of sophisticated, new medical technologies might be suggested for future studies:

1. The primary diffusion pathway of a new technology or procedure is likely to be from a specialty group in a tertiary medical center to members of that same specialty in other tertiary centers.
2. The secondary diffusion pathway, occurring at some time after the primary pattern has begun, is probably still within the same medical specialty but is from tertiary centers to community hospitals.
3. If a technology or procedure has an expanding range of applications beyond its initial use, another diffusion pattern, or set of patterns, is likely to be from the primary specialty group to an adjacent specialty group, initially within the tertiary center, then among tertiary centers, and finally from tertiary centers to community hospitals.

4. The diffusion of a technology from use in the hospital setting to use in other settings will depend on the physical characteristics of the technology and the incentives to move it out of the hospital setting (especially those incentives created by reimbursement policies of third-party insurers).

Other hypotheses can be suggested; however, to understand the dynamics of change within medicine, it is necessary to recognize certain conceptual limits and the correlative need for enriching future diffusion studies by a more detailed specification of the nature of innovations, the communication channels, and the social system of medicine.[31]

Utilization

Underutilization of hospitals is one factor contributing to higher costs of medical care. Excess capacity leads to either longer hospital stays (thus increasing the cost per hospitalization) or higher charges (thus increasing the cost per day).

End-stage renal disease has never confronted this problem. Before Medicare, existing facilities were usually operating at capacity. Indeed, home dialysis grew in the 1960s largely because it extended access to life-saving treatment without increasing the capacity of highly utilized facilities.

The Medicare ESRD program created strong incentives for high utilization of facilities[11] by establishing minimum utilization rates for both dialysis and transplantation. Per treatment reimbursement for both hospital and nonhospital outpatient dialysis was set at $150, with adjustments for physician fees (reimbursed on a cost basis) and laboratory tests (reimbursed on a charge basis).

Facilities under this ceiling were given an incentive to spread fixed costs over a larger number of treatments by using machines (or stations) at their maximum practical capacity. Held and Pauly presented data suggesting that a declining cost curve existed for such facilities as a consequence of increasing the number of treatments.[32] Thus high utilization was driven by a desire to maintain revenues above expenses by taking advantage of economies of scale.

The minimum utilization rate, although superfluous for dialysis, has had quite a different history for transplants. Since transplants were performed solely in Medicare-certified hospitals, they were reimbursed on a cost basis from the Hospital Insurance Trust Fund (Part A).* The minimum utilization rate, therefore, had no economic justification.

---

* Outpatient dialysis is reimbursed from the Supplementary Medical Insurance Trust Fund (Part B).

The primary justification was an alleged relationship between the volume of transplants performed and the outcome in the survival of the patient and transplanted kidney.[33] The process of formulating minimum utilization rate policy, however, revealed that such a relationship could not be supported by data.[11] Early policy discussions in 1973 were dominated by large-volume transplant surgeons who spoke of 50 operations annually as a reasonable basis for certification. In the Notice of Proposed Rulemaking of mid-1975, however, unconditional certification was set at 25 transplants per year and conditional at 15 to 24 annually.[34] Public comment attacked even these levels as too high, and the final rule published in mid-1976 stipulated 15 operations annually as the basis for unconditional certification and 7 to 14 operations for conditional status.[35] Data could not support the volume-outcome relationship.

One contemporary issue intersects the utilization question. In order to be certified for reimbursement purposes, ESRD dialysis facilities must obtain Certificate-of-Need (CON) approval from the health planning system. Critics charge that this system creates barriers to entry, resulting in regional monopoly power to certified facilities.[32] If CON is eliminated, they argue, facilities will be allowed to compete for patients, greater amenities will be provided to patients, and fewer rents (or excess profits) will be extracted. Costs will be controlled by something comparable to the screen. Movement in this direction may occur within the next few years, but during the interim, discussion is very likely to intensify on the prospects for competition within the government-made market for the treatment of end-stage renal disease.

Costs

The ESRD program remains an issue because of large expenditures for relatively few beneficiaries. Table 1 shows the average annual enrollment in the program and the annual benefit payments for the period from 1974 (the first full year of the program) through 1980.* These data include beneficiaries receiving both dialysis and trans-

---

* The data in Table 1 and Table 2 were generated by the Office of Financial and Actuarial Analysis, Division of Medicare Cost Estimates, Health Care Financing Administration. They are used because they are consistent over the duration of the program, are updated and revised constantly, and are consistent with HCFA's reporting of data on the entire Medicare program. The data from the Actuary's Office do not correspond exactly to those data published by the Office of Special Programs, Office of End-Stage Renal Disease, HCFA. There has been no published effort to reconcile discrepancies between the two HCFA data sources.

Table 1   Medicare ESRD Patient Population and Annual ESRD Benefit Payments, 1974–1980

Calendar Year	Average Annual Enrollment*	Benefit Payments ($ million)†
1974	19,000	$ 283
1975	27,000	450
1976	35,000	606
1977	41,000	787
1978	48,000	977
1979	56,000	1,198
1980	64,000	1,453

SOURCE: Office of Financial and Actuarial Analysis, Division of Medicare Cost Estimates, Health Care Financing Administration, May 1980.

* Estimated.
† Incurred basis; data are calculated according to the date the services were rendered.

plantation services, and all the Medicare costs—renal and other—incurred by these beneficiaries.

The projected beneficiary population and benefit payments are shown in Table 2. These estimates suggest that the patient population may reach an equilibrium 10 to 20 years from now, assuming no change in ability to prevent end-stage renal disease. The cost estimates are questionable since they are quite sensitive to the following assumptions: that ESRD costs will increase at the same rate that overall Medicare costs are predicted to increase; that no economies of scale, learning in production, or cost-reducing technical change will occur for dialysis equipment and supplies; and that no significant preventive measures for end-stage renal disease will be found. If cost projections, for example, allow no increase in per beneficiary expenditures, using 1980 figures, the cost burden for 1985, 1990, 1995, and 2000 will be $1,930 million, $2,111 million, $2,157 million, and $2,202 million, respectively.

It is often alleged that the ESRD program costs are out of control. The response to this charge involves several points. First, entitlement programs have long been referred to as "uncontrollables" because the formula for expenditures is established in statute law, not by the annual appropriations process. In this respect the ESRD program, as an open-ended entitlement program within Medicare, is—by deliberate decision—part of the "uncontrollable" portion of the federal budget.

Second, the original Congressional "estimates" were wrong; they were absurdly low.[27] The fourth year expenditure projection used in the 1972 Senate debate was exceeded in the first full year of the pro-

Table 2   Projected Medicare ESRD Patient Population and Annual ESRD
           Benefit Payments*

Calendar Year	Average Annual Enrollment	Benefit Payments ($ million)†
1985	85,000	$2,910
1990	93,000	4,647
1995	95,000	6,694
2000	97,000	9,252

SOURCE: Office of Financial and Actuarial Analysis, HCFA.

* The population projections assume an increase of new entrants of less than 1 percent per year with a slight increase in the numbers of transplants performed. Cost projections assume an inflation factor of about 10 percent per year for hospital costs and about 4 percent per year for nonhospital costs. These assumptions are consistent with the 1979 Annual Reports of the Board of Trustees of the Federal Hospital Insurance Trust Fund and the Federal Supplementary Medical Insurance Trust Fund.

† Incurred basis.

gram. These low initial estimates do not mean, however, that a program is out of control.

Third, the program is expensive and its size, relative to the few beneficiaries, lead many to conclude that costs are out of control.

Fourth, with the exception of a recent paper by Held and Pauly,[32] no systematic cost studies have been conducted and published by the program. This absence of data merely reinforces the view that costs are not being controlled. In fact, if the program expenditures are deflated to 1972 dollars, the cost growth is much less than Medicare and relatively stable in benefit payments per beneficiary (an admittedly crude indicator).[11]

The costs of the ESRD program do raise two disturbing questions related to policy and ethics. First, the program does not include costs to the government incurred as a result of patients receiving both ESRD benefits, and monthly disability income support payments from Social Security. The burden on the disability program is not known; historically, Congress expected that patients receiving treatment would be rehabilitated, but reality has differed from that initial, unexamined expectation.

The second question that deserves consideration relates to "opportunity costs" to Medicare that are imposed by the ESRD program —missed opportunities for funding other programs because resources used for one purpose, in this case the ESRD program, cannot be used for another. The ESRD program now requires nearly 5 percent of total Medicare expenditures for one fourth of 1 percent of the Medicare beneficiary population. The nation, while willing and able to finance care for end-stage renal disease, might be faced with the decision to forego the financing of other expensive, life-saving medical technology because the impact on general health care benefits to the elderly and disabled was deemed unjustifiable.

Conclusion

In the current context of heightened concern for the allocation of scarce medical resources, the emergence of expensive, life-saving medical technologies poses very difficult societal choices. Does a society, acting through its national government, respond to the existence of such procedures by paying for treatment for all who can possibly benefit? Does it limit or ration access to treatment on the measure of social worth? Does it refuse to make any public resources available for treatment on the calculated basis that the social costs of such action exceed the social benefits?*

This discussion raises the problem of "tragic choice"—a dilemma that is particularly painful because scarcity limits society's ability to affirm fundamental values. In the case of expensive, life-saving medical technologies, the societal choice either affirms that life is beyond price or acknowledges that scarcity forces us to compromise that value.[36] According to Calabresi, tragic choices "are situations for which there is no right decision."[37]

The societal dilemma could be resolved by making access to treatment a universal entitlement, which was done in 1972 for victims of end-stage renal disease. That decision, however, only highlighted other equity considerations such as the justification of alternative uses of the resources spent for end-stage renal disease.

Assessment of end-stage renal disease raises the following questions: What is the value of saving a few lives when resources could be deployed for the benefit of a much larger number of individuals? What relationship should exist between medical care and rehabilitation? What role, if any, should quality-of-life considerations play in the formulation and implementation of public policy? The future will require policymakers to deal with these issues in relation to use and assessment of other medical procedures.

Endnotes

1. Wing, A.J., Brunner, F.P., Brynger, H., Chantler, C., Donkerwolcke, R.A., Gurland, H.J., Hathway, R.A., and Jacobs, C. (1978). Combined report on

---

* The polar case at the "no-treatment" end of this continuum is a societal decision that bars any resources, public or private, being used for treatment. Such a policy would be justified by decision makers on the grounds of equal denial of access, but since it would also require vast government coercion of individuals, it is regarded as infeasible in the United States.

regular dialysis and transplantation in Europe, VIII, 1977. In *Dialysis Transplantation Nephrology: Proceedings of the Fifteenth Congress of the European Dialysis and Transplant Association*, eds. B.H.B. Robinson and J.B. Hawkins, p. 23. Kent, England: Pitman Medical Publishing Co. Ltd.

2. Lacke, C., Senekjian, H.O., Knight, T.F., Frazier, M., Hatlelid, R., Kozak, M., Baker, P., and Weinman, E.J. (1981). Twelve months' experience with continuous ambulatory and intermittent peritoneal dialysis. *Arch. Intern. Med.* 141:187–190.

3. Simmons, R.G., Klein, S.D., Simmons, R.L. (1977). *Gift of Life: The Social and Psychological Impact of Organ Transplantation*. New York: John Wiley & Sons.

4. Rettig, R. (1980). The politics of health cost containment: End-state renal disease. *Bull. N.Y. Acad. Med.* 56:115–138.

5. Advisory Committee to the Renal Transplant Registry (1977). The 13th report of the human renal transplant registry. *Transplantation Proceedings* IX:9–26.

6. Terasaki, P.I., Opelz, G., Mickey, M.R. (1976). Analysis of yearly transplant kidney transplant survival rates. *Transplantation Proceedings* VIII: 139–144.

7. Tilney, N.L., Strom, T.B., Vineyard, G.C., and Merrill, J.P. (1978). Factors contributing to the declining mortality rate in renal transplantation. *N. Engl. J. Med.* 299:1321–1325.

8. McDonald, J.C., Vaughn, W., Filo, R.S., Piron, G.M., Niblack, G., Spees, E.K., and Williams, G.M. (1981). Cadaver donor renal transplantation by centers of the Southeastern Organ Procurement Foundation. *Ann. Surgery* 193:1–8.

9. Gotch, F.A., and Krueger, K.K., eds. (1975). Adequacy of dialysis. *Kidney Int.* 7(Suppl. 2), January.

10. Gutman, R.A., Stead, W.W., and Robinson, R.R. (1981). Physical activity and employment status of patients on maintenance dialysis. *N. Engl. J. Med.* 304:309–313.

11. Rettig, R.A. (1980). Implementing the end-stage renal disease program of Medicare. R-2505-HCFA/HEW. Santa Monica: The Rand Corporation.

12. Rennie, D. (1981). Renal rehabilitation—Where are the data? *N. Engl. J. Med.* 304:351–352.

13. Moore, F.D. (1972). *Transplant: The Give and Take of Tissue Transplantation*. New York: Simon and Schuster.

14. Fox, R.C., and Swazey, J.P. (1978). *The Courage to Fail: A Social View of Organ Transplants and Dialysis*, 2nd ed. rev. Chicago: University of Chicago Press.

15. Kolff, W.J., and Berk, H.T.J. (1944). The artificial kidney: Dialyzer with a great area. *Acta Med Scand.* 117:121–134.

16. Scribner, B.H., Buri, R., Caner, J.E.Z., Hegstrom, R., and Burnell, J.M. (1960). The treatment of chronic uremia by means of intermittent hemodialysis: A preliminary report. *Trans. Amer. Soc. Artif. Intern. Organs* 6:114–121.

17. Brescia, M.J., Cimio, J.E., Appel, K., and Hurwich, B.J. (1966). Chronic

hemodialysis using venipuncture and a surgically created arteriovenous fistula. *N. Engl. J. Med.* 275:1089ff.

18. Board of Directors, National Association of Patients on Hemodialysis and Transplantation (1979). Dialyzer reuse: NAPHT's statement of position. *NAPHT News* August, p. 23.

19. Willingmyre, G.T. (1979). *Reuse of Single-Use Hemodialyzers: Technical, Legal, and Economic Implications.* Washington, D.C.: Health Industries Manufacturers Association.

20. Hollander, S. (1965). *The Sources of Increased Efficiency: A Study of DuPont Rayon Plants.* Cambridge, Mass.: The MIT Press.

21. Abernathy, W.J., and Utterback, J.M. (1978). Patterns of industrial innovation. *Tech. Review* June-July, pp. 41–47.

22. *Research Needs in Nephrology and Urology,* Vol. 1. Washington, D.C.: Department of Health, Education, and Welfare, p. viii.

23. Rogers, E.M. (1962). *Diffusion of Innovations.* New York: The Free Press.

24. Rogers, E.M., and Shoemaker, F.F. (1971). *Communication of Innovations: A Cross-Cultural Approach.* New York: The Free Press.

25. Coleman, J.S., Katz, E., and Menzel, H. (1966). *Medical Innovation: A Diffusion Study.* Indianapolis: The Bobbs-Merrill Co., Inc.

26. Eveland, J.D., Rogers, E.M., and Klepper, C. (1977). *The Innovation Process in Public Organizations.* Ann Arbor: University of Michigan.

27. Rettig, R.A. (1976). The policy debate on patient care financing for victims of end-stage renal disease. *Law and Contemporary Problems* 40:196–230.

28. Petersdorf, R.G. (1980). The evolution of departments of medicine. *N. Engl. J. Med.* 303:489–496.

29. Relman, A.S. (1980). The new medical-industrial complex. *N. Engl. J. Med.* 303:963–970.

30. Russell, L.B. (1979). *Technology in Hospitals: Medical Advances and Their Diffusion.* Washington, D.C.: The Brookings Institution.

31. For a detailed discussion on this point, see Young, D.A. (1980). The communication network linking clinical research and clinical practice. Conference on the Development and Dissemination of Biomedical Innovations: Foundations for Program Development. National Institutes of Health, Mt. Poconos, Pennsylvania, March 17–19.

32. Held, P.J., and Pauly, M.V. (1980). An economic analysis of the production and cost of renal dialysis treatments. Center for Health Services and Policy Research, Northwestern University, June 19.

33. Luft, H.S., Bunker, J.P., and Enthoven, A.C. (1979). Should operations be regionalized? The empirical relation between surgical volume and mortality. *N. Engl. J. Med.* 301:1364–1369.

34. 40 *Federal Register* 27782, July 1, 1975.

35. 41 *Federal Register* 22502, June 3, 1976.

36. Calabresi, G., and Bobbitt, P. (1978). *Tragic Choices.* New York: W. W. Norton.

37. Calabresi, G. (1974). Commentary on Kenneth J. Arrow. In *Ethics of Health Care,* ed. L.R. Tancredi, pp. 53–55. Washington, D.C.: National Academy of Sciences.

END-STAGE KIDNEY DISEASE: TRENDS IN STATISTICS

by Robert J. Wineman, Ph.D.

This chapter provides a summary of trends in the patient population and other statistics for patients undergoing treatment for end-stage kidney disease (ESRD) by either dialysis or transplantation. The primary sources of data utilized are the reports of the Dialysis Registry from 1972 through 1976[1] and the ESRD–Medical Information System (ESRD-MIS) of the Health Care Financing Administration.[2-6] Additional data concerning the number of transplants are derived from reports of the National Institute of Allergy and Infectious Diseases.[7] An earlier analysis[8] has been revised and brought up to date.

Patient Population

The United States dialysis population is displayed in Figure 1.[1-4] By the end of 1979, 45,565 patients were undergoing dialysis. According to Medicare figures from the year 1978, 92.2 percent of all dialysis patients were Medicare patients. The remaining 7.8 percent were non-Medicare patients whose treatment was covered by the Veterans Administration or Public Health Service hospitals, and those whose medical costs were supported entirely by insurance or private resources. From the plot of the population changes with time, a slight lowering of the rate of increase can be seen for the entire population; however, this may be an artifact of data sources and is too small a change in the trend to be very certain. The effective date of the ESRD modification of the Medicare law is noted in Figure 1 at mid-1973. On this date the federal government assumed the responsibility for payment of 80 percent of the costs of dialysis and transplantation for the ESRD population eligible for Medicare.

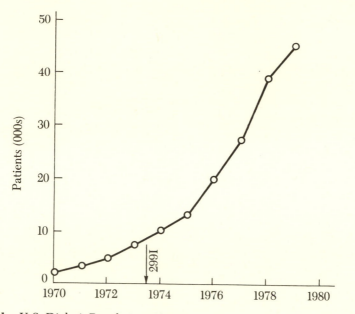

Figure 1   U.S. Dialysis Population, 1970–1980. The growth of the patient population in thousands (ordinate) is shown as a function of time in years (abscissa).

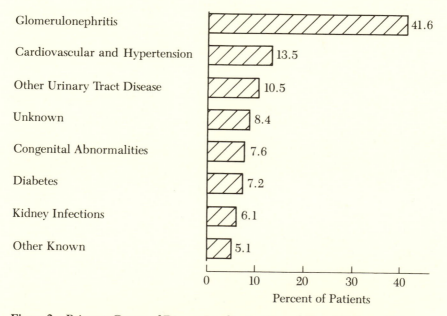

Figure 2   Primary Causes of Disease Leading to Chronic Renal Failure. The percentage of patients whose renal disease is due to various causes is given as a histogram. (*SOURCE: Derived from the National Dialysis Registry data*).

Primary Causes of Disease and Possible Prevention

A histogram of the causes of the disease leading to chronic renal failure is displayed in Figure 2. An examination of all causes of disease that might be preventable with currently available information indicates several facts. First, a fraction of the cardiovascular and hypertension cases may be prevented in the future by improved hypertension control, a measure which has gained more patient acceptance in the last few years. Second, certain congenital abnormalities, if detected and corrected at an early age, *may* permit the stricken individual to live a normal life with reasonable kidney function. Third, vigorous follow up of renal tract infections may prevent loss of kidney function in a fraction of the affected individuals. Fourth, a small fraction of patients exhibiting rapidly progressive glomerulonephritis might possibly be helped by the application of a new technique known as plasmapheresis, although the efficacy of this approach has not yet been proven. The prevention of diseases resulting in end-stage kidney disease is dependent on information derived from current and future research.

Demographics

The average age of the dialysis population[1] and the ESRD population is continuing to increase, as indicated in Figure 3. Many dialysis centers report an average age of 50 years or more. The increase in mean age among dialysis patients has been attributed in part to the gradual broadening of the patient acceptance criteria, so that more older patients are treated. A second reason for this increase has been the preference given to transplantation for the younger patients.

The age distribution of ESRD patients for 1978 is shown in Figure 4.[5] Compared to a similar analysis in 1976,[6] the 1978 data show a lower proportion of the total population in the below-20 age group, which is due to an increase in the greater-than-65 age group. The latter was approximately 15 percent of the population in 1976 and approximately 22 percent in 1978. The proportion of patients in the 20 to 44 and the 45 to 64 age groups has remained approximately the same.

Figure 5 shows the percentage of female patients. From 1970 to 1978 the percentage of female patients increased from approximately 32 percent in 1972 to approximately 45 percent in 1978.[1,5,6] No analyses have been done to explain this trend. Whether it represents a sex difference in the incidence of the disease or a sex bias in the treatment of patients has not been determined.

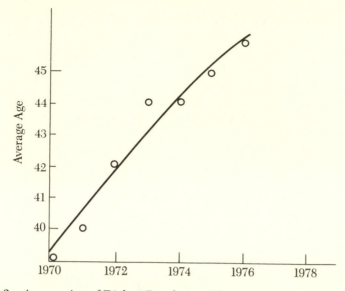

Figure 3 Average Age of Dialysis Population. The average age of the population (ordinate) is shown as a function of time in years (abscissa).

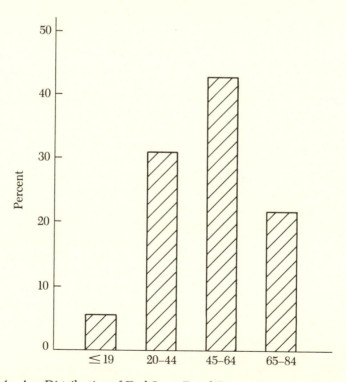

Figure 4 Age Distribution of End-Stage Renal Disease Patients, 1978. This histogram shows the distribution of patients in various age groups (abscissa) as a percentage of the total population (ordinate).

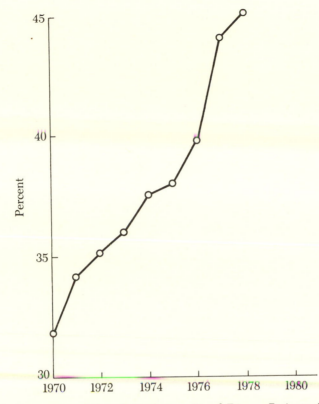

Figure 5 Percentage of Female End-Stage Renal Disease Patients, 1970–1978. The growth of the number of female patients (ordinate) as a percentage of the total population is shown as a function of time (abscissa).

The race distribution of the ESRD population is given in Figure 6. The 1978 data[5] show no appreciable change since 1976.[6] A higher representation of blacks can be seen among ESRD patients compared to their proportion in the national population. This is similarly true of races classified as "other." Approximately 25 percent of the overall ESRD population is black and 3 percent is represented by other races. The prevalence of ESRD by race is given in Figure 7.[6] Overall, there are approximately 14 patients per 100,000. Among whites, the prevalence is 10 per 100,000, while in the black and other races, the prevalence is a factor three times that of the white population, or approximately 30 per 100,000. One of the reasons for this difference is that the black population has a higher percentage of hypertension than whites.

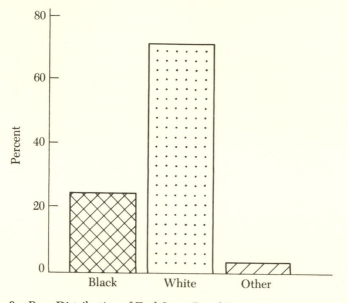

Figure 6   Race Distribution of End-Stage Renal Disease Patients, 1978. This histogram shows the distribution of patients in three race groups (abscissa) as a percentage of the total population (ordinate).

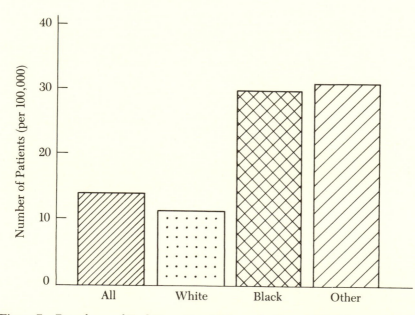

Figure 7   Prevalence of End-Stage Renal Disease by Race, 1978. The prevalence of end-stage renal disease by race (abscissa) is shown in this histogram in terms of patients per 100,000 of the general population (ordinate).

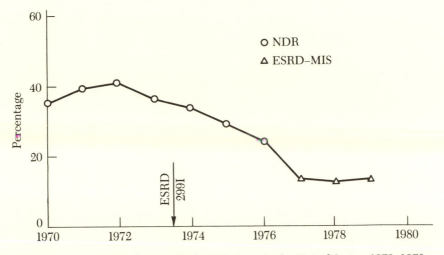

Figure 8 Percentage of Home Dialysis Patients in the United States, 1970–1979. The percentage of home or self-care patients (ordinate) is shown as a function of time (abscissa). Points with circles are from National Dialysis Registry Data, while triangles are from the ESRD Medical Information System.

Utilization of Self-Care or Home Dialysis

Home or self-care dialysis generally costs less than in-center dialysis, but a controversy exists concerning the amount of savings attributable to home patient care. In 1978 the General Accounting Office reported that the cost of in-center dialysis per patient year was approximately $16,520, compared to about $6,500 for home dialysis after the first year.[11] The first-year costs for home care are higher since that usually includes purchase of equipment. Figure 8 shows the percentage of home dialysis patients in the U.S. dialysis population over time.[1,5] In 1972, prior to the passage of the Medicare ESRD amendment, the percentage of home patients peaked at approximately 40 percent. By 1978 this figure had fallen to about 12 percent of the dialysis population, and in the last year for which data was available, 1979, the percentage increased slightly up to 13 percent. In 1979, however, 30.8 percent of new patients selected home dialysis compared to 24 percent in 1978.[3,4] In 1978 and 1979 the average outpatient dialysis cost per treatment was $149.[3,4]

Transplantation

In 1972 transplantation as a therapeutic modality for the end-stage kidney patient reached a level of approximately 3,000 per year in the

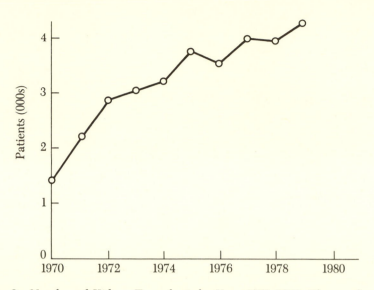

Figure 9 Number of Kidney Transplants by Year, 1970–1979. The number of patients in thousands (ordinate) undergoing kidney transplantation as a function of time (abscissa) is shown.

Figure 10 Annual Number of Transplants as a Percentage of the U.S. Dialysis Population, 1970–1979. The number of transplants per year as a percentage of the dialysis population of that year (ordinate) is shown as a function of time (abscissa).

United States as indicated in Figure 9.[7] Seven years passed before transplantation reached a level of approximately 4,000 per year. In 1979 4,271 renal transplants were performed in the United States, an increase of 8.2 percent from the preceding year. The number of living related transplants was 1,203, or 28.2 percent of this total.[4]

In Figure 10, the number of transplants as a percentage of the dialysis population of each year is displayed. This proportion reached a peak in 1971 and has been in decline since that time. In the last three years, approximately 10 percent of dialysis patients received transplants each year.

In 1979 the average cost for a kidney transplantation was $19,300.[4] Overall transplant rate and kidney graft survival sharply influence the national cost of end-stage renal disease therapy. An increase in the rates of transplantation, or gains in graft survival, are approaches to reduce the national costs for this patient group. As commented earlier, advances in research which might lead to better preventive measures would be preferable.

Rehabilitation

No broad-based studies are available in regard to patient rehabilitation within the United States ESRD population. A measure of rehabilitation frequently utilized in the absence of such data is the hospitalization rate of patients. A five-center study of dialysis patients was conducted in 1976, utilizing as a data base 1,100 patients or 5 percent of the dialysis population at the time.[9] In Figure 11, a

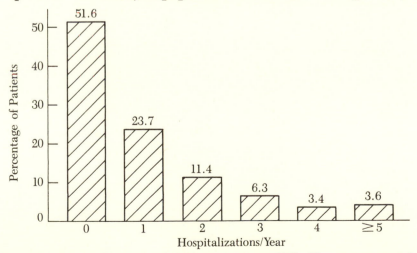

Figure 11 Hospitalizations of Dialysis Patients, 1976. The percentage of patients (ordinate) experiencing frequencies of hospitalizations (abscissa) in the study year is displayed as a histogram.

histogram indicates the percentage of patients who required varying numbers of hospitalizations during that year. From this data it is evident that one half of the patients were never hospitalized and, of the other half, about 50 percent of these were hospitalized once, and the remainder at increasing frequencies. In 1979 the average annual per capita cost of hospitalization for inpatient services was $2,795 for all ESRD patients.

Survival

The U.S. ESRD Medical Information System data is not yet available for analyzing patient survival. Therefore comparative survival data from Europe were utilized in Figure 12, which shows survival data of the European Dialysis and Transplant Association (EDTA).[10] Data are displayed for an eight-year time period. A sharp difference in survival rates can be seen, depending on patient age and other medical risk factors. The EDTA data presented do not correct for any of these factors, and are overall average figures.

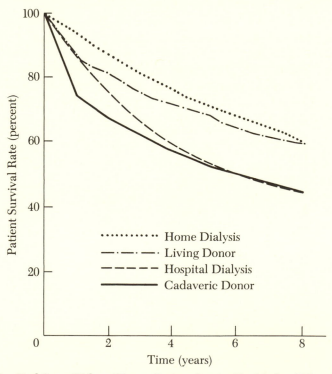

Figure 12   End-Stage Kidney Disease Patient Survival by Mode of Therapy. Survivals of ESRD patients (ordinate) according to therapy are shown as a function of time (abscissa).

Future Trends

Future changes in the U.S. dialysis population will depend on a variety of factors. Policy-related issues such as patient acceptance criteria or availability of funding might influence the shape of the curve (see Figure 1). Continued broad acceptance criteria will lead to a substantial, additional growth. Conversely, if acceptance of patients were restricted based on multiple disease states or other risk factors, a reduced rate of growth would occur, similar to the situation that occurred in the United Kingdom.[10]

Other factors relate to technology. As commented previously, advances in research may lead to better methods for either arresting or preventing kidney disease. Improved therapeutic techniques, in turn, could lead to a larger patient pool because of longer average survival of patients. Similarly, transplantation as a mode of therapy might influence the size of the future dialysis population, depending on such variables as the availability of cadaveric kidneys and graft survival.

In the 1980 report to the Congress, the Health Care Financing Administration[5] estimated that the average annual enrollment in the ESRD program would increase from 68,200 in 1981 to 83,700 in 1985. During that time period the total benefit payments could be expected to rise from approximately $1.5 billion in 1981 to $2.6 billion in 1985.

Endnotes

1. Bryan, F.A. (1976). Final Report, National Dialysis Registry. Artificial Kidney-Chronic Uremic Program, National Institutes of Health, Report No. AK-8-1387-F, PB269174/AS.
2. Health Care Financing Administration, Medicare Bureau, Office of Financial and Actuarial Analysis, May 8, 1978.
3. End Stage Renal Disease Program, Annual Report to Congress 1979, P.L. 95-292, 1801 (g), DHEW, Health Care Financing Administration.
4. Office of Special Programs, Health Care Financing Administration, Second Annual Report to Congress, FY 1980, Department of Health and Human Services.
5. Renal Disease Program (1979). Report of Medical Information System, April. Health Care Financing Administration.
6. Profile of the ESRD Patient Population (1976). Report of the End-Stage Renal Disease Medical Information System, Contract HSA-240-75-0123, DHEW, Health Care Financing Administration.
7. National Institute of Allergy and Infectious Disease (1977). Report: "Research Resources for Microbiologic and Immunologic Studies." National Institutes of Health, DHEW.

8. Wineman, R.J. (1978). End-stage renal disease: 1978. *Dialysis and Transplantation* 7:1034.
9. Unpublished data. Artificial Kidney-Chronic Uremic Program, Dialysis Complication Study. A further analysis is Hirschman, G.H., Wolfson, M., Mosimann, J.E., Clark, C.B., Dante, M.L., and Wineman, R.J. (1981). Complications of dialysis, *Clin. Nephrology* 15:66–74.
10. Brunner, F.P., Brynger, H., Chantler, C., Donckewolcke, R.A., Hatheway, R.A., Jacobs, C., Selwood, N.H., and Wing, A.J. (1979). Combined report on regular dialysis and transplantation in Europe, IX. In *Proceedings of the European Dialysis and Transplant Association*, ed. B.H.B. Robinson, pp. 1–85. London: Pitman.
11. Report of the General Accounting Office. Treatment of chronic kidney failure, dialysis, transplant costs and the need for more vigorous efforts. HRD 78-17, p. 3.

RECENT ADVANCES IN CLINICAL HEMODIALYSIS TECHNOLOGY

by Edmund G. Lowrie, M.D.

In recent years technological advances have increased the clinical effectiveness of hemodialysis to such an extent that lives now depend upon this therapy.[1] Socially and politically, hemodialysis has been the subject of much study and debate, and may now represent the first test of universal, catastrophic health insurance in the United States.

The significant and progressive changes that have occurred during the past few years in the development of the artificial kidney have resulted in a better understanding of the mechanisms by which dialysis affects the uremic syndrome, and are the combined efforts of individuals from various professional disciplines. Figure 1 illustrates this cooperative process and shows the interaction between different fields of technical study. First, nontoxic materials which are compatible with human blood and which perform certain designed functions must be developed (see Figure 1). Next, materials must be assembled into artificial kidney systems; here engineering is critical for the design and testing. As artificial kidneys function only to remove water and solutes from the blood, techniques that predict and describe these effects quantitatively are germane to the

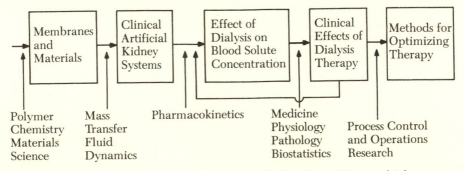

Figure 1 Schematic Representation of the Multidisciplinary Nature of Advances in Dialysis Treatment.

301

prescription of therapy. Clinical pharmacologists have applied principles of pharmacokinetics to intermittent dialysis. The intensity of exposure or "dose" of dialysis might therefore be viewed in terms of the relative rates of solute generation and removal which, in turn, affect the solute concentration or burden to which the body is exposed.[2-6] Finally, dialysis hopefully improves the patient's condition and well-being, as shown by the transition from the third to the fourth panel of Figure 1.

I would like to focus my discussion on two recent advances in the Figure 1 continuum which have enhanced our understanding of dialysis therapy. First, until recently, no specific method has been accepted for prescribing dialysis on an individualized basis. Pharmacokinetics provides a potential tool for achieving this goal, and the principles, as applied clinically in the dialysis setting, will be discussed. Second, evidence is accumulating which suggests that artificial kidneys are not simply inert products in an extracorporeal circuit; instead, they appear to interact with the formed and humoral elements of blood—perhaps to the detriment of the patient. Certain biomaterials seem less effective than others, and new artificial kidneys seem to have more deleterious effects than those which are reused.

Pharmacokinetic and Metabolic Modeling for the Individualized Prescription of Dialysis

Figure 2 simulates the blood urea nitrogen (BUN) concentration profile for an average patient receiving routine, three times weekly

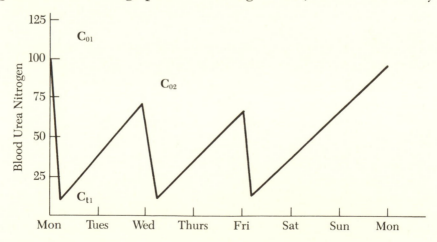

Figure 2   Simulation of a One-Week Dialysis Cycle for an Average Patient. C_{01} and C_{02} are predialysis concentrations. C_{t1} is a postdialysis concentration.

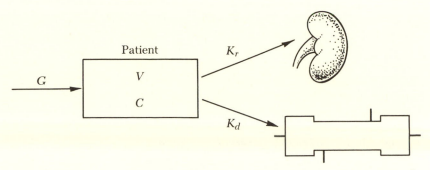

Figure 3 Schematic Diagram Showing the Flow of Material (Urea) in the Patient Dialyzer System. The balance equation describing the concentration *(C)* profile in the body of volume *(V)* is simply $V \, dC/dt = G \, C \, (K_r + K_d)$. (See text for explanation.)

hemodialysis. The BUN level falls during dialysis and increases during the next succeeding interdialysis interval, resulting in a saw-toothlike concentration across the time profile.

Figure 3 illustrates the theoretic model used to describe dialysis therapy. The human body is assumed to be a volume of water in which urea is dissolved, which, in fact, is quite accurate. Urea is manufactured within the body as a major product of protein metabolism and accumulates within the body unless excreted. Urea may be removed by two potential routes—the biologic kidney, which functions continuously, and the artificial kidney, which functions only during dialysis treatments. During the interval between dialyses, the rate of urea generation exceeds its rate of removal by the diseased kidney, resulting in an increase in concentration in the body and the blood. During dialysis the rate of urea removal (due to the combined effects of the diseased, biologic kidney and artificial kidney) exceeds the rate of generation, causing the BUN concentration to fall.

Sargent and Gotch[3-6] have described the mathematics of urea kinetics. Conceptually, however, urea nitrogen concentration at any time during dialysis reduces to a function of six variables:

$$C = f \, (K_r, \quad V, \quad G \, : \, TD, \quad \theta, \quad K_d)$$
$$\text{Patient-dependent} \qquad \text{Prescription-dependent}$$

C represents urea nitrogen concentration. K_r is the residual urea clearance of the diseased kidneys; *V* is the volume of urea distribution which approximates total body water; and *G* is urea nitrogen generation rate, which is determined by the intake of dietary protein. These variables are termed "patient-dependent." *TD* is the length of each treatment, θ is the interdialysis interval, and K_d is the urea clearance of the dialyzer. K_d is, in turn, a function of the blood flow rate to the dialyzing membrane, the dialysate flow rate from the dialyzing membrane, dialyzer surface area, and the dialyzer's overall permeability to urea.[7,8] TD, θ, and K_d are prescribed by the medical team guiding the therapy.

Residual kidney function can be measured, and K_d is determined by the dialyzer and the conditions under which it is operated. TD and θ are prescribed and are well controlled. V and G are more difficult to determine, however. The kinetic equations can be rearranged to estimate volume and generation rate so that these parameters can be calculated from a knowledge of other variables and the blood urea nitrogen concentration at various times in the dialysis cycle:[2-4]

$$V = f\ (K_r, G, TD, K_d : C_{01}, C_{t1})$$

and

$$G = f\ (K_r, V, \theta : C_{t1}, C_{02})$$

Here, C_{01}, C_{t1}, and C_{02} are the blood urea nitrogen concentrations measured prior to and after one dialysis and prior to the next succeeding treatment. Once values for V and G are determined, they can be used to prescribe future therapies. For example, the equations are simply rearranged so that

$$K_d = f\ (K_r, V, G : TD, \theta, \text{desired } C)$$

and clearance may be prescribed by choosing an appropriate dialyzer type and manipulating the dialysate and blood flow rates. The aim of prescribing therapy, then, is to choose a TD, θ, and K_d combination to achieve a desired C subject to the patient's K_r, V, and G. An additional restriction is that patients must consume a nutritionally complete diet.

Figure 4 shows the clinical course of an individual whose therapy was altered using these techniques to increase blood urea nitrogen from 70 mg/kl predialysis to approximately 120 mg/dl. The patient's protein catabolic rate (PCR, which determines G) in grams per day is also shown. Note that PCR remained constant and the patient's blood urea nitrogen level rose promptly into the target range. Figure 5, on the other hand, shows an individual whose therapy was shortened from five to three hours, thereby providing less treatment exposure. K_d, however, was increased in an effort to maintain BUN unchanged. Note again that PCR remained constant, as did predialysis BUN, even though the time spent for dialysis was shortened. Figure 6 shows a patient whose therapy was altered in a fashion similar to that of the patient shown in Figure 4. Treatment was altered so that predialysis BUN should increase to about 120 mg/dl to test the kinetic system in preparation for a larger scale clinical trial.[2] Note, however, that while BUN increased promptly, it did not

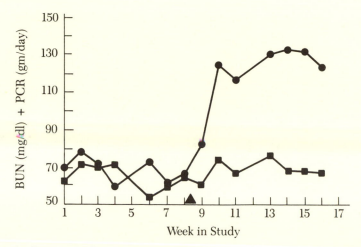

Figure 4 Patient Dialysis History Showing Mid-Week Predialysis BUN (●) and Protein Catabolic Rate (PCR) (■). Dialysis was changed at ▲ by reducing K_d to achieve a predialysis target BUN of about 120 mg/dl while maintaining *TD* constant. (See text.)

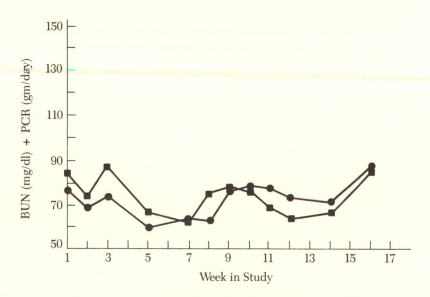

Figure 5 Patient Dialysis History Showing Mid-Week Predialysis BUN (●) and Protein Catabolic Rate (PCR) (■). Dialysis was changed at week 8 by reducing *TD* from five to three hours while making a compensatory increase in K_d to maintain BUN constant. (See text.)

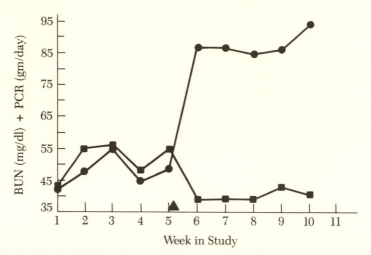

Figure 6 Patient Dialysis History Showing Mid-Week Predialysis BUN (●) and
Protein Catabolic Rate (PCR) (■). Dialysis was changed at ▲ by reducing K_d to
achieve a predialysis target BUN of about 120 mg/dl, while maintaining TD constant. Note the reduction in PCR and the short fall in BUN.

reach the target range. A prompt fall in PCR was associated with the
early rise of BUN. This suggests that patients may attempt to defend
against *relative* underdialysis by reducing spontaneously their intake
of dietary protein.

Figure 7 illustrates the relationship between PCR and the dialysis
prescription. Patients' therapies were changed to produce a calculated alteration in their predialysis BUN. The ordinate of the figure
shows the deviation from the estimated target value for predialysis
BUN which the dialysis prescription was designed to achieve. The
abscissa shows the change in PCR that occurred after therapy was
altered. A close relationship is observed.

The biological toxicity of a substance such as urea should be questioned if it is to be used to control therapy. By itself, urea is thought
to be relatively non-toxic—at least at the blood concentrations usually
encountered in dialysis patients. In practice, however, a substance
need not be toxic in and of itself to be used successfully as a clinical
marker to guide therapy. Instead, it should be easy to measure,
should be understood in terms of its metabolism and toxicity, and
should be clinically relevant. Not only does urea meet these criteria,
but it is the major byproduct of protein metabolism and, as such, its
relative rate of generation, elimination, and accumulation might be
reasonably expected to co-vary with other products of protein metabolism. Such products have long been suspected of being important
in causing the uremic syndrome, and protein restriction is the therapeutic first defense for treating uremia.

We conclude from these and other experiences[2] that urea nitrogen

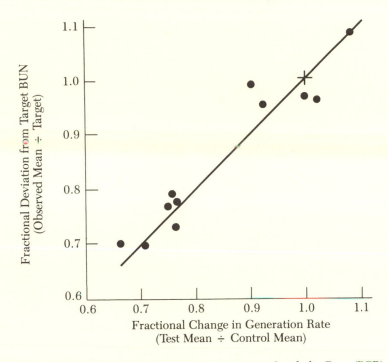

Figure 7   Relationship between Changes in Protein Catabolic Rate (PCR) and the Deviation of Measured Predialysis BUN from Target. Patients were dialyzed during a control period in which PCR was measured. Therapy was then altered to achieve a target BUN from 70 to 120 mg/dl, given specified changes in *TD*. The data suggest a close relationship between PCR change and deviation from target.

concentration can be controlled adequately using pharmacokinetic tools. The techniques permit physicians to predictably manipulate blood solute concentration in terms of important patient attributes by changing elements of the treatment protocol. Proceeding in this way, physicians can prescribe an appropriate dialysis or time-dialyzer clearance combination to achieve a desired or target concentration of blood urea nitrogen. Thus the necessary tools are provided to prescribe dialysis on an individualized basis. Studies which will evaluate the clinical relevance of pharmacokinetically guided therapy in relation to medical outcome are in progress.[2]

Biocompatibility of Materials and Reprocessed Artificial Kidneys

Relatively few studies have been published on the *in vivo* biological effect of the materials used in the manufacture of artificial kidneys. Their effect on the formed and humoral elements of blood is largely unknown. Knowledge is beginning to expand in this area, and further efforts are warranted and required.

Figure 8 shows the effect of dialysis on white blood cell count, which falls to very low levels early in dialysis. This phenomenon has been recognized for a number of years, and white blood cells are thought to be sequestered in the lung.[9] Curves are shown for cellophane and cuprophan dialyzers, both of which are cellulosic materials. The third frame shows the neutrophil curves for a dialyzer manufactured from a noncellulosic material—polymethylmethacrylate (PMMA). Curves for both new (first use) and reprocessed (third use) artificial kidneys are shown. White blood cell count falls much more with new cellulosic dialyzers than with either used cellulosic or noncellulosic artificial kidneys.[10]

The clinical consequences of this periodic plummeting in white blood cell count are not clear. Dialysis-induced leukopenia, however, also is associated with significant functional and structural changes

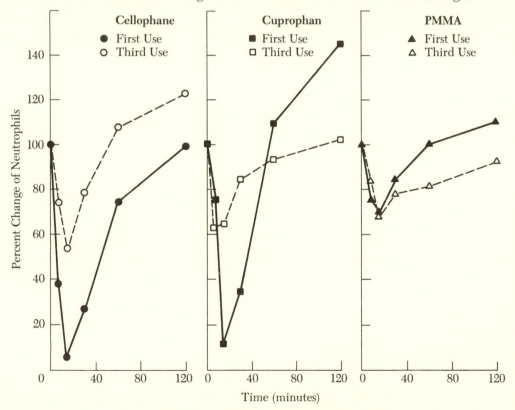

Figure 8   White Blood Cell Count Shown as a Function of Time for Dialyzers Using Regenerated Cellulose, Cuprophan, and Polymethylmethacrylate (PMMA) Membranes during First and Third Use.

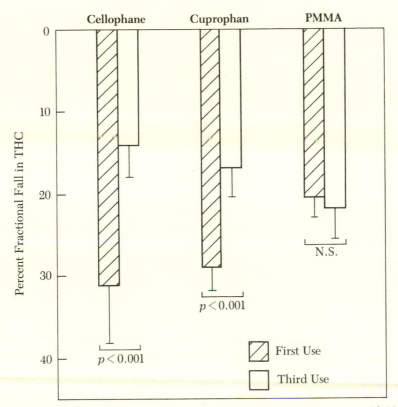

Figure 9 Change in Serum Complement Observed During Treatment with New and Reused Dialyzers Made from Regenerated Cellulose, Cuprophan, and Polymethylmethacrylate (PMMA) Membranes.

in neutrophils.[11-14] Both migration and chemotaxis, important processes in fighting infection, are severely impaired. Patients undergoing dialysis are prone to bacterial infection, which is the principal cause of hospitalization in chronic dialysis patients.[15] This apparent injury to white cells may contribute to infection and the morbidity experienced by patients who must receive artificial kidney treatments. The effect may be blunted dramatically either by using reprocessed artificial kidneys and/or by developing membranes which are less injurious to white blood cells.

Figure 9 is taken from the same series of experiments shown in Figure 8,[10] and illustrates the fractional fall in the total hemolytic complement in serum which occurs with new and reprocessed cellulosic and noncellulosic artificial kidneys. Again, serum complement falls significantly during the first use of cellulosic dialyzers. The decrement is much less pronounced when either reused cellulosic or noncellulosic kidneys are employed. The clinical sequelae of this

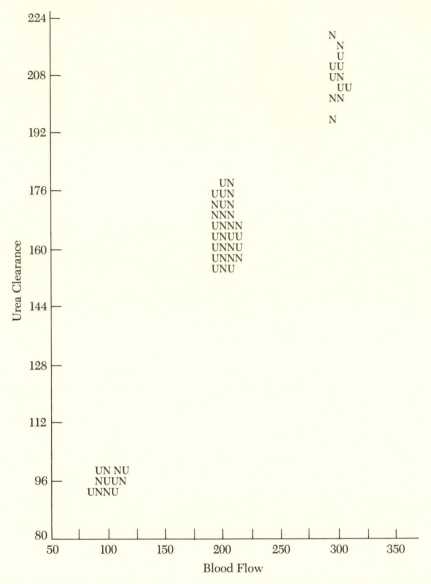

Figure 10   Urea Clearance Shown as a Function of Blood Flow Rate for New (N) and Reused (U) Artificial Kidneys, Using a Cuprophan Membrane (HPF-200).

periodic activation of the complement pathway is not clear, however. Data suggest that the alternate system for complement activation is initiated by dialysis membranes.[9] Phylogenetically, the system is old and can be activated by several polysaccharides such as bacterial cell walls, inulin, zymosan, and endotoxin. Therefore, it is not surprising that the polysaccharide structure of cellulosic membranes may initiate complement activity. The activation of complement may be only one aspect of blood materials interactions, and other enzymes systems in the blood may be activated by exposure to dialysis membranes. For example, the activation of Hageman factor—the first substance in a cascade of reactions which characterize the blood clotting pathway—also occurs as a result of exposure to foreign surfaces such as dialysis membranes. A host of subsidiary reactions may play important roles in dialysis-related morbidity.[16]

Observations concerning the apparent enhanced biocompatibility of reused artificial kidneys are especially relevant. The data suggests that using reprocessed artificial kidneys is not only safe but may provide medical benefits for patients because they appear to cause less damage to the formed and humoral elements of the blood. Reusing a device several times on the same patient could provide a contributing technique for controlling the cost of dialysis therapy. To be reused, however, a unit should not only be safe, but its function should be reasonably preserved. Figure 10 plots urea clearance by blood flow rate, and clearance increases with flow in classical fashion. Dialyzers were tested when new (shown as N) and during a third use (shown as U), and the two curves can be superimposed.[16] If the reprocessing procedure is strictly controlled, the multiple use of artificial kidneys may provide measurable benefits for patients while permitting significant cost reductions.

Summary

Two important fields of technical endeavor relevant to clinical dialysis have recently emerged. The first involves developments designed to permit an understanding of methods by which dialysis may be prescribed quantitatively on an individualized basis. The second involves our current embryonic understanding of the interaction of humans with biomaterials.

Endnotes

1. Kolff, W.J. (1965). First clinical experiences with the artificial kidney. *Ann. Intern. Med.* 62:608.

2. Lowrie, E.G. and Sargent, J.A. (1980). Clinical example of pharmaco-kinetic and metabolic modeling: Quantitative and individualized prescription of dialysis therapy. *Kidney Int.* 18 (Suppl. 10):S11–S16.

3. Gotch, F.A., Sargent, J.A., Keen, M.L., Seid, M., and Foster, R. (1973). Comparative treatment time with Kiil, Gambro, and Cordis-Dow kidneys. *Proc. Dial. Transpl. Forum* 3:217–228.

4. Gotch, F.A., Sargent, J.A., Keen, M.L., and Lee, M. (1974). Individualized, quantified dialysis therapy of uremia. *Proc. Dial. Transpl. Forum* 4:27–35.

5. Sargent, J.A. and Gotch, F.A. (1974). The study of uremia by manipulation of blood concentrations using combinations of hollow fiber devices. *Trans. Amer. Soc. Artif. Intern. Organs* 20:395–401.

6. Sargent, J.A. and Gotch, F.A. (1975). The analysis of concentration dependence of uremia lesions in clinical studies. *Kidney Int.* 7:S35–S44.

7. Henderson, L.W. (1976). Hemodialysis: Rationale and physical principles. In *The Kidney*, eds. B.M. Brenner and F.C. Rector, pp. 1643–1671. Philadelphia: W.B. Saunders.

8. Hampers, C.L., Schupak, E., Lowrie, E.G., and Lazarus, J.M. (1973). *Long Term Hemodialysis*, 2nd ed. New York: Grune & Stratton.

9. Craddock, P.R., Fehr, J., Dalmasso, A.P., Brigham, K.L., and Jacob, H.S. (1977). Hemodialysis leukopenia: Pulmonary vascular leukostasis resulting from complement activation by dialyzer cellophane membranes. *J. Clin. Invest.* 59:879.

10. Hakim, R.M., and Lowrie, E.G. (1980). Effect of dialyzer reuse on leukopenia, hypoxemia and total hemolytic complement system. *Trans. Amer. Soc. Artif. Intern. Organs* 26:159–163.

11. Bjorksten, B., Mauer, S.M., Mills, E.L., and Quie, P.E. (1978). The effect of neutrophil chemotactic responsiveness. *Acta. Med. Scand.* 203:67.

12. Geotzl, E. Private communication.

13. Gellin, J., and Wolf, S. (1975). Leukocyte chemotaxis: Physiological considerations and abnormalities. *Clin. Hemat.* 4:567.

14. Miller, M.E. (1975). Pathology of chemotaxis and random migration. *Hematology* 12:59.

15. Nsouli, K., Lazarus, J.M., Schoenbaum, S., Gottlieb, M., Lowrie, E.G., and Shocair, M. (1979). Bacteremic infections in hemodialysis. *Arch. Intern. Med.* 139:1255.

16. Lowrie, E.G., and Hakim, R.N. The effect on patient health of using reprocessed artificial kidneys. *Clin. Dial. Transpl. Forum*, in press.

WORKSHOP ON PRENATAL DIAGNOSIS: AN OVERVIEW

by Stephen G. Pauker

Technologies capable of identifying abnormalities in the unborn fetus are now available both in the research laboratory and in day-to-day medical practice. Many of these diagnostic tools are "technologies in evolution"—each month tests move from the laboratory into routine availability and new diagnoses become possible. Parents who embark on an accepted course of prenatal diagnosis sometimes find themselves in uncharted waters when an unexpected observation raises the specter of a disease that has been prenatally identified only on a case report basis.

Furthermore, the very concept of antenatal diagnosis abounds with moral and ethical dilemmas. Good outcomes for the prospective parents, for the fetus, and for society are often at variance with one another. The ultimate goal of these technologies is to identify pregnancies affected by certain abnormalities so those pregnancies can be terminated—in other words, selective abortion in the hope of replacing abnormal fetuses with more normal ones. Of course, selective abortion is among the crudest forms of genetic manipulation. As medicine becomes more sophisticated, *in utero* treatment of the fetus and, eventually, "treatment" of the gamete itself will surely follow. Since the burden of raising genetically defective children falls largely on the parents, many couples might desire prenatal diagnosis even though such testing might not be "cost-effective" and might thereby increase, rather than lessen, the "burden" of genetic disease on society. Even if such testing were desired and "economically justified," society might not have sufficient resources to support large-scale diagnostic programs; also the allocation of resources to such programs might preclude other equally "cost-effective" benefits to society. Moreover, certain members of society might find selective abortion to be morally wrong in any circumstances and might vigorously object to allocation of any of the "public commons" for this purpose.

It is in this milieu of evolving technology, moral conflict, and limited resources that we conduct a workshop to discuss prenatal diagnosis. Four papers are presented here—those presented in the original workshop and a fourth which uses the first three as a "take-off" point for considering the moral and ethical issues that surround prenatal diagnostic technologies.

From the perspective of a clinical geneticist who serves as a primary pediatrician to many of the couples whom she counsels, Susan Pauker provides an overview of prenatal diagnosis and the social pressures that it places on prospective parents. Her arguments are illustrated by case studies of several couples. The basic point of her paper is that the ultimate decision about prenatal diagnosis—whether or not to utilize the technology—lies with the individual couple. Although there are many pressures, the final choice is, and must be, very personal, as that couple must live with the consequences of their choice. What is not obvious is the unfortunate fact that most genetic counseling today provides information only: Prospective parents are rarely given a logical framework for making a well-reasoned choice.

J. Michael Swint provides an economic perspective on prenatal diagnosis, contrasting the cost-effectiveness and cost-benefit approaches. Swint correctly points out that, if the issue of finite resource availability is to be addressed, some value must be placed on various morbidities and, in fact, on life itself. If life cannot be valued, then all of society's resources might (or maybe even should) be applied to saving a single life. Swint also addresses the impact of "out-of-pocket" costs (that is, those not reimbursed by third-party payers) and resource limitations on individual demand for prenatal diagnostic services. He projects usage levels for the next decade on the basis of present experience. One point in his paper deserves some amplification—the distinction between the "replacement" and the "nonreplacement" case, with the former providing greater economic benefits. Recall that the logic of selective abortion is to allow prospective parents to replace a defective pregnancy with a normal one. Of course, not all couples will "try again" after terminating an affected pregnancy. Since normal individuals contribute a net gain (that is, productivity) to society, the replacement of a defective fetus by a normal one should provide a net gain (rather than less loss) to society and should therefore be counted as either additional benefit or diminished cost to a program of prenatal diagnosis.

Pauker, Pauker, and McNeil address individual demand for prenatal diagnosis services from a somewhat different perspective. In prior publications, they have demonstrated the feasibility of quantitating patients' attitudes toward various health outcomes and how such attitudes can be used to choose optimal therapeutic and diagnostic strategies. In this paper, they show how the distribution of such attitudes, if they were available, might be used to project the acceptability of and individual demand for antenatal diagnosis. Their theoretical paper must be viewed as a model and not as an actual projection for several reasons. The attitudinal data (that is, the distribution of utilities) must be gathered from a representative population, and such utilities must be assessed in a manner that insures that they truly reflect the ideas of the population. Such projections assume that, on average, people will act in a logical, unbiased way. The model they offer assumes that the only reason for prenatal diagnosis is in preparation for selective abortion—it ignores motivations such as "reassurance". Furthermore, the assumed sensitivity, specificity, and risk of prenatal diagnosis are based on relatively small reported series and may not truly be representative. Nevertheless, the decision analytic model of technology assessment, particularly with the use of subjective utility distributions, may become increasingly important and deserves careful consideration and refinement.

Finally, John C. Fletcher address the ethical and moral dilemmas posed by prenatal diagnosis. He considers why selective abortion might be considered different from neonatal euthanasia but points out that the line is becoming increasingly blurred. He touches on the related issues of supply, demand, limited resources, and access to prenatal diagnosis. He concludes by reviewing the historical evolution of our society's attitudes toward sexuality, reproduction, and individual choice.

In summary, this section of the book should provide the reader with a broad, nontechnical overview of prenatal diagnosis as it presently is practiced and as it may well evolve in the future.

PRENATAL GENETIC DIAGNOSIS BY AMNIOCENTESIS: THE HUMAN PROBLEM

by Susan Perlmutter Pauker

Although prenatal diagnosis can provide parents with important information concerning the outcome of pregnancies, the ultimate decision about undergoing these procedures, based on the information supplied, must lie with the parents. This paper will illustrate the various dilemmas couples must face in making these choices and will present actual case studies in which prenatal diagnosis has been offered.

The prenatal detection of serious hereditary disorders can be accomplished in the mid-trimester of pregnancy by a variety of technologies aimed at identifying specific abnormalities in the fetus. This procedure is available when selective interruption of the pregnancy is possible. Ultrasound studies can date the gestation, detect twins, and identify gross abnormalities in fetal anatomy such as anencephaly. Samples of maternal blood can reveal levels of normal and abnormal fetal proteins. More directly, these levels can be measured by a sample of amniotic fluid, from which the chromosomes of the fetus can be analyzed. Fetoscopy can provide, at an increased risk, an actual visualization of the fetus and an analysis of blood samples.

The potential value of these tests varies widely, depending on the feelings of the parents. The costs and psychological burdens of these procedures are substantial; long-term and short-term risks of many of these techniques, such as ultrasound examination, are not yet completely known. Examining the issues involved in selecting prenatal diagnostic procedures thus seems worthwhile.

Amniocentesis is the removal of a small amount of the amniotic fluid surrounding the growing fetus by transabdominal needle aspiration. The fluid can then be examined biochemically for proteins and enzymes which can reflect the health of the fetus, or the fetal fibroblast cells, normally "shed" into that fluid, can be cultured and the chromosomal composition (the karyotype) of those cells can be determined. With this technique Down's syndrome can be detected through cytogenetic studies, and Tay-Sachs disease can be identified in the 16th to 17th week of gestation. In sex-linked disorders, such as Duchenne's muscular dystrophy, the carrier status of the prospective mother can be estimated by a Bayesian analysis of the familial history (pedigree) and measurement of maternal serum enzyme levels. Determination of the sex of the fetus by amniocentesis can then allow prospective parents to terminate the pregnancy if the probability of disease is high. Multifactorially inherited disorders of the neural tube, such as anencephaly or meningomyelocele, can be detected by maternal serum alpha fetoprotein (AFP) levels, sometimes visualized by ultrasound, and confirmed by elevated amniotic fluid levels of AFP or by direct fetoscopic visualization.

Nevertheless, most amniocenteses provide the parents with reassurance. Mid-trimester amniocentesis is performed most often to detect Down's syndrome in pregnant women over age 35 years who, at the time of delivery, have at least a 1/365 chance of having an affected child. Most mothers seeking amniocentesis at that age will be reassured that their fetus has a normal chromosomal complement. Even at age 45 years, there is a 97 percent chance that the fetus will be unaffected by Down's syndrome. However, amniocentesis carries an 0.5 percent risk of abortion, a smaller risk of damaging an otherwise normal viable fetus, and a false-positive and false-negative reporting rate perhaps as high as 0.5 percent. The test is performed at 16 or 17 weeks gestation, and a delay of up to four weeks occurs before the results of cytogenetic analysis are available. Approximately one to ten percent of first amniocentesis cell cultures do not grow well in the laboratory; in such cases, a second procedure is requested at 18 to 22 weeks gestation, repeating the risk of abortion.

When deciding about most currently available prenatal diagnostic techniques, parents must consider elective abortion of a pregnancy that appears to be affected by a given disorder. A mid-trimester abortion is an in-hospital procedure with roughly twice the mortality rate to the mother of a term delivery (20/100,000 as opposed to 10/100,000). The injection of a salt or prostaglandin solution into the uterus is followed by induction of labor with Pitocin®. At the time that the abortion would need to be performed, the mother has often been aware of fetal movement for several weeks. Some parents who have experienced an elective mid-trimester abortion relate a great deal of depression during and after the labor and delivery. Prophylactic psychological support services must be extended to these couples to assist them in coping with their experience.

The Procedure

Amniocentesis is often preceded by an ultrasound examination to locate the fetus, placenta, and umbilical cord in the womb to minimize the chance of damaging these tissues during the procedure. Ultrasound can usually detect twins (which occur in approximately one in 80 pregnancies), can identify some major congenital anomalies, and can be used to estimate gestational age by measurement of the fetal size. Expectant mothers can see the visual image of their fetuses formed electronically from the echoes of high-frequency soundwaves. Parents have reported an intense sense of bonding to their unborn child; this sense is increased by being able to see a visual image of the child. A loss of objectivity often accompanies this feeling, confounding the parents' ability to think clearly about a possible elective abortion pending the results of the amniocentesis.

The amniocentesis takes about fifteen minutes to perform either in an outpatient area of a hospital or in the physician's office. Using sterile technique, the physician introduces a needle through the mother's abdominal wall into the uterus. The procedure involves some physical discomfort, with a pinching sensation as the needle enters the uterus, as the fluid is withdrawn, and as the needle is removed from the abdominal wall. The procedure may need to be repeated if a sample of amniotic fluid could not be obtained or if that fluid were contaminated by blood.

The procedure costs approximately $550, including the ultrasound examination, the obstetrician's fee, genetic counseling, and cytogenetic studies. Third-party insurers vary in the coverage arrangements, so parents must often weigh the short-term costs of the procedure versus the long-term costs of raising a handicapped or retarded child. Those long-term costs include lost parental income, possible institutionalization, special education, equipment, skilled child care, and so forth.

Problems and Pressures

Chromosome abnormalities other than Down's syndrome can be detected through amniocentesis; these findings can pose difficult decisions regarding abortion. Parents may be prepared to abort electively fetuses found to have extra chromosomes 21 (Down's syndrome), 13 (Patau's syndrome), or 18 (Edward's syndrome), with their retardation and accompanying birth defects. However, when the clinical significance of a chromosomal abnormality is unclear, such as in the controversial XYY syndrome, parents need to reconsider the possible outcomes. In addition, the sex of the fetus will be determined in the search for extra or deficient chromosomal material in the fetal cells. Parents can opt to receive this information about their child's gender or they may elect to have it withheld until delivery.

Major magazines in the United States have publicized the "miracle" of amniocentesis. This exposure puts a great strain on couples who have delayed their reproduction for career planning or because of late or second marriages and who feel impelled to avail themselves of prenatal diagnosis of potential retardation problems. Previously parents of a Down's syndrome child could count on family and neighbors for support and sympathy. Since this disorder is now widely considered preventable (by mid-trimester abortion), advanced-age parents in the later segment of the reproductive cycle might find themselves isolated by the society now burdened by the lifelong care of their child with Down's syndrome.

In the past retarded children were often placed in state-supported institutions. Today most of these children are raised at home, placed for adoption via foster homes, or institutionalized privately at substantial cost. In the past children with major birth defects were allowed to die in the newborn nurseries. Today many birth-defective children are resuscitated in the newborn period and are taken to specialized tertiary-care institutions where, either with parental consent or by court order, an attempt is made to repair their defects. These children often require extensive multispecialty medical follow-up, and the eventual outcomes of this intensive and expensive therapy are quite variable.

All of these facts, and many more, need to be assimilated by parents before they weigh the risk of having a child with a particular birth defect against the complications of prenatal diagnosis. Religious, cultural, intellectual, and social biases abound in this difficult area. The incidence of birth defects among liveborns not amenable to prenatal diagnosis in the general population is 2–4 percent; the decision to undergo prenatal diagnosis to detect anomalies must be viewed in this context.

Case Studies

To understand the implications of the technology available for prenatal diagnosis, it is helpful to consider the histories of actual patients who have been offered these procedures.

Case 1: Down's Syndrome.

> *Mrs. G held her newly born daughter with Down's syndrome (Mongolism), trying to decide whether to nurse the baby or to place her for adoption. Mrs. G feared the effect of her daughter's lifelong*

dependency on her two healthy sons. She turned to her husband, who was not with her for the genetic counseling session six months before, and said, "I made the right decision then. I knew the risks and I didn't need the test because I knew this would never happen to me."

Age 35 years at the time of delivery, Mrs. G had faced a 1/365 risk of having a child with Down's syndrome, classical Trisomy 21. That nondisjunctional chromosomal abnormality occurs in 1/1000 liveborns to 28-year old women, and in 1/100 liveborns in 40-year-old women. Her daughter may eventually learn a small income-producing skill but, with an IQ of 55 to 75, probably will not be able to achieve independence in daily living. Because Mrs. G perceived her risk to be low and her luck to be high, she chose not to avail herself of mid-trimester amniocentesis to detect this chromosomal problem in utero. She understood that the amniocentesis would increase the mid-trimester miscarriage risk by approximately 1/200, but she knew she would not elect to abort the fetus even if it were abnormal. Thus, for her, this risk was not worth taking. Her decision pivoted on her sense of invulnerability—she "knew this could never happen" to her.

Although the outcome of her pregnancy was far from ideal, Mrs. G's amniocentesis decision was appropriate—the best choice *at that time* and under those circumstances—a choice with an outcome she could accept. Her daughter now walks with support at age one year and has been well integrated into her family.

Case 2: Hemophilia. Hemophilia is an X-linked recessive disorder for which women, presumed to be carriers on the basis of their family histories, would, in the past, often choose adoption rather than risk having an affected son. With the availability of prenatal fetal sex detection by amniocentesis, they now could selectively abort males, although only 50 percent of these fetuses would actually be affected by hemophilia. The advent of fetoscopy changed the future for some of these women:

Mrs. C and Mrs. R were both presumed carriers for classical hemophilia A. Both had strongly positive family histories and maternal Factor VIII antigen levels which confirmed that they were carriers. Both opted to undergo prenatal diagnosis to detect male fetuses, which they would then elect to abort. Each woman had an ultrasound examination which excluded twins and gross fetal malformations, verified the gestational age, and located the placenta to facilitate the amniocentesis.

Both women underwent amniocentesis at 16 weeks gestation and were found to have male fetuses with a 50/50 chance of being affected by this debilitating disorder. The next step in prenatal detection was fetoscopy—the insertion of a fiber-optic instrument into the uterus to allow removal of placental blood for analysis of Factor

VIII levels. This test, which increases the risk of abortion by approximately 4 percent and causes an approximately 8 percent risk of very premature delivery, was discussed extensively with both couples. Mrs. C accepted fetoscopy and, at 29 weeks gestation, delivered a normal male who required assisted ventilation to survive. Mrs. R suddenly refused fetoscopy as the procedure was underway, and delivered a hemophiliac male at term.

These cases illustrate the problems and emotional anguish which the availability of this complex testing can involve. Nevertheless, these women appreciated the opportunity to learn more about their pregnancies, having seen the medical hardships endured by their relatives.

Case 3: Neural Tube Defects.   Defects in development of the neural tube—a spectrum of anomalies from anencephaly to meningomyelocele (openings of the spine and spinal cord linings)—are inherited in a multifactorial or polygenic fashion: environmental and developmental influences, as well as the genetic constitution of the parents, can contribute to these birth defects. Neural tube defects occur in approximately 1/1000 live births. Usually, this birth defect is not evident in a family history.

Mr. and Mrs. H had refused genetic counseling when referred by their obstetrician because Mrs. H would not be 35 years of age until just after delivery, and she did not expect to consider amniocentesis until her next pregnancy. Their family histories did not include hereditary disorders. They were healthy and avoided all drugs, smoking, alcohol, and reputed teratogens.

At the beginning of the third trimester of an otherwise normal pregnancy, the fetal head was found to be disproportionately small. Ultrasound examination verified anencephaly, or failure of brain development. The pregnancy was electively terminated. The parents were anxious for an earlier detection of any recurrence of such a problem in a future pregnancy.

The recurrence risk for this couple is less than 5 percent, but because the parents were interested in prenatal detection, a series of tests could be performed. Even if they did not electively abort an affected fetus, prenatal detection could still assist the obstetrician in planning the delivery to minimize birth trauma to an affected fetus.

Anencephalic children die at or before birth or within a few days. Children who survive with open spinal cord defects may be unable to be toilet trained, unable to walk, and could be at increased risk for meningitis, encephalitis, or hydrocephaly (rapidly increasing head size due to excess cerebral spinal fluid in the ventricles). Such complications require hospitalizations, repeat surgical procedures, and extensive newborn and early childhood care. Properly managed children can sometimes have normal intelligence.

The first step in the prenatal diagnosis of neural tube defects, performed at approximately 16 weeks gestation, is the determination of the level of alpha-fetoprotein (AFP) in the mother's blood. That protein is found in increased concentration in the amniotic fluid surrounding fetuses with the 90 percent of neural tube defects which are open (that is, not covered by fetal skin). The protein leaks from the fetus into the amniotic fluid, is circulated through the placenta, and passes into maternal blood where it can be measured. Other causes of elevated amniotic AFP levels include the presence of twins, blood contamination of the amniotic fluid sample, gastroschisis (open abdominal wall), and fetal demise.

If maternal serum AFP levels are elevated, ultrasound examination is performed to exclude the possibility of twins and to detect the presence of some cases of spina bifida, hydrocephaly, and anencephaly. Amniocentesis to allow direct measurement of the AFP level in the amniotic fluid would be the next step. This test allows a more accurate means of detection but, as mentioned earlier, carries a 0.5 percent increased risk of mid-trimester abortion. With direct amniotic fluid AFP levels, an additional 10 percent of open neural tube defects can be detected. New techniques, including rapidly adhering cell analysis and measurement of acetylcholinesterase activity, may further increase the accuracy of prenatal detection of neural tube defects. Amniography, injection of radio-opaque dye into the amniotic fluid, can outline the fetal back on x-ray examination to help confirm the diagnosis. Finally, fetoscopy can provide visual confirmation of the diagnosis. Both of these procedures involve higher risk, however, and are sometimes followed by either miscarriage or premature birth of otherwise normal fetuses.

Perspective

Many parents are encouraged by the new medical technology available for prenatal diagnosis, but the risks involved and the possibility of the elective abortion of affected pregnancies have a sobering effect. Parents' attitudes about these possible outcomes must be obtained and appropriately considered before diagnostic procedures can be suggested.

Physicians should not recommend prenatal diagnosis to parents but, rather, should inform them of the risks and assist them in incorporating their own values into the decision. Parents should work closely with their obstetricians and geneticists as they consider the alternatives and should take risks only when weighed against their values and desire for information.

Prenatal testing may be desirable for many parents, but the burden of risk to a normal fetus may be more than some parents can handle. Parents should be given a structure and logic for planning their decision. They should have the best possible information, should be supported in their choice, and should be helped to process the relative risks and benefits of being pregnant in the 1980s.

ANTENATAL DIAGNOSIS OF GENETIC DISEASE: ECONOMIC CONSIDERATIONS

by J. Michael Swint

The extension of health services and inflation in the cost of individual care are causing decision makers to scrutinize existing and proposed health programs more closely. Health related expenditures in the United States comprise approximately 9 percent of the GNP and have exceeded $250 billion per year. Unfortunately adequate funding is not available for all health programs with desirable objectives. Economic considerations in expanding the use of available technology are important in decision making, and we will examine antenatal diagnosis of genetic disease in this context. The application of economic analysis to genetic disease intervention is quite new. Thus delineation of the costs involved in the use of available technologies in this area is intended to help in the assessment of extending these services, as opposed to alternative health- and nonhealth-related expenditures.

The aggregate economic impact of genetic disease is substantial. Based on a study of 43,558 infants, the overall frequency of chromosomal abnormalities is 5.6 per 1,000 births.[1] The frequency of major congenital anomalies is about 3 percent among newborn infants. While these figures are evidence of relatively rare outcomes, the aggregate impact is not negligible; that is, an estimated 20,800 infants in 1975 were born with chromosomal abnormalities in the United States.[2,3]

The net future commitment of society to the maintenance and care of individuals affected with chromosomal abnormalities alone is in excess of $3 billion per year (in excess of $2 billion in 1971 prices).[2] This estimate omits costs for research and for psychological and psychiatric counseling.

With respect to lost productivity, the 1970 present values of lifetime earnings of workers (using 1970 prices) have been estimated to be $185,000 for men in the general population; $164,000 for mildly retarded men (IQ 50 to 70), and $37,000 for moderately retarded men (IQ 40 to 49).[3] Comparable figures for women were $117,000, $106,000 and $23,000. In general Conley found that the earnings of mild retardates were similar to those of the general population up to age 25, but then declined to about 86 percent of normal. The earnings of moderate retardates were found to be about 20 percent of the general population. While these relative differences are probably about the same today, the absolute differences should be inflated.

Institutionalization and/or special education must also be considered. If an infant affected with a genetic disease was institutionalized in 1972 for a period of 70 years, the discounted present value of this cost (in 1972 prices) was $149,000.[4] Since the survival rate usually is not that long and a significant amount of time would be spent away from an institution, the cost is overstated. If we assume that a Down's or Hunter's syndrome case is institutionalized for an average of 20 years, between the ages of 10 and 30, the current cost of care would be $65,000.

Finally, medical care costs must be incurred for the care of affected individuals. The variation in medical costs between diseases and (often) between cases of a particular disease is great, ranging from about $1,600* (exclusive of physician fees and clinic costs) per hospitalization for sickle-cell patients;[5] to $20,000 per year for patients with hemophilia;[6] to as high as $60,300 to $120,600 (discounted present value) for just two and one half years of care for patients with Tay-Sachs disease, in 1976.[7]

Cost-benefit analysis (CBA) is utilized in the health field to examine the net economic impact of intervention alternatives—a comparision of the benefits derived from intervention with the benefits society must forego by not using these resources for alternative purposes (opportunity costs). This is economic desirability from the social perspective and includes economic consequences borne by society as well as those borne by individual couples making the decisions in the case of genetic disease intervention.

The application of CBA to antenatal diagnosis of genetic disease does *not* imply that economic consequences form the decision criterion for program approval. Rather, CBA provides information designed to improve the ability to make rational decisions; it functions as one source of information that must be combined with ethical, sociopolitical, and other (intangible) information for net evaluation. Prest and Turvey[8] illustrate this point by stating that CBA "is only a technique for making decisions within a framework which has to be decided upon in advance and which involves a wide range of con-

---

* Figures adjusted to 1978 prices.

siderations, many of them of a political or social character." Intangibles are explicitly omitted; they cannot be weighed by the analyst. Thus CBA does not place a monetary value on a human life—that would imply monetization of intangibles—but only measures the economic consequences associated with changes in life expectancies, mortality rates, and so forth. Methods to incorporate intangible considerations, such as the willingness-to-pay approach to CBA, are still in the early research stages.[9] Whenever program evaluation decisions are made without the benefit of net economic impact data, an implicit economic value has been assumed nevertheless.

An alternative technique, cost-effectiveness analysis (CEA), can be very useful but supplies less information than CBA. It compares programs on the basis of the cost required to achieve a given output or, alternatively, on the basis of the amount of output attained for a given cost. The outcome measures are either health service inputs (medical services provided, for example, because health impact is undetermined) or health outcomes. They are not monetized. Thus CEA can only help find the most efficient method to achieve a given objective; it cannot help determine whether the objective is worth achieving in the first place. If CEA is used, the social desirability must be taken as predetermined. The following statement from Herbert Klarman illustrates this point:[10]

Retreating from the valuation of benefits to their mere measurement [going from CBA to CEA] entails a substantial loss: the analysis no longer assists in determining priorities among several fields of public activity. The reason is simple. While cost-benefit analysis cuts across diverse objects of public expenditure, cost-effectiveness analysis can only help in choosing among alternative ways of achieving a given, presumably desired, outcome. [bracketed material added]

Essentially, in using CEA the economic impact of a health outcome is determined by decision makers' often implicit assumption, while with CBA it is the result of an explicit analytical exercise.

One criticism of CBA comes from what Grosse[11] calls miguided humanitarianism—the belief that where human life is involved the benefits are infinite and the costs, therefore, irrelevant. According to Grosse, this view neglects the fact that the amount of resources available for disease control programs is limited. Hence we may use an extraordinary amount of resources to save a single life, but in doing so we have cost many more lives by denying resources to other uses. Thus the "truly moral problem is not to distinguish between good and evil, but rather to select appropriately among alternative goods."

With regard to antenatal intervention in genetic disease, while the decision to undertake intervention is highly individualistic, collective concerns that remain as the consequences of such decisions have impacts not borne by the couples in question. Even if a couple can afford to pay the monetary costs of caring for an affected child, the medical (and other) resources involved are precluded from alternative uses. Thus, from the social perspective, the *measurable* economic benefits of diagnosis and intervention, if indicated, include reduction of medical care costs; avoidance of productivity loss (shorter life expectancy, lower productivity if able to work, and reduced labor force participation rates for mothers as some remain home to care for affected children); the addition of productivity in the "replacement case" (that is, the birth of an unaffected child subsequent to the therapeutic abortion of an affected fetus); and avoidance of such costs as institutionalization, special education, and training. In addition, there is avoidance of risks to the mother (such as toxemia associated with carrying a fetus affected with Barts hydrops fetalis), and avoidance of interruption of normal pregnancies (that is, pregnancies that previously may have been interrupted due to the high-risk group of the mother and the prior absence of antenatal diagnostic techniques). The economic costs of intervention include program (or direct) costs (counseling, diagnostic procedures, intervention procedures); indirect costs to patients (time loss, travel, and so forth); misdiagnoses of procedures and their economic consequences; and risks of procedures to the mother and to the fetus and their economic consequences.

In general, for a genetic disease to be suitable for antenatal diagnosis and intervention from the economic perspective, it should: (1) be confined to a well-defined population (identified by demographic characteristics, including maternal age, race, carrier screening, family history, or previous pregnancy history) such that the risk of occurrence is relatively high; (2) lack curative treatment for affected infants (except where antenatal diagnosis increases the probability of successful postnatal treatment); and (3) represent a substantial economic burden to society. While many genetic diseases meet these criteria, intervention activities are currently very limited in view of what is technically possible, particularly with regard to amniocentesis for advanced maternal age. As such, evidence of its economic impact, while generally favorable, is limited.

An Economic Assessment of Current Amniocentesis Use

Trisomy 21

Current efforts at antenatal intervention in Down's syndrome detect about 50 to 60 of the 3,000 affected births per year in the United

States. Approximately 750 of these births are to women ≥35 years, such that the *potential* impact of programs focused on this risk group alone is about one quarter of the annual incidence. However, each year in the United States another 700 to 800 cases of serious chromosomal abnormality could be detected with the same test. Down's syndrome and other trisomies do meet the economic criteria for intervention in genetic disease, and evidence indicates the economic desirability of expanded efforts in this direction.

Conley and Milunsky[12] studied 526 cases of amniocentesis with detection of 16 affected fetuses and found a benefit-cost ratio for Down's syndrome and trisomy 18 interventions of 2.9 for the replacement case and 1.9 for the nonreplacement case. However, these calculations do not reflect patient and physician education costs that would be a necessary part of a large expansion of such services. The estimated discounted present value of benefits of preventing a case of Down's syndrome is $100,000 for the replacement case and $65,000 for the nonreplacement case (1972 prices), while the cost estimate of detecting one case is approximately $35,000. For the current perspective, the benefit figures would have to be adjusted upward for several years' worth of inflation. Also, with technical change, laboratory costs may decline; for example, while somewhat lower than other labs, the New York Megalab[13] reports that the prospective cost estimate of detecting one case of Down's syndrome is $20,000. The general indications of these results are reinforced by a recent study by the Center for Disease Control (CDC)[14] in which the present value of benefits for preventing a (nonreplacement) case of Down's syndrome is $124,380 (1977 terms); the nonreplacement benefit-cost ratio is 1.50; while the replacement benefit-cost ratio is 1.65 (for maternal age ≥35). Recent evidence[15] indicates that the spontaneous fetal death rate for Down's syndrome fetuses detected during the second trimester (for which the mother elected not to abort) may be higher than incorporated in these studies. Thus program benefits may be slightly overestimated.

Cohort analysis has established evidence to support the economic viability of amniocentesis for women ≥35 years. However, women aged 35 through 39 are in the same cohort risk group and a more precise measure is needed. Hook and Chambers[16] estimated the incidence of liveborn Down's syndrome infants by yearly maternal age for mothers aged 20 to 49.* Evidence from their cost-benefit analysis indicates that maternal age as low as 33 or 34 years may be appropriate for amniocentesis from the economic perspective. At a minimum, this supports previous findings of the economic viability

---

* For a comparison of three similar studies recently conducted, see Hook and colleagues.[15–17]

of amniocentesis for women ≥35 years and also lends support to the inclusion of women aged 33 and 34. The issue deserves further examination. The study states that "graphed reported rates (Down's syndrome by maternal age) suggested a linear increase in rates between 20 and 30 (or just over 30) and a logarithmic increase in rates from 33 to 49 with a transitional region between these intervals."[16]

Tay-Sachs Disease

Until recently Tay-Sachs disease (TSD) was the only single-gene disorder detectable by amniocentesis for which there have been widespread carrier detection efforts. In the past decade more than 100,000 U.S. Jews have been screened for carrier status, and cost-benefit analyses have shown that these programs, in combination with diagnostic and therapeutic abortion services for couples at risk, have substantial economic returns. In evaluations of two separate programs, Nelson et al.[18,19] and Swint et al.[20] found benefit-cost ratios ranging from 3.2 to 10.0 for community screening efforts (with community organization and publicity) and from 1.6 to 3.2 for on-demand hospital screening.

Using 1970 data O'Brien[21] estimated the economic impact of screening for TSD on a nation-wide basis (this presumed 100 percent compliance). For antenatal diagnosis and pregnancy interruption, if indicated in families with a previous TSD child (18 percent of cases are in such families), the benefit-cost ratio was 53.9; for heterozygote detection in relatives of affected persons who are ≥30 years, with amniocentesis and interruption where indicated, the benefit-cost ratio was 13.2; and for screening and intervening for all U.S. Jews ≥30, the benefit-cost ratio was 11.1. While certain organizational, publicity, and administrative program costs are omitted from these estimates and cause them to be somewhat optimistic, the results leave a considerable margin for error in meeting the economic criteria. Thus while the number of TSD cases per year is small (approximately 50 in the United States, of which approximately 80 percent occur in Jewish families),[21] the risk group is sufficiently well defined to justify expansion of such activities in terms of economic criteria. As more screening occurs, families in which the trait occurs are identified, the carrier detection rate (per number screened) increases, and costs per TSD case detected decline further, resulting in an increased benefit-cost ratio.

Other Diseases

A technique more complicated than amniocentesis—fetoscopy—can be used to detect sickle cell anemia (in about 65 percent of cases) and alpha thalassemia, but these applications are new and are not currently in widespread use. Amniocentesis has recently been used in combination with an amniotic fluid alpha-fetoprotein test to detect neural tube defects. In all, about 15,000 amniocenteses were performed in the United States in 1978 at a cost of approximately $7.5 million. The economic cost of an amniocentesis with laboratory services is about $500.[22] Pregnancies falling in medically indicated risk groups for such tests totaled 150,000 to 200,000.[23]

Implications of Expanded Use of Amniocentesis

Economic considerations involved in the potential widespread expansion of antenatal diagnostic techniques are:

1. The level of demand by individuals for such services. As the use of diagnostic and intervention services is voluntary, individual demand is required for expansion of such services, regardless of other factors.
2. The feasibility of attracting and organizing adequate resources to supply large increases in the availability of services at costs that do not overwhelm the (collective) economic desirability or excessively deter individual demand.
3. The net economic impact of intervention alternatives, that is, comparison of the benefits derived from intervention in genetic disease with the costs of that intervention.

Demand

The demand for an antenatal diagnostic procedure (given its accuracy/risk profile), like other commodities, is a function of information, individual preferences, price, income, and certain aggregate considerations—in this case, demographic factors that may alter the size of risk groups for some genetic diseases. For example, the risk of Down's syndrome increases with maternal age, and if some women postpone having children, these births will be shifted to a higher risk group. Also, as physicians influence the individual's demand for services in the health care system, the changing legal requirements of physicians may become important[24] in terms of requirements to provide, on an unsolicited basis, certain types of genetic risk-group information and genetic diagnostic or referral services. This also underscores the importance of educating physicians and patients to the benefits and risks of individual genetic disease interventions.

Changing individual preferences have facilitated increases in demand for antenatal diagnostic services. Two cases in particular are worth noting. The first is increasing social acceptance of therapeutic abortions.[25,26,27] Second, as birth rates and family size have declined in the United States, there are indications that couples are more concerned with the "quality" of their children: with fewer children, they are more concerned with rearing the "perfect child."[28] This would be manifested in increased investments in the child,[29] such as increased parental care and attention and opportunities for formal education. This might also be true in terms of increased demand for antenatal diagnostic procedures in an effort to avoid having a child affected by genetic disease.

Demand and Price.   Empirical evidence indicates that price does have a negative impact on the demand for preventive and non-preventive medical care services,[30-33] but the extent of impact varies with the type of care, the population in question, and the institutional arrangements within which care is delivered. Price does form a barrier for the consumption of medical services. It is doubtful that antenatal diagnostic services are greatly different, but we cannot be certain without better data. At a minimum, available evidence from the medical care literature indicates that the price elasticity issue (for antenatal diagnostic services) merits more careful attention and further research.

The price (out-of-pocket cost) to an individual is the residual after payments by third parties. The availability of third-party reimbursement thus determines the price individuals must pay. The level of third-party coverage for genetic disease intervention varies between states and between types of coverage within states. As a general rule, a procedure is covered by private insurance only if it is medically necessary for the insured—the mother in this case. Thus the use of any antenatal diagnostic technology for the detection of genetic disease in a fetus but not for a medically indicated need for the mother is usually not covered.

Medicaid coverage also varies among states, but with respect to antenatal diagnosis of genetic disease, there is near uniformity in excluding all diagnostic technologies for genetic disease. The criterion of medical necessity for the mother exists with Medicaid coverage as well, and it is strictly interpreted with regard to genetic disease intervention.

Current coverages are based on relatively low levels of utilization of genetic disease diagnostic services and have not been given careful attention—that is, guidelines were established for medical care situations in which some genetic procedures are not easily classified. The expanded use of these services may force a reexamination of coverage guidelines for genetic disease diagnosis and intervention and thus increase the likelihood for expansion of third-party cover-

Table 1   Projected Age-Specific Births in the United States, 1978–1988

Year	Maternal Age				Cumulative	
	30 to 34	*35 to 39*	*40 to 44*	*>44*	*≥35*	*≥30*
1978	398,611	113,548	23,766	1,845	139,159	537,770
1979	410,392	116,341	23,653	1,811	141,805	552,737
1980	448,619	123,180	24,548	1,781	149,509	598,128
1981	487,977	137,471	26,473	1,786	165,730	653,707
1982	495,399	138,418	26,546	1,790	167,754	663,153
1983	507,465	143,227	27,948	1,808	172,983	680,448
1984	518,811	151,090	29,025	1,838	181,953	700,764
1985	532,472	158,799	29,843	1,877	190,499	711,971
1986	544,257	166,681	30,242	1,912	198,835	743,092
1987	556,599	166,279	32,532	1,978	200,789	757,246
1988	566,457	168,923	33,538	2,057	204,548	770,975

Percent Change In Annual Levels:
1978 vs. 1988:   42.1   48.8   41.1   11.5   47.0   43.4

Absolute Change in Annual Levels:
1978 vs. 1988: 167,846   55,375   9,772   212   65,389   233,205

SOURCE: Bureau of the Census, 1977.

age. Presumably this is less likely for private carriers since they do not bear the social costs of loss productivity, special education, institutionalization, and so forth. While the decision to undertake antenatal diagnosis and intervention must remain an individual one, if an action results in a *net* reduction of social costs, it is in society's interest to remove economic barriers to utilization. This would argue for consideration of Medicaid coverage.

Demand and Maternal Age. In 1976 approximately 50 percent of amniocenteses performed in the United States were indicated by advanced maternal age; by 1979 this percentage increased to 75 percent. Thus age-specific birth rates are important determinants of the demand for genetic disease diagnostic services. Table 1 provides projections of age-specific numbers of births expected in the United States over the next ten years. The major assumption underlying these SERIES II estimates—the intermediate series from the Bureau of the Census[34]—is a cohort fertility rate of 2.1 (the level of fertility required for the population to replace itself, given expected mortality rates and 400,000 net immigrations per year). The projections suggest that the rate of growth will be below the current rate of 0.8 percent per year. The assumed level of fertility corresponds closely to that suggested by recent survey data related to birth expectations.[35]

From 1972 to 1976 the fertility rate among wives aged 18 to 24

Table 2   Projected Amniocenteses for Advanced Maternal Age by Utilization Rate, 1979–1988 (in 1000s)

Year	Utilization by Women ≥ 35				Utilization by Women ≥ 30			
	3.6%*	15%	35%	50%	3.6%	15%	35%	50%
1979	5.1	21.3	49.6	49.6	19.9	82.9	193.5	276.4
1980	5.4	22.4	59.3	74.8	21.5	89.7	209.3	299.1
1981	6.0	24.9	58.0	82.9	23.5	98.1	228.8	326.9
1982	6.0	25.2	58.7	83.9	23.9	99.5	232.1	331.6
1983	6.2	25.9	60.5	86.5	24.5	102.1	238.2	340.2
1984	6.6	27.3	63.7	91.0	25.2	105.1	245.3	350.4
1985	6.6	28.6	66.7	95.3	26.0	108.4	253.0	361.5
1986	7.2	29.8	69.6	99.4	26.8	111.5	260.1	371.5
1987	7.2	30.1	70.2	100.3	27.3	113.6	265.0	378.6
1988	7.4	30.7	71.5	102.3	27.8	115.7	169.9	385.5

* Approximate utilization rate, 1975 to 1976.

was 2.0, with an expected lifetime fertility of 2.1. The 1977 fertility level of 1.8 and lifetime expectancy of 2.0 for this age group suggest not necessarily that women are "living up to" expectations, but possibly that they may be postponing having children until later in life.[36] Thus it is possible that the estimates of births to women of advanced age are conservative (Table 1).

Using these projections, Table 2 shows that the number of amniocenteses for advanced maternal age alone was projected for utilization rates ranging from 3.6 percent (the approximate national utilization rate in 1975 to 1976) to 50 percent (Table 2). Women aged 30 to 34 have apparently increased their demand for amniocentesis.[37,38] As the projected age-specific birth rates in Table 1 underscore the importance of this apparent trend, Table 2 also includes the ≥30 group.

National utilization of amniocentesis services for women ≥35 increased to approximately 10 percent by 1980. In certain medically sophisticated communities, in terms of health-care consumers and medical personnel, the rate of utilization is already much higher than this. For example, in 1979 the utilization rate (for pregnant women ≥35) in the San Francisco Bay Area was nearly 50 percent.[39]

Supply

Many factors which influence demand will simultaneously influence supply (third-party reimbursement, physician education to the risks of genetic disease, legal requirements, and so forth). Beyond these, however, certain bottlenecks may preclude some demand from being serviced. The provision of laboratory services for processing amniotic fluid (karyotyping) and alpha-fetoprotein, and biochemical tests involving fetal or maternal blood are of particular concern. In 1979 most of the approximately 15,000 to 20,000 amniocenteses undertaken in the United States were conducted in university hospitals, ranging up to 1,000 to 1,200 tests per year each. For this type of activity, indications are that capacities larger than this become administratively inefficient, resulting in increased error.[39,40] Also, laboratories apparently are nearing current capacity.

Megalabs and commercial labs are adjuncts to the current system that should be considered. The megalab is designed to take advantage of high-population density, permitting high-volume testing where transportation time and costs are not prohibitive. For the first megalab, which recently opened in New York City, the projected capacity is 4,000 tests per year, and estimates show costs to be $20,000 per case of Down's syndrome detected.[13] The experience of this lab will be followed with interest; however, there are relatively few cities with sufficient population density to support a megalab without processing the amniotic taps on a regional basis. Conceivably, a regional transportation system could be designed which would allow rapid and inexpensive transport of test materials from outlying areas to a regional (mega) lab. This has not yet been demonstrated; however, there does exist some precedent in the organization of regional labs for other types of tests.[41]

Currently several commercial labs do karyotyping. However, published data are not available to indicate that these labs maintain high levels of quality control. A major issue is thus whether expansion of lab capacity without appreciable reduction in quality control can be achieved. To help ensure quality control, efforts to certify various types of lab personnel and labs themselves are currently under way.

If third-party reimbursement for lab services becomes more extensive, these developments imply the decline of commercial labs with low-quality control; third-party payment to labs without certification is unlikely. By the same token, if demand and the availability of third-party payment increases, the larger commercial labs may be able to invest in better quality control (personnel), gain certification, and still earn an adequate rate of return.

In summary, the current system is not capable of handling a large expansion of amniocentesis activity—such as an increase from 15,000 annually to 100,000—in the short run. Although a lack of facilities plays a part, a lack of qualified personnel to administer and, to some extent, to staff the labs is more important. In the event of a rapid expansion of demand, a portion of it will go unfilled—an unfortunate but unavoidable occurrence. The consequent rationing in such a situation raises some serious ethical considerations that need to be given more careful attention. For example: What criteria would be used to discriminate between cases? Who would establish those criteria, and how would they be enforced, if at all?

The feasibility of alternative approaches to increase capacity over time is one issue to be considered. Do we simply train more cytogeneticists and supporting personnel (a lengthy process), hire them away from alternative activities (the pool of such individuals is small), or encourage expansion of urban or regional megalabs, university hospital lab services, and/or large certified commercial labs? Choosing among these alternatives and implementing them will take time; major capacity changes therefore cannot be expected immediately. If currently available, computerized karyotyping alone, which can do the complete analysis for 90 to 95 percent of tests, might prevent supply shortages in the short-to-medium run. Unfortunately, the introduction of this technology is several years away.

Considerations involved in attracting adequate manpower to perform amniotic taps are quite different. Whereas there is a physical shortage of cytogeneticists to administer a large-scale expansion of cytogenetic lab services, a large pool of obstetricians exists that could be drawn upon to perform amniocenteses, and the period necessary to train them to draw amniotic fluid is relatively short (several days if only the less sophisticated uses of ultrasonography are included).

Quality control is of major concern in cases where community physicians are used to perform the amniotic tap. In this regard, faculty of the Baylor College of Medicine cytogenetic program currently perform about 50 percent of the approximately 500 amniotic taps per year that are diagnosed in their lab; the remainder are done by private physicians in the Houston area. The re-tap rate for community physicians is approximately the same as for Baylor faculty.[40] A learning curve is involved, and the rate of necessary re-taps takes time to fall to these levels. If this experience can be generalized, and if the community physicians in question are given adequate incentives and careful training, manpower to perform amniotic taps should not be a constraint. Professionalism in the face of increased demand by the physicians' own patients may provide the greatest incentive; legal requirements may provide further incentive, and the extension of third-party reimbursement would remove a current disincentive.

Net Economic Impact of Expanded Use

As previously discussed, economic considerations indicate the expansion of amniocentesis for diagnosis of Down's syndrome and other trisomies, and Tay-Sachs disease. Serious ethical and social considerations may preclude widespread use of detection for sickle cell disease; however, from an economic perspective, the nature of this disease makes it not unlikely that the benefits of prevention would outweigh the costs of detection and intervention. In addition to premature death, as was previously noted, the average cost per hospitalization of a sickle cell patient is approximately $1,600. In their analysis of hospital admissions, Tetrault and Scott[6] found that costs per hospitalization of sickle cell patients ranged from approximately $300 to $19,000 (1978 prices).

At the present time in the United States, approximately 50,000 individuals are affected with sickle cell disease. For antenatal diagnosis, amniocentesis costs would be approximately $2,000 per case detected.* While carrier screening costs also have to be considered, it is not unreasonable to expect positive net economic benefits. Sickle cell disease is about six times more frequent in black populations than Tay-Sachs disease is in Jewish populations, and intervention programs for the latter have been shown to have positive economic impacts.

Other Technologies

Carrier Screening.   Carrier status screening is intended to define more narrowly the risk groups for selected diseases, resulting in lowered diagnostic costs and improved economic returns. As previously discussed, this has resulted in positive economic returns for Tay-Sachs screening and prevention programs.[18,19,20] Unless screening results are followed by diagnosis *and* intervention, however, there is no basis for economic benefits. Thus, with the new potential for diagnosis by amniocentesis, sickle cell disease should now be considered for increased screening. However, as just mentioned, the economic perspective is dependent upon ethical and social considerations that will determine diagnostic and intervention utilization on the part of the at-risk populations. Many of these issues are discussed by Fletcher in this volume. Research into these issues and the prospective economic consequences of screening-diagnosis-intervention programs for sickle cell disease would be useful.

Ultrasonography.   While ultrasonography has become the primary noninvasive method of fetal visualization,[42] evidence of its economic consequences is limited. The primary reason is that ultra-

---

* There are approximately 3,000 at-risk pregnancies for sickle cell disease in the United States each year and about one quarter as many births; approximately 65 percent of these are detectable.

sound is normally used in conjunction with other diagnostic technologies, and program evaluations assess the consequences of their joint impact on disease detection and intervention, making the individual contribution of ultrasonography difficult to isolate. Thus it is used as an adjunct to amniocentesis in programs (Down's syndrome and Tay-Sachs disease) which have been shown to yield positive economic returns.[3,4,14,18,20,21,48] To the extent that ultrasonography is judged medically necessary in such programs, the economic returns certainly justify expanded use to keep pace with program growth.

The routine use of ultrasonography in all pregnancies is another issue. The significant economic costs associated with such widespread use underscore the need for a careful accounting of the potential health benefits involved and the economic consequences. Data indicating the positive economic consequences of such a step are not currently available.

A third issue is the use of ultrasonography to detect certain inborn errors of metabolism (dwarfism, for example) and neural tube defects. As this technology is beyond the research and experimental stage (in terms of risk levels and diagnostic accuracy), the economics of such intervention appear promising for indicated risk groups. The economic burden of these diseases is considerable, and the risk groups are well defined by family histories. A prospective evaluation of the economic consequences of such interventions would be valuable for future consideration of expansion or contraction of this intervention.

Fetoscopy.   While advances in fetoscopy or placental aspiration may now permit detection of severe hemophilia and beta thalassemia *in utero*, these techniques remain in the research and experimental stages due to risk levels involved in their use. Fetal loss associated with placental aspiration has declined but is still high at approximately 5 percent.[43] In a study of 54 fetoscopies, Mahoney and Hobbins[8] found three miscarriages and six prematurities. Substantive advances permitting widespread use of fetoscopy must wait for improvements in instrumentation; until then, it is rather conjectural to consider economic issues. All couples undergoing these procedures have a 25 percent chance of having an affected offspring, and the economic burden of these diseases is substantial. However, the economic costs of the risks are also large, and possibilities of misdiagnosis (particularly with expanded use) exist as well. Procedural costs per case detected are approximately $5,000, although changing laboratory technology will likely lower this over time.[43] This cost is compared to annual treatment costs of about $4,500 in the United Kingdom[44] for individuals affected with beta thalassemia and $20,000 in treatment costs per year for a case of hemophilia in the United States.[7]

Alpha-Fetoprotein. The technology exists for alpha-fetoprotein (AFP) tests to detect, *in utero*, 80 percent of the approximately 5,000 neural tube defects (NTDs) each year in the United States. In aggregate terms, and assuming 100 percent utilization, if all U.S. pregnancies were screened with maternal serum AFP tests, about 2 percent (60,000) would have levels high enough to indicate a backup sonogram, and 50 percent of these would be indicated for amniocentesis (30,000). The detection of roughly 4,000 NTDs would result.[45]

While the economic consequences of expanded use of AFP need to be explored, neural tube defects pose a substantial economic burden. Spina bifida cystica can result in paralysis, urinary tract infections, and other illnesses, and 25 percent of these cases result in mental retardation; encephalocele results in serious neurologic deficit and mental retardation for nonfatal cases. Anencephaly is lethal, but in a study by Layde and associates,[46] 13 percent of the patients survived from three days to one month with high medical care costs.

Layde and associates[46] estimated the economic consequences of a multi-tiered program utilizing maternal serum AFP screening, ultrasonography, and amniocentesis (where indicated) for a theoretical cohort of 100,000 women in the United States. Assuming 100 percent compliance and program costs of $2,047,780, the benefit-cost ratios ranged from 1.98 to 2.37 for nonreplacement and replacement cases, respectively. If these results can be generalized, it would cost over $60 million to test all pregnancies in the United States in this manner (these figures omit the costs of initial serum screening, follow-up testing, and extra physician office visits), with expected economic benefits approximately twice that amount.

This does not imply that such a policy should be undertaken. This research is an initial effort in the area and many concerns remain, particularly with respect to the possibility of both unacceptably high rates of misdiagnosis and procedural risks.[47]

A prior study conducted in Scotland by Hagard and Carter[48] did not find favorable economic results, possibly because of lower costs of care and institutionalization in Scotland, and the fact that their study was followed by clearer evidence of the sensitivity of maternal serum to AFP screening.

The use of amniotic fluid AFP to detect neural tube defects *in utero* at present is primarily limited to women with previously affected pregnancies, although it is used for those with positive serum screens. These women have a 2 to 5 percent recurrence rate. Of fetuses with NTDs, 50 percent are either stillborn or die within three days of birth; 25 percent do not live beyond the age of five years; and those who do live beyond five years have physical ailments, and

50 percent of this group are mentally retarded. The economic costs are high, and given the favorable findings of Layde, et al.,[46] for a program with greater costs and lower incidence, the net economic impact of the current testing for neural tube defects is clearly positive.

Concluding Comment

With all indications for amniocenteses included, the total at-risk population is 150,000 to 200,000 pregnancies per year in the United States versus 15,000 to 20,000 amniocenteses currently performed per year. The aggregate economic cost of providing such services would exceed $150 million. These figures use ≥ 35 as the criterion for advanced maternal age; if women aged 30 to 34 increase their demand, the numbers involved would be correspondingly higher. While the available evidence discussed indicates a positive *net* economic impact for certain types of intervention, the difficulties in achieving adequate funding are large, particularly in light of the poor third-party coverage for services and a low level of funding under the National Genetics Disease Act. As the private insurance carriers bear only a small fraction of the full social cost of genetic disease, it is unlikely that they will be the first to initiate expansion in coverage.

Endnotes

1. Lubs, H. and Ruddle, F. (1970). Chromosomal abnormalities in the human population: Estimation of rates based on New Haven newborn. *Science* 169:495.
2. Milunsky, A. (1973). *Prenatal Diagnosis of Hereditary Disorders*. Springfield, Ill.: Charles C. Thomas.
3. Conley, R. (1973). *The Economics of Mental Retardation*. Baltimore: The Johns Hopkins University Press.
4. Swanson, T.E. (1971). Economics of mongolism. *Ann. N.Y. Acad. Sci.* 171:679.
5. Tetrault, S. and Scott, R. (1974). Urban hospitalization cost analysis of patients with sickle cell disease. Unpublished paper, Howard University; abstracted in the Proceedings of the First National Symposium on Sickle Cell Disease, DHEW, NIH Publication No. 75–723.
6. National System of Hemophilia Treatment Centers (1979). Personal communication.
7. Mahoney, M. and Hobbins, J. (1978). Fetoscopy: Procedures and Problems. Paper presented at the Workshop on Prenatal Approaches to the Diagnosis of Fetal Hemoglobinopathies, Los Angeles.

8. Prest, A. and Turvey, R. (1965). Cost-benefit analysis: A survey. *Econ. J.* 75:683.
9. Acton, J.P. (1976). Measuring the monetary value of lifesaving programs. Unpublished paper presented at the Law and Contemporary Problems Conference on the Value of Human Life.
10. Klarman, H. (1967). Present status of cost-benefit analysis in the health field. *Am. J. Pub. Health* 57(11):1948.
11. Grosse, R. (1972). Cost-benefit analysis of health services. *Ann. Am. Acad.* 339:89.
12. Conley, R. and Milunsky, A. (1975). The economics of prenatal genetic diagnosis. In *The Prevention of Genetic Disease and Mental Retardation*, ed. A. Milunsky. Philadelphia: W.B. Saunders Co.
13. Hsu, L. (1978). Personal communication.
14. Center for Disease Control (1978). Mental retardation, birth defects and genetic disease control programs: A cost-benefit analysis. Atlanta, Ga.: CDC.
15. Hook, E. (1978). Spontaneous deaths of fetuses with chromosomal abnormalities diagnosed prenatally. *N. Engl. J. Med.* 299(19):1036.
16. Hook, E. and Chambers, G. (1977). Estimated rates of Down's syndrome in live births by one-year maternal age intervals for mothers aged 20–39 in a New York state study: Implications of the risk figures for genetic counseling and cost-benefit analysis of prenatal diagnosis programs. *Birth Defects: Original Article Series*, Vol. XII, No. 3A.
17. Hook, E. and Fabia, J. (1978). Frequency of Down's syndrome in live births by single year maternal age interval: Results of a Massachusetts study. *Teratology* 17(3):223.
18. Nelson, W., Swint, J., Caskey, C. (1978a). An economic evaluation of a genetic screening program for Tay-Sachs disease. *Am. J. Hum. Genet.* 30(2):160.
19. Nelson, W., Swint, J., and Caskey, C. (1978b). A comment on the benefits and costs of a genetic screening. *Am. J. Hum. Genet.* 30:219.
20. Swint, J., Shapiro, J., Corson, V., Reynolds, L., Thomas, G., and Kazazian, H. Jr. (1979). The economic returns to community and hospital screening programs for genetic disease. *Prev. Med.* 8:463.
21. O'Brien, J. (1970). Discussion of Massachusetts metabolic disorders screening program. In: *Early Diagnosis of Human Genetic Defects: Scientific and Ethical Considerations*, Fogarty International Center Proceedings, No. 6, M. Harris (ed.) National Institutes of Health, DHEW Publication No. (NIH) 72–25, Washington, D.C.: U.S. Government Printing Office.
22. Block, E. (1977). Cost accounting of amniocentesis for prenatal diagnosis. U.C. Medical Center, San Francisco (unpublished).
Population Growth, Vol. 1, Part 4.
23. National Institutes of Health (NICHD) (1979). *Antenatal Diagnosis: Predictors of Hereditary Disease and Congenital Defects*, DHEW Pub. no. 79–1973.
24. *New York Times* (1978). Doctor held liable in abnormal births.
25. Jones, E. and Westoff, C. (1971). Attitudes toward abortion in the U.S.

in 1970 and the time trend since 1965. *Demographic and Social Aspects of Population Growth*, Vol. I, Part 4.

26. Jones E. and Westoff, C. (1978). How attitudes toward abortion are changing. *J. Pop.* 1:5.
27. Blake, J. (1971). Abortion and public opinion: The 1960–1970 decade. *Science* 171:540.
28. De Tray, D. (1973). Child quality and the demand for children. *J. Polit. Econ.* 81:S70.
29. Becker, G. (with Lewis, H.) (1976). On the interaction between the quantity and quality of children. In *The Economic Approach to Human Behavior*, ed. G. Becker. Chicago: University of Chicago Press.
30. Newhouse, J. and Phelps, C. (1974). Price and income elasticities of medical care services. In *The Economics of Health and Medical Care*, ed. M. Perlman. New York: Wiley.
31. Slesinger, D., Tessler, R., and Mechanic, D. (1976). The effects of social characteristics on the utilization of preventive medical service in contrasting health care programs. *Med. Care* 14(5):392.
32. Lairson, D. and Swint, J. (1978). A multivariate analysis of the likelihood and volume of preventive visit demand in a prepaid group practice. *Med. Care* 16(9):730.
33. Lairson, D. and Swint, J. (1979). Estimates of preventive versus non-preventive medical care demand in an HMO. *Health Serv. Res.* 14(1):33.
34. Bureau of the Census (1977). Population characteristics, Current Population Reports. Series P-20, No. 308.
35. Bureau of the Census. 1976. Fertility and American women, Current Population Reports. Series P-20, No. 308.
36. Bureau of the Census. 1977. Population estimates and projections of the population of the U.S., Current Population Reports. Series P-25, No. 704.
37. Golbus, M. (1979). Personal communication.
38. Kaback, M. (1979). Personal communication.
39. Golbus, M. (1978). Personal communication.
40. Riccardi, V. (1978). Personal communication.
41. Bailey, R. (1977). From professional monopoly to corporate monopoly: The clinical laboratory in transition. *Med. Care* 15(2):129.
42. Miles, J. and Kaback, M. (1978). Prenatal diagnosis of hereditary disorders. *Ped. Clin. N. Am.* 25:593.
43. Leonard, C. and Kazazian, H. (1978). Prenatal diagnosis of hemoglobinopathies. *Ped. Clin. N. Am.* 25:631.
44. Modell, B. (1978). Paper presented at the Conference of Parental Diagnosis of Hemoglobinopathies, Los Angeles, California.
45. Oakley, G. (1978). Personal communication..
46. Layde, P., Von Allamen, S., and Oakley, G. (1979). Maternal serum alpha fetoprotein screening: A cost-benefit analysis. *Am. J. Pub. Health* 69:566.
47. Chamberlin, J. (1978). Human benefits and costs of a national screening programme for neural-tube defects. *Lancet* 8093–8105:1293.
48. Hagard, S. and Carter, F. (1976). Preventing the birth of infants with Down's syndrome. *Br. Med. J.* 1(6012):753.

IMPLICATIONS OF PARENTS' ATTITUDES ON ALTERNATIVE POLICIES FOR PRENATAL DIAGNOSIS

by Stephen G. Pauker, Susan P. Pauker,
and Barbara J. McNeil

Although prenatal diagnostic procedures can offer prospective parents the opportunity to identify and terminate abnormal pregnancies, society and the medical care system must determine which potentially available prenatal diagnostic procedures should be made available. For prospective parents, the decision to employ prenatal diagnosis rests on two types of variables: the first concerns the frequency of various potential outcomes of a pregnancy; the second, the couple's attitudes toward these potential outcomes. These potential outcomes range from the birth of a normal child to the birth of an affected child and include elective abortion and miscarriage as a result of prenatal diagnostic procedures. Thus far analyses of societal decisions regarding the allocation of resources for prenatal diagnosis have considered mainly the economic costs of affected children, the cost of prenatal diagnosis, and disease frequency. Little attention has been directed toward attitudes of prospective parents regarding pregnancy outcomes.

This study was supported in part by Research Career Development Awards KO4 GM 00349 (Dr. Stephen Pauker) and KO4 GM 00194 (Dr. Barbara McNeil) from the National Institute of General Medical Sciences, by research grant PO1 LM 03374 and training grant T15 LM 07027 from the National Library of Medicine, and by research grant P41 RR 01096 from the Division of Research Resources, National Institutes of Health, Bethesda, Maryland.

Goal and Methods of Study

We shall illustrate the impact of prospective parents' attitudes toward pregnancy outcomes on the evaluation of a prenatal screening program. For this purpose we shall study amniocentesis as a means of identifying pregnancies affected by Down's syndrome. We shall not address economic issues concerning the cost of such a prenatal diagnostic program nor questions of availability of sufficient resources for it. Instead we shall assume that the prenatal diagnostic program is both economically justified and feasible, but that implementation and planning depend on estimates of the acceptability of prenatal diagnosis (and selective abortion) to the target population.

The Amniocentesis Decision for Down's Syndrome

To assess the value of amniocentesis for the prenatal detection of a child with Down's syndrome, a couple must balance the risks of the procedure against the information it might provide. They must also weigh the relative burdens of miscarriage or elective abortion, incurred as a result of the procedure, against the life-long burden of raising a child affected by Down's syndrome.

This choice can be represented by the decision tree in Figure 1.[1,2] The decision is denoted by the square node on the left; that is, the couple may elect amniocentesis (upper branch) or decline the procedure (lower branch). Each circular node denotes a chance event under the control of neither the couple nor the physician. For example, amniocentesis might be followed by a miscarriage (with probability r). If a miscarriage does not occur, the fetus may (with probability d) or may not (with probability $1-d$) have Trisomy 21 (Down's syndrome). In either case, a positive or a negative test after amniocentesis is performed might result. A test that is negative in the presence of Trisomy 21 is a false-positive result. If amniocentesis is not performed, the fetus may or may not have Trisomy 21. The terminal branches of the tree summarize the four possible outcomes associated with the screening program: miscarriage, abortion, a child with Down's syndrome, or an unaffected child. To compare the relative values of performing or not performing amniocentesis, the relative values of these four outcomes must be determined. On a scale where the cost* of an unaffected child is zero and a child with Down's syndrome is 100, the relative costs of elective abortion

---

* Cost refers to a decrement in utility and not specifically to economic cost. Thus it is a measure of the burden of an outcome to the couple involved.

and a miscarriage are critical in determining the value of amniocentesis in the search for Down's syndrome.

Parental Attitudes

A variety of technologies could be employed to assess relative values of the four outcomes described in Figure 1. These include the lottery technique, direct estimation, and decomposition into utility functions for separate attributes. We shall not focus on these mechanisms but, for illustrative purposes, shall presuppose that the attitudes of a population of prospective parents have been assessed. Thus, each of the four potential outcomes will have been assigned utility or cost by the prospective parents.

Since we defined the cost of an unaffected child to be zero and an affected child to be 100, we must assign a cost to the remaining

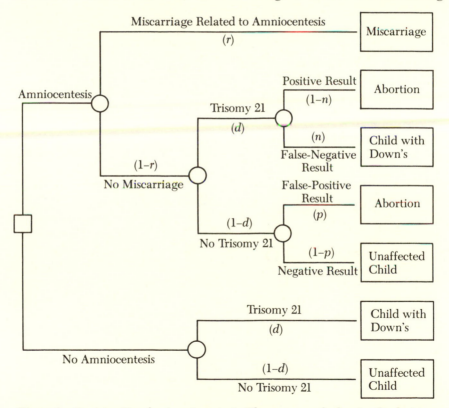

Figure 1   Decision Tree for Amniocentesis. The square node denotes the decision each prospective couple faces. Each circular node describes a chance event. The boxes at the right edge denote each outcome state. The expressions on each branch denote the probability of that event occurring.

two outcomes. If we assume that most parents feel that miscarriage is no less desirable an outcome than selective abortion, we can make the conservative assumption in this analysis that the cost of both of these outcomes is the same, thereby eliminating one variable. Since we may be overestimating the cost of miscarriage, this assumption implies that we will be calculating a *lower bound* on the proportion

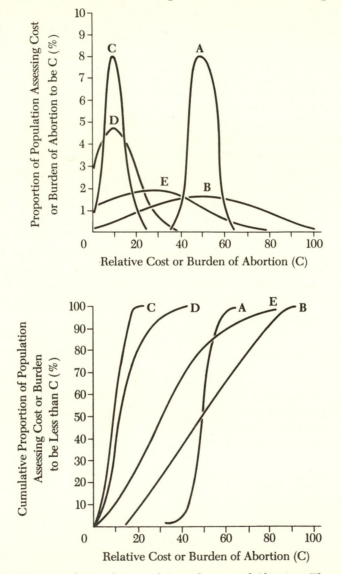

Figure 2 Theoretical Distribution of Attitudes toward Abortion. The panel on the top shows the proportion of five populations (A–E) which would assign the relative cost or burden of abortion to be the value on the horizontal axis. The panel on the bottom displays the cumulative distributions.

Table 1 Summary of Hypothetical Distributions of Attitudes* Toward Elective Abortion

	Perceived Cost of Abortion	
Distribution of Attitudes	*Mean*	*Standard Deviation*
Distribution A	50	5
Distribution B	50	25
Distribution C	10	5
Distribution D	10	10
Distribution E	25	25

* On a scale where the cost of a child with Down's syndrome is 100 and the cost of an unaffected child is zero.

of prospective parents who would be expected to accept amniocentesis if the choice were offered to them.

In this analysis we will present five theoretical distributions of attitudes toward an elective abortion, illustrated in Figure 2. The density distributions can be found on the left and the cumulative distributions on the right. Distributions C and D place a relatively small burden on an elective abortion, whereas curves A and B place an intermediate burden on that outcome. Distributions A and C show little variation in attitudes, whereas curves B and D show significantly more variation. Curve E shows the largest spread in attitudes; this distribution approximates parental attitudes described in a previous study.[2] The means and standard deviations of these five distributions are summarized in Table 1.

The Basic Analysis

Table 2 summarizes the definitions of symbols used in this analysis. To calculate the expected costs of each alternative in Figure 1, we first multiplied the probability of each branch of a chance node by its cost and then totaled all such products. Recalling that the cost of an unaffected child was zero, we saw that the expected cost

Table 2 Definition of Symbols

Symbol	*Definition*
d	Probability of Trisomy 21 (*disease*)
r	Probability of miscarriage after amniocentesis (*risk*)
n	False-*negative* rate (probability of negative karyotype in the face of Trisomy 21)
p	False-*positive* rate (probability of positive karyotype in the absence of Trisomy 21)
C	Cost of elective abortion or miscarriage

of performing amniocentesis was $rC + (1-r)(d(1-n)C + 100dn + (1-d)pC)$; similarly, the expected cost of refusing amniocentesis was $100d$. The optimal decision was the one with the lower expected cost. Hence, we calculated the cost of an elective abortion which just made amniocentesis the optimal decision. This threshold cost (C^*) can be shown to be

$$\frac{100d(1-n(1-r))}{r + (1-r)(p(1-d) + d(1-n))}$$

If the cost of abortion assigned by an individual couple were below C^*, they would benefit from amniocentesis; if their assigned cost of abortion exceeded C^*, they would not benefit from amniocentesis.

In this investigation, assuming several levels of risk and accuracy of amniocentesis, we calculated the threshold cost of elective abortion as a function of maternal age. For this purpose, we used data[3] indicating that the age of the mother has no known effect on the risk or accuracy of amniocentesis but sharply affects the incidence (d) of Trisomy 21 (Figure 3, heavy line). We also calculated the expected proportion of the U.S. population expected to benefit from the availability of prenatal screening programs if certain distributions of

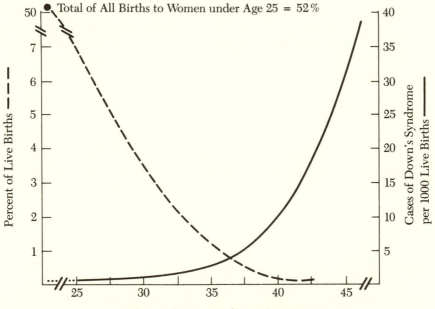

Figure 3   Frequency of Live Births and Down's Syndrome Children by Maternal Age. The broken line shows the percent of live births and is to be read on the left scale. The solid line displays the frequency of Down's Syndrome and is read on the right scale.

attitudes (Figure 2) were to hold. For this purpose we used data on the percentages of live births as a function of age (Figure 3, broken line).[4]

Results

The results of this analysis are summarized in Figures 4 through 6 and Table 3.

The Threshold Cost of Abortion as a Function of Age

For a miscarriage risk (r) of 1/200, and false-positive (p) and false-negative (n) rates of 1/200, the calculated threshold cost of abortion (C^*) varies markedly with age. Thus, as shown by the solid line in Figure 4, for women who are 30 years of age at delivery, the couple would have to assess the cost of abortion to be below ten for amniocentesis to be appropriate; for women who are 40 years of age, the assessed cost of abortion would have to be below 47. Therefore,

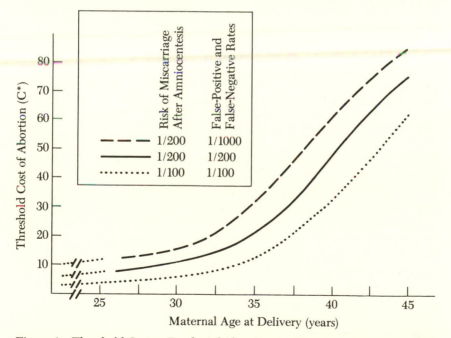

Figure 4   Threshold Cost or Burden of Abortion as a Function of Maternal Age. If the couples assessed the cost to be *below* the threshold cost, then amniocentesis would be appropriate for them. The solid line corresponds to baseline assumptions about risk and accuracy; the lower line corresponds to lower accuracy and greater risk; the upper line corresponds to higher accuracy and smaller risk.

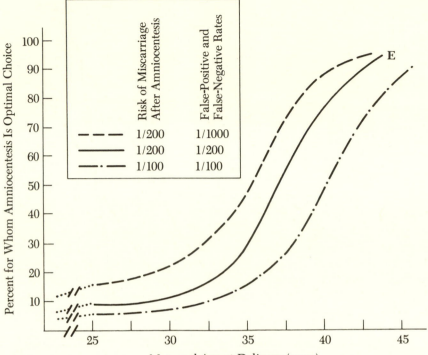

Figure 5   Proportion of a Population for Which Amniocentesis Would Be Appropriate. Attitudes toward abortion correspond to Distribution E in Figure 2. This proportion is shown as a function of maternal age. The upper line corresponds to greater accuracy than baseline; the solid line corresponds to baseline assumptions; the lower line corresponds to assumptions of lower accuracy and higher risk.

at older ages, the cost of an abortion can be considerably higher than at younger ages, and amniocentesis can still be indicated.

The dashed line in Figure 4 shows the results of similar calculations under the assumption that amniocentesis carries a higher risk and is associated with more errors (that is, r, n, and p are all $1/100$). This increase in risk and decrease in accuracy might occur as the procedure became more widespread, placing some strain on laboratory facilities and encouraging obstetricians with less experience to perform the procedure. In contrast, the dotted line in Figure 4 summarizes calculations based on the baseline risk of miscarriage ($1/200$), but with an improvement in test accuracy such that the false-positive and false-negative rates drop considerably to only $1/1000$. The effect of increased risk and decreased accuracy is to lower the threshold cost of abortion—that is, to require the couple to view elective abortion and miscarriage as less severe outcomes (relative to the outcome of having a child with Down's syndrome)

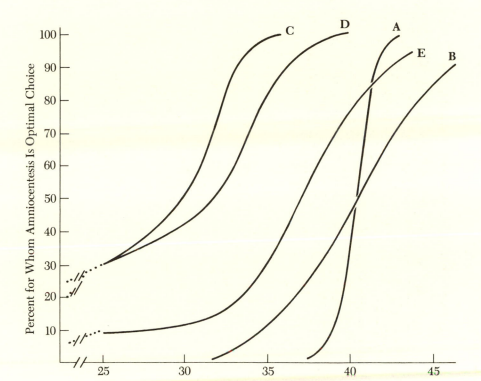

Figure 6 Proportion of Any Population, as a Function of Maternal Age, for Whom Amniocentesis Would Be the Optimal Choice. Each labeled line corresponds to a particular distribution of attitudes toward abortion as summarized in Figure 2. The baseline assumptions about the risk and accuracy of amniocentesis are used.

for amniocentesis to remain appropriate. Conversely, an increase in test accuracy has the effect of raising the threshold cost of abortion.

Maternal Age and the Acceptability of Amniocentesis

Having derived the threshold cost of abortion for various maternal ages, we can compare those thresholds to the distribution of attitudes toward abortion (Figure 2) for the population to be offered amniocentesis. That comparison can provide an estimate of the proportion of couples for whom amniocentesis would be the optimal decision. Consider, for example, a maternal age of 35 years, under the baseline assumptions about risk and accuracy. As seen in Figure 4, the threshold cost of abortion would be 22. Assuming that the couples' attitudes follow distribution E and using the cumulative distribution curve in Figure 2, 30 percent of population E would

Table 3 Projections of Proportion of Pregnancies for Which Amniocentesis Would Be Optimal Choice

Label on curve	Distribution of Attitudes among Prospective Parents				
	A	B	C	D	E
Mean cost of abortion	50	50	10	10	25
Standard deviation of cost of abortion	5	25	5	10	25

Conditions of Testing			Proportion of Pregnancies for which Amniocentesis Would Be Optimal Choice (%)				
Risk of Miscarriage after Amniocentesis	False-Positive Rate	False-Negative Rate	A	B	C	D	E
0.005	0.005	0.005	1.0	1.7	35.9	34.3	9.7
0.01	0.01	0.01	0.5	0.7	19.5	15.4	6.5
0.005	0.001	0.001	1.7	4.7	65.7	52.1	17.8

have assessed the cost of abortion to be less than 22. Thus amniocentesis would be the optimal decision for 30 percent of those couples in population E for which the woman's age would be 35 years at the time of delivery. In Figure 5 we have displayed the results of such calculations for all maternal ages below 45 years: under the baseline assumptions of risk and accuracy (solid line); under the assumption of increased risk and decreased accuracy (dashed line); and under the assumption of baseline risk and increased accuracy (dotted line). Clearly, the proportion of the population for which amniocentesis would be appropriate is affected markedly by maternal age but is also affected substantially by test accuracy and risk.

Under the baseline assumptions of risk and accuracy, we repeated the above calculations for each of the five distributions of attitudes shown in Figure 2. The results are displayed in Figure 6. From these acceptance distributions, the policymaker could calculate the overall acceptance rate in any population or subpopulation in which the distribution of live births as a function of maternal age is known. Population A has the narrowest distribution of attitudes (mean 50, standard deviation 5) and shows the sharpest curve in Figure 6, rising from 10 percent of the population at age 39 years to 95 percent of the population by age 43 years. Population B had the identical mean attitude, but a far greater standard deviation (that is, 25). The corresponding curve in Figure 6 has a more gentle slope, rising from 10 percent at age 35 years to only 85 percent by age 45 years. Both curves have medians falling between years 40 and 41. A similar relation is seen between the curves corresponding to populations C and D which have the same mean cost of elective abortion (that is, ten) but different standard deviations. In this case, the two curves intersect at the 30th percentile (rather than the 50th) because, as seen in Figure 1, the attitudes of population D are more skewed to the right that are those of population C. Obviously amniocentesis would be appropriate for a greater proportion of these latter two populations than of the former two.

For example, at a maternal age of 35 years, zero and 10 percent of populations A and B respectively, should choose amniocentesis, whereas 98 percent of population C and 80 percent population D should elect the procedure. Population E is most interesting since it corresponds roughly to the distribution of attitudes seen in 90 individuals counseled about amniocentesis through the Harvard Community Health Plan.[2] Below a maternal age of 30, we would expect amniocentesis to be appropriate for only 10 percent of that population; by age 35, the procedure should be appropriate for 30 percent of couples; by age 40, the procedure should be appropriate for 77 percent of couples; by age 45 (where the risk of Trisomy 21 is 1/32), virtually all couples should find amniocentesis the appropriate choice.

Projections Across All Ages

Although Figure 6 provides some estimates of the proportion of couples who would benefit from amniocentesis as a function of maternal age, the policymaker might want to project that estimate across the entire population, asking, in effect, if amniocentesis were generally available, what proportion of couples might benefit from the procedure. The results of these calculations for each of the five populations and for each of the three assumed conditions of testing are given in Table 3. Estimates of the proportion of couples benefiting from amniocentesis vary from 0.5 percent (high risk, low accuracy, with population A) to 65.7 percent (low risk, high accuracy, with population C). For the baseline condition of risk and errors (.005, .005, .005) and for the only population whose attitudes have been studied to date (couples enrolled in the Harvard Community Health Plan as depicted in curve E, Figure 2), amniocentesis would be the optimal choice in 10 percent of the pregnancies. The effect of variations in the attitudes of the population to be tested is greater than the effect of changes in test risk or accuracy.

Discussion

We have presented a model of the interactions between the attitudes of prospective parents toward various health outcomes and the number of such parents for whom a particular diagnostic procedure, amniocentesis for the prenatal diagnosis of maternal age–related Down's syndrome, would be beneficial. We have demonstrated, for various theoretical distributions of patient attitudes, that the number of people expected to benefit from the availability of testing is strongly influenced by the projected accuracy and risks of those tests (see Table 3).

The importance of patient attitudes in judging the efficacy of various diagnostic and therapeutic procedures has been underscored. Classic measures of efficacy which depend on objective measures of outcome benefit (such as, dollars saved and cases of Down's syndrome averted) and cost (such as dollars spent and miscarriages induced) do not include patients' attitudes about health outcomes. Indeed, it is very likely that patients' attitudes will show wide variation and that the distributions of attitudes may vary from population to population (Catholic blue-collar workers contrasted with Protestant professionals, for example) and may contribute to observed variation in the utilization of resources.[5] Even within a relatively homogenous population, variation may be great from individual to individual. Thus, as is the case with Down's syndrome, all couples should be offered the opportunity to decide for themselves.

Some situations do exist in which the ability to make an individual decision depends upon the availability of resources or is determined by a previous societal decision. The decision to offer screening for neural tube defects by measuring alpha fetoprotein levels is an example. Currently, resources of this program are being withheld until more data are available; thus even if an individual should want to be tested for a pregnancy potentially afflicted with a neural tube defect, the resources are not generally available. In cases involving policy decisions which affect the group at large, the health planner might find a projection of average benefit over a given population useful in determining the overall benefit of testing.

This paper provides one model for making such projections and estimating the proportion of a given population that would benefit; we calculated the benefit rate as a function of maternal age, assuming that the distributions of attitudes were age-invariant. We then integrated those distributions across the entire population in the United States. In other situations the integration process would involve the maternal age-birth rate distribution of other subpopulations.

The readily observable effect of variations in the distribution of attitudes on the proportion of parents who would benefit from the availability of prenatal diagnosis should serve to establish both the importance of developing techniques to acquire such attitudes and of collecting samples of different populations to provide estimates of the distributions. Techniques of utility acquisition for this purpose are still in the nascent stages.

Finally, prospective parents are not ideal decision makers and their actual choices about the utilization of a diagnostic or therapeutic technology might not agree with what "should" be the logical choice. We would thus expect that our projections of the number of parents who should benefit from the availability of amniocentesis might not coincide with the actual frequency with which parents opt to utilize the procedure. A systematic discrepancy might be shown to exist between our projections and actual use—either because of errors in our model or in our utility distributions, or because prospective parents are making systematically suboptimal choices. Thus research must be directed toward the comparison of expected and actual decisions and the understanding of any discrepancies. Such resolution might involve changing the model or improving patient decisions through better education or the use of decision aids, or both. The development of these techniques to assess the efficacy of medical technologies should improve our understanding of the proper place of these technologies in our diagnostic and therapeutic armamentarium and should thereby improve the quality, and maybe, even lower the cost, of medical care.

Summary

If prospective parents' attitudes concerning the outcome of a pregnancy can be quantified, then those attitudes can be evaluated in the light of the risks and accuracies of a prenatal diagnostic procedure to determine whether that particular parent should undergo the procedure. By assessing the attitudes of a spectrum of prospective parents, the policymaker might estimate what proportion of parents would benefit from the procedure. Since the accuracies and risk of prenatal diagnosis might differ under different policy plans (for example, changes in volume and resources, or who performs the procedure), the policymaker might then estimate the usefulness of utilizing different management strategies for a prenatal diagnostic procedure applied to various populations.

Endnotes

1. Pauker, S.P. and Pauker, S.G. (1977). Prenatal diagnosis: A directive approach to genetic counseling using decision analysis. *Yale J. Bio. Med.* 50:275–289.
2. Pauker, S.P. and Pauker, S.G. (1978). The amniocentesis decision: An explicit guide for parents. In *Risk, Communication, and Decision-Making in Genetic Counseling*, eds. C.J. Epstein et al., Part C of *Ann. Review of Birth Defects*. National Foundation—March of Dimes, Birth Defects Original Article Series XV:5C. New York: Alan R. Liss, 1979, pp. 289–324.
3. Hook, E.B. and Chambers, G.M. (1977). Estimated rates of Down's syndrome in live births by one year maternal age intervals for mothers aged 20–49 in a New York State study—Implications of the risk figures for genetic counseling and cost-benefit analysis of prenatal diagnosis programs. In *Numerical Taxonomy of Birth Defects and Polygenic Disorders*, eds. D. Bergsma and R.B. Lowry. New York: Alan R. Liss. For the National Foundation—March of Dimes, BD:OAS XIII (3A):123.
4. Hook, E.B. (1979). Genetic counseling and prenatal cytogenetic services: Policy implications and detailed cost-benefit analysis of programs for the

prevention of Down's syndrome. In *Service and Education in Medical Genetics*, eds. I.H. Porter and E.B. Hook, pp. 29–52. New York: Academic Press.

5. Sokal, D.C., Byrd, J.R., Chen, A.T.L., Goldberg, M.F., and G.P. Oakley Jr. (1980). Prenatal chromosomal diagnosis: Racial and geographic variation for older women in Georgia. *JAMA* 244:1355–1358.

ETHICAL ISSUES IN PRENATAL DIAGNOSIS: PAST, PRESENT, AND FUTURE

by John C. Fletcher

Prenatal diagnosis of disease and congenital malformations in the fetus is one of four activities that comprise applied human genetics. A full presentation of the ethical issues raised by applied human genetics would show the significant discussion that has been recorded in the literature between 1967 and 1980.[1-3] Only prenatal diagnosis will be discussed in this paper, but this technology is influenced by the goals of genetic screening and genetic counseling, and by the development of therapies for genetic diseases.

The thesis of this discussion is that the moral problems experienced by the first generation of patients, their families, and their physicians in the use of prenatal diagnosis are expanding. First, these problems are being experienced by many more persons than was the case in the period from 1970 to 1975. Second, they are being transmuted into larger socioethical issues that will be important in resolving public policy disputes. Further, I believe that a "moral policy" does function in this society to guide conflicts on both the moral and socioethical levels—a moral policy that responds largely to the claims of the ethical principles of freedom and fairness that guide the increasingly voluntaristic activities of human sexuality and reproduction.[4]

Historical Background

The distinction between moral problems and socioethical issues is found in discussions in the philosophical literature of differences in the *moral* and *ethical* levels or dimensions of human experience.[5] A moral problem is defined as a concrete dilemma that produces con-

359

flict between individuals or groups, especially when the dilemma involves the conflict of moral rules or beliefs. Moral problems are reported in the literature, or in case reports at meetings, usually with the question: "What ought I or we to have done in this situation?" The answers are practical but under the rubric of moral rules rather than expediency.

Ethics is widely described in the literature as reflective activity on the relevance and meaning of moral rules and principles to human decisions and human fulfillment. This reflective activity normally takes the form of reasoned argument that moves from the best possible description of whole classes of moral problems to an examination of recurrent conflicts in the light of ethical principles. Arguments based on ethical principles help to move these conflicts to a level of impartiality and rationality at which the "best interests" of the society and the individual can be ideally considered. This is the level of universality that is presupposed by the term "ethics." At this level, the question is usually raised in the following ways: "Can the reasons that I or we used to give practical guidance in these recurrent moral dilemmas really be defended when we are arguing in the framework of the entire society?" "Would I or we hold up our moral advice as an example for everyone to follow?" A socioethical issue embodies matters that are in dispute for large sectors of society as a consequence of questions about human responsibility raised by the frequency and intensity of moral dilemmas in individuals and groups.

Public policy expressions in the form of legal, constitutional, or administrative remedies are also often necessary to provide guidance for the resolution of socioethical issues, in addition to ethical debate on both sides of a question. "Moral policy" is used, following Callahan,[6] to mean the way a particular community chooses to order the moral rules about a particular area of social life. The term "community" is inclusive of a wide variety of human groups and collectives. Moral policies mediate the moral experience of human communities to moral judgments in the unfolding problems of everyday life. Moral policies evolve and change over time. In the area of social life that involves human sexuality and reproduction, a moral policy has been constructed that has its roots in cultural changes of the 16th and 17th centuries in the West. These areas are increasingly identified as structured by a voluntaristic ethos.[7] The moral policy of freedom with fairness will be more thoroughly discussed in the final section; however, an appreciation of the historical background is necessary for the resolution of problems on either moral or ethical grounds.

Moral Problems in Prenatal Diagnosis, 1970 to 1980

From 1968, when the first reported abnormal fetal karyotypes were cultured from cells derived from amniotic fluid, to the present, the most frequently experienced moral problems clustered around four considerations of prenatal diagnosis. I have arranged these in order of frequency and intensity of experienced moral dilemmas. This typology is based mainly on research in the literature and case reports.

Selective Abortion Following Prenatal Diagnosis

Susan Pauker's research illustrates with case examples how the moral dilemma about abortion appears in the clinical setting. Parents are caught between consideration of two good but irreconcilable outcomes. They ought to be concerned about the welfare of their wanted and expected child, and they ought to be concerned about protecting themselves and their families against disaster. If the diagnosis is positive, they cannot obtain both goods.

Physicians are in a similar moral bind. They ought to want to cure disease, and they ought to want to prevent incurable disease. A living, affected, and aware child in the same family whose mother undergoes prenatal diagnosis will also be in a dilemma. He or she will feel threatened because of the likelihood of nonexistence if prenatal diagnosis had been available or used during early fetal life. Yet, there will also be an opposing sense of obligation to spare another person from the pain and suffering that have been endemic to living with an affected child. All of these conflicting moral sentiments can be seen at work in meetings of national organizations to aid handicapped persons whose disease is now open to antenatal diagnosis. The most controversial indications for amniocentesis—sex choice and maternal anxiety—are primarily problems that reflect the abortion issue.

Risk Factors Related to Techniques and Their Sequelae

The risk of fetal death and/or maternal complications due to amniocentesis was historically a source of moral concern, especially in the period from 1968 to 1974, but this concern was greatly reduced by widely accepted controlled studies in the United States, Canada, and the United Kingdom.[8,9,10] The risks of fetal loss and maternal complications, as Susan Pauker stated, are significantly higher with the use of fetoscopy. For this reason, prenatal diagnosis for hemoglobinopathies that involves fetoscopy should be presented as applied research rather than standard practice and should be adopted in new centers with utmost care and under careful supervision.[11] The risks of sonography, if any, are unknown because the question has not been studied.

Patients should be informed about other technical risks as well. False-positive and false-negative rates due to human error and to the limitations of various laboratory procedures are small but real risks. In the general use of amniocentesis, the accuracy rate is at least 99.4 percent. In screening maternal serum and amniotic fluid for elevated alpha-fetoprotein (AFP), the rate of false-positive and false-negative is directly related to the cut-off points established by screening programs. Programs which set a lower cut-off point for screening normal serum AFP values have a higher detection rate for neural tube defects but carry a higher false-positive rate. Those that set high cut-off points detect fewer defects with a lower false-positive rate. These possibilities should also be communicated to the patient.

The moral content of the risk issue derives from the obligations to avoid harm and to increase benefits. When the risks of amniocentesis were unknown in the early 1970s, some physicians[12,13] argued that parents who were not willing to accept abortion should not be allowed to have amniocentesis, because assuming the unknown risks was justified only by the "benefit" of abortion when necessary. Other physicians vigorously opposed this view of the risk/benefit dilemma.

Link to Neonatal Decision Making

The third moral problem in decisions about prenatal diagnosis occurs less frequently in practice but has been intensely debated in public. Is there a morally relevant difference between the fetus with a diagnosed disease and a newborn with the same degree of impairment? Does permitting abortion for the fetus warrant pediatric euthanasia? The opposite question is relevant: If one would not permit euthanasia in an active sense with the newborn, how can one morally abort a fetus at an earlier stage for reason of disease? This moral problem contains the same polarities as does the abortion problem cited earlier, with an additional feature of precedent for decisions with the newborn.

I have argued[14] that the only morally relevant difference between the fetus and the newborn is that the latter is separate from the mother and that the "separateness criterion" provides physicians and others with a reason to consider the best interests of the newborn even when parents disagree with the best medical strategy that is indicated. Since the mother's best interests cannot be finally considered independently from those of the fetus, and vice versa, before delivery, the mother's interests on occasion should override those of the fetus. The moral status of the fetus prior to "separateness" can be, in my view, compromised in the interests of living persons with serious moral dilemmas. Since the successful outcome of *in vitro* fertilization of ova and embryo transfer, the "separateness argument" could be questioned. Further, low birthweight infants and other prematurely delivered infants are supported technologically at a stage of gestation not much removed from the mid-trimester point at which abortions of the type under consideration here are normally done. Advancing technology may render the separateness argument a moot point.

The apparent moral problem here is that one who approves of aborting a fetus with a severe disease condition does not necessarily approve of the same reasoning being applied to justify pediatric euthanasia. The opportunities for pediatric euthanasia increase as neonatal age decreases. Also, the use of sonography to trace the growth and development of the fetus will enable prenatal assessment of therapy before and after delivery. Obviously the fetus will be a patient under such conditions, and the interests of the mother and fetus cannot be separated. Conflicts will undoubtedly arise between physicians and parents in such circumstances. More work needs to be done to define the moral status of the fetus at all stages of gestation and in various conditions of technological support. The moral relevance of additional knowledge about therapeutic potentials for the fetus and technically supported newborn should enliven and change our concepts of viability, potentiality, and quality of life. These three concepts are important contributors to a larger concept of the moral status of the fetus.

Access to Prenatal Diagnosis

Studies of prenatal diagnosis indicate a higher use rate by more highly educated and economically affluent people.[15,16] Although the exact role of costs of services as a barrier to potential low-income users is not well known, costs may pose a significant problem to many whose pregnancies are at risk. Third-party payment for prenatal diagnosis varies widely. Medicaid payment for prenatal diagnosis is practically nonexistent. Legal barriers also exist to using Medicaid funds to reimburse for abortion for a genetic disease. The moral problem here is the offense to distributive justice caused by an inadequate distribution of medical benefits, especially when Medicaid funds are used to reimburse for conditions that cause far less damage to an infant and family than many of the genetic diseases that can be diagnosed.

A delay of more than one year in making alpha-fetoprotein testing available for public use, due to disagreement about whether the medical community can justly provide the collateral services needed to optimize such tests, also constitutes a serious moral problem of access for those whose pregnancies are at risk for neural tube defects. Reports from the United Kingdom[17] and pilot programs in the United States[18] show that this form of testing is safe and effective. When used with a straightforward approach to educating the patient and giving her the choice of informed participation from the outset, there is a minimum of anxiety about the unknown. I believe that AFP test kits should be quickly approved by the Food and Drug Administration[19] for distribution with restrictions applicable only to laboratory performance and communication between laboratory and physician.

Socioethical Issues of Prenatal Diagnosis in the 1980s

Swint's paper ("Antenatal Diagnosis of Genetic Disease") provides reasons why the demand for prenatal diagnostic services will grow rapidly in the 1980s. Three such reasons are: (1) increased numbers of women who plan pregnancies later in life and incur the risk of maternal age; (2) more success in diagnostic techniques with additional diseases; and (3) evidence of the economic benefits of prevention. If demands for services are actually met, one can expect each of the previously discussed moral problems to occur with greater frequency in physician-patient interactions. Because some forms of prenatal diagnosis, such as amniocentesis and sonography, will be performed in the context of the general practice of medicine, one should also expect that other practitioners besides specialists will be more frequently exposed to these moral problems.

In addition to the wider experience of moral problems in prenatal diagnosis, more questions of the socioethical type will be raised in public forums. These questions will concern the justification for vastly expanding the system of genetic services, the social priority of genetic services, and issues related to the early possibilities of genetic therapies.

Supply and Demand

If only 35 percent of the women over 35 years of age in 1988 utilize amniocentesis, one can expect a demand for over 70,000 procedures in that year alone for this single indication. By contrast, in 1978 probably only 15,000 procedures were done for all reasons. A significant public issue exists now, and will be subject to more debate, about meeting the need for medically indicated requests for prenatal diagnosis.

In the face of obvious cost-benefit ratios of prenatal diagnosis, again indicated by Swint, one might expect to see little public debate about the wisdom of supporting increased programs. Since this area of medical practice and research involves fetal life, concepts in genetics, and potential for dramatic changes in the process of visualizing and treating the fetus, one should expect that many other value-laden conflicts about the proper goals and direction of such activities will occur. The abortion debate will not disappear but will provide a constant background to the discussion of meeting the demand for diagnosis annually of 150,000 to 200,000 pregnancies at risk for conditions that can be detected. The entire area of human genetics is in process of demystification, and one should expect religious explanations of the functioning of genes and the direction of natural selection to hover beneath the surface of public debate. The politics of masculinity and femininity also might play a role as it is tied into part of the sex choice issue and into the abortion debate itself.

Priority of Genetic Services

No extensive public debate has occurred on the priority of genetic services in the context of the overall needs of public health in the United States. The National Genetic Diseases Act, passed by Congress in 1975, had its origin in screening for a small number of particular genetic diseases.

As the role of genetics is better understood in many more diseases, and as screening techniques become more available, the public will need to know what weight to assign to screening, prevention, and treatment of genetic diseases. Notions of genetic health could be oversimplified or the importance of genetic services could be overestimated in contrast to other health measures that could improve the quality of life in far more direct ways, such as nutrition and environmental health. On the other side of the problem, there is a danger that the priorities question will not be addressed because of fear of controversy about abortion or the lingering eugenic fallacy. However, because of unavoidable needs in the supply and demand sides of the problem, there will be an equivalent need to assess the relative priority of these services.

Potential for Advances in Therapies for Genetic Diseases

A future debate on the priority of genetic services will be partly influenced by the empirical possibilities of therapies for genetic diseases and other malformations. There is a great deal of current interest in the use of purified genetic material as a therapeutic agent in humans.[20] The outcomes of research with animals ought to determine whether and how soon human clinical trials should begin to correct, replace, or compensate for deleterious genes. If research in animals is successful, the future could brighten for gene therapy trials. On the other hand, if animal research shows that new genes do not remain in target cells, cannot be regulated appropriately, or cause harm to the target call or other cells, no sound ethical basis would exist for designing tests with humans, even at a Phase I level. More work with animals would be needed to solve these problems, if they can be solved, and the differences between human and other animal genetic models would need further analysis.

The potential of gene therapy is relevant to the debate on the priority of genetic services in terms of purposes and goals. Today the only reliable means to prevent genetic diseases are abortion, adoption, and voluntary infertility. Even with these means, prevention must be considered only a modest and limited goal because of the continued births of new carriers. If effective gene therapies did result from current and new research, these measures would be regarded as primitive but necessary forerunners of more desirable goals of cure and permanent prevention, assuming that corrected genetic material could be transmitted in the next generation. Additionally, if genetic therapy were possible in an affected individual, why would not genetic correction be possible in a carrier? Under these circumstances, great support could be marshaled for screening and diagnosis if therapy were also a clear option and goal.

The lack of therapeutic options, in the event that genetic therapies are scientifically impossible, will play a role in the process of assigning public priorities to genetic services. There will be less incentive to use public funds to make genetic screening significantly more available to the population if abortion alone continues to be the major means of responding to a positive diagnosis. The great benefits from the potentials of gene therapies should be a source of support for continuing research in human genetics, as well as future research in the human embryo and earliest stage of gestation. Those on both sides of the abortion debate could agree that the success of gene therapy would reduce, albeit in a small way, the number of abortions.

Moral Policy in Human Sexuality and Reproduction

Survey research shows that 83 percent of respondents to a 1980 national study agree with the statement that abortion for "serious defect in the fetus" ought to be an acceptable reason for a legal abortion.[21] The paper in this volume by Pauker, Pauker, and McNeil recommends a method of health policy planning for prenatal diagnosis based on attitudinal survey results gathered from prospective parents. They assume that the preferences of prospective parents about abortion fall roughly into the proportions indicated by national surveys. These findings and assumptions relate not only to attitudes but, in my view, to a deeper moral policy on which these attitudes rest. This moral policy presently functions to guide conflicts on both the individual and social levels of decision making in the broad social areas of human sexuality and reproduction.

The historical background of a moral policy of freedom with fairness lies in the origins of changes in practices of marriage and family relations in 16th and 17th century European nations, following the Protestant Reformation and the rise of the modern state. Gradually the freedom of the individual to choose his or her own marriage partner, to pursue the self-fulfillment possible through marriage, and even to dissolve the marriage was enjoyed by more and more members of society. The principle of fairness thus became increasingly relevant to the resolution of conflicts of interests between men and women, as well as between children and their parents.

Since this earlier period sexuality and reproduction have increasingly been defined as voluntaristic social practices that respond largely to the principles of freedom and fairness. Freedom is defined both negatively and positively. On the one hand, it is freedom from external restriction, harm, and the disabling of voluntaristic activity. On the other hand, freedom is self-realization and the satisfaction of basic needs. Fairness is defined as impartial and equal treatment of those who participate in voluntaristic activity, and as the obligation to protect all who participate in these activities equally.

As the scientific and medical possibilities for controlling reproduction emerged in the 20th century, these means were adapted (not without great conflict) within the emergent moral policy guiding sexuality and reproduction. An overriding moral policy has developed that protects the freedom of parents and physicians to apply knowledge gained from research and technology to avoid or achieve reproduction, even while treating with fairness those who would not themselves, on moral grounds, use such freedom. Knowledge and technology have been diffused through a public filter of individual choice. Persons who want to practice contraception may do so; those who do not are not extrinsically punished. Persons at risk for genetic disease are counseled but not prohibited from reproduction. Abortion decisions are overwhelmingly the choice of the woman. Mentally retarded persons are given sex education and in some instances equipped with contraception. It is considered increasingly blameworthy to sterilize another person involuntarily. The injunctions of the moral policy as it bears upon these techniques, including prenatal diagnosis, are twofold: "I will respect your freedom to choose to avoid or achieve reproduction with or without technological intervention, if you will respect my freedom to do likewise. Further, I will work for a society in which neither of us will suffer harm or punishment because of the differences that exist when freedom is cherished."

This moral policy provides a common ground for the coexistence of differing moral policies, as long as each community assents to the principles inherent in the policy. A strong similarity exists between the function of this moral policy and the dynamics of religious toleration. The practical application of the policy depends upon the lack of coercion in public or private spheres and the readiness of society to prohibit punishment for those whose moral traditions differ from the prevailing view. Our society, in my view, must continue to provide genetic services to those who freely choose to control their reproduction, as well as to provide support for those who freely choose to take the risks of reproduction, even when they are aware of potential dangers.

My conclusion is that individuals in this society who make choices about using the benefits of science and technology to resolve problems that affect their identities, as well as their health, will continue to recognize tools such as prenatal diagnosis as a benefit only if the moral policy of freedom with fairness continues to guide these choices. Science and medicine are dependent for their support and wider use on a deeper range of values than knowledge and healing alone.

Endnotes

1. Shinn, R.L. (1978). "Gene Therapy: Ethical Issues." In *Encyclopedia of Bioethics*, ed. W.R. Reich, pp. 520–527. New York: Free Press.
2. Institute of Society, Ethics, and the Life Sciences: Research Group on Ethical, Social and Legal Issues in Genetic Counseling and Genetic Engineering (1972). Ethical and social issues in screening or genetic disease. *N. Engl. J. Med.* 286:1129–1132.
3. Hilton, B., Callahan, D., Harris, M., Condliffe, P., and Berkeley, B. (1973). *Ethical Issues in Human Genetics*. New York: Plenum Press.
4. Fletcher, J.C. (1978). Prenatal diagnosis: Ethical issues. In *Encyclopedia*

of Bioethics, ed. W.R. Reich, pp. 1336–1346. New York: Free Press.

5. Aiken, H.D. (1962). *Reason and Conduct*. New York: Knopf.

6. Callahan, D. (1970). *Abortion: Law, Choice, and Morality*. New York: Macmillan.

7. Stone, L. (1975). The rise of the nuclear family in early modern England: The patriarchal stage. In *The Family in History*, ed. C.E. Rosenberg. Philadelphia: University of Pennsylvania Press.

8. NICHD National Registry for Amniocentesis Study Group (1976). Mid-trimester amniocentesis for prenatal diagnosis: Safety and accuracy. *JAMA* 236:1471–1476.

9. Simpson, N.E., Dallaire, L., Miller, J.R., Siminovich, L., Hamerton, J.L., and McKeen, C. (1976). Prenatal diagnosis of genetic disease in Canada: Report of a collaborative group. *Can. Med. Assoc. J.* 115:739–748.

10. Report to the Medical Research Council by their Working Party on Amniocentesis (1978). An assessment of the hazards of amniocentesis. *Brit. J. Ob. Gynae.* 85: Supp 2.

11. Fletcher, J.C. (1979). Prenatal diagnosis of the hemoglobinopathies: Ethical issues. *Am. J. Ob./Gyn.* 135:53–56.

12. Fuchs, F. (1971). Amniocentesis: Techniques and complications. In *Early Diagnosis of Human Genetic Defects: Scientific and Ethical Considerations*, ed. M. Harris. Fogarty International Center, Proceedings No. 6 HEW Pub. No. NIH 72-25.

13. Littlefield, J. (1970). The pregnancy at risk for a genetic disorder. *N. Engl. J. Med.* 282:627–628.

14. Fletcher, J.C. (1979). Prenatal diagnosis, selective abortion, and the ethics of withholding treatment from the defective newborn. *Birth Defects: Original Article Series* 15:239–254.

15. Bannerman, R.M., Gillick, D., and Van Coevering, R. (1977). Amniocentesis and educational attainment. *N. Engl. J. Med.* 297:449.

16. Sokal, D.C., Byrd, J.R., Chen, A.T.L., Goldberg, M.F., and Oakley, G.P. (1980). Prenatal chromosomal diagnosis: Racial and geographic variation for older women in Georgia. *JAMA* 244:1355–1357.

17. Report of the Working Group on Screening for Neural Tube Defects (1979). London: Department of Health and Social Security.

18. Haddow, J.F., and Macri, J.W. (1977). *Screening for Neural Tube Defects in the United States*. Portland, Me.: Pilot Press.

19. Food and Drug Administration (1980). Proposed rule on alpha-fetoprotein test kits. *Fed. Regis.* 45:74160.

20. Anderson, W.F., and Fletcher, J.C. (1980). Gene therapy in human beings: When is it ethical to begin? *N. Engl. J. Med.* 303:1293–1297.

21. Granberg, D., and Granberg, B.W. (1980). Abortion attitudes, 1965–1980: Trends and determinants. *Fam. Plan. Persp.* 12:250–261.

THE ARTIFICIAL HEART: COSTS, RISKS, AND BENEFITS

by Deborah P. Lubeck and John P. Bunker

Since 1964 the National Heart, Lung and Blood Institute (NHLBI) has funded a research program for the development of a permanently implantable artificial heart. At the program's inception, researchers were optimistic that its successful development would provide a means of treating serious cardiac disease by 1970, well before biomedical advances were expected to produce effective preventative treatment. But after fifteen years, a totally implantable artificial heart is still a distant goal.

Cardiac disease kills over 800,000 persons yearly. The number that might benefit from total cardiac replacement depends on the severity of concomitant illness, age restrictions, access to emergency coronary care, and the nature of the device itself. An estimated 34,500 people yearly may be helped, based on the assumption that a prospective patient's death is imminent, that circulation can be supported long enough for transportation to an institution with appropriate facilities, that the patient does not suffer from serious and chronic noncardiac disease, and that she/he is under 65 years of age. A lower estimate of 16,000 patients per year may be realistic in some parts of the country where inadequate mobile coronary care and surgical facilities exist.

The perfection of the artificial heart will have a far-reaching impact on cardiac disease, now and in the future. Life expectancy may be increased by 0.6 of a year. We estimate that the life expectancy for a randomly chosen member of the population might increase by only 35 days with the availability of the artificial heart. Optimistically, 60 percent of artificial heart recipients employed prior to

This paper is adapted from the summary of "The Artificial Heart: Costs, Risks, and Benefits," a case study conducted as part of the Office of Technology Assessment project, The Implications of Cost-Effectiveness Analysis of Medical Technology. (Washington, D.C.: Office of Technology Assessment, 1981.)

implantation might return to work. The experience of coronary artery by-pass surgery indicates that as few as 50 percent may return to work. However, the range of estimates varies with the reliability of the device and the adequacy of rehabilitative care.

As the procedure becomes available, the demand for its widespread use will undoubtedly be as great as was demonstrated for hemodialysis. Minimum cost estimates for the artificial heart involve an amount that would be prohibitive for most families. Estimates for the cost of manufacturing and surgically implanting an electrically powered device (not including previous development costs) range from $24,000 to almost $75,000 per patient. These are only initial costs; continuing medical and technological care could range from $1,500 to $8,800 per patient per year. If insurance companies prove unwilling to cover the high costs of this treatment without special premiums or other incentives, the federal government will be faced with a serious dilemma—to allow those who cannot afford to pay privately to go without a lifesaving device or, alternatively, to devote up to an additional $2 to $3 billion annually to this new medical technology. Such a commitment dwarfs the funds spent to date on the development of the artificial heart and other circulatory assist devices.

Federal funding for artificial heart implantation involves additional costs and planning for adequate facilities, trained personnel, and a strong program to rehabilitate patients who must deal with the inconvenience and anxiety related to daily recharging of batteries, potential mechanical or electrical failure, and total reliance on an implanted machine. Cost considerations must also take into account potential loss of other worthwhile programs displaced by the development of the artificial heart. The artificial heart may proportionately raise social expenditures financed through Medicare and Social Security from other social programs. Funds that support the training of heart surgeons and technicians for a large-scale implantation program may deter research on cardiac disease prevention or alternative treatments. Recent work in cardiac disease prevention at Stanford University and in Finland indicates that an effective prevention program may have a greater potential to reduce death from cardiac disease, and at less cost, than the artificial heart.

While artificial heart research has led to useful therapeutic inventions, as well as substantial advances in understanding, the question still remains whether an artificial heart will be realized in the near future. As yet neither a hemocompatible material nor a portable power source that can meet the specifications for a long-term, implantable heart in laboratory testing has been developed. Current prototypes of two-year and five-year partial assist devices use electrical battery systems which still have mechanical and operational liabilities. In clinical trials (projected for the mid-1980s), these devices will provide an experimental model to assess the reliability of the engine under conditions of extended use, as well as the quality of life that might be expected from implantation of an artificial heart. Production and implantation also will result in a more accurate picture of total economic costs of the device and surgical procedure.

In addition to battery powered devices, several million dollars of Energy Research and Development Administration (ERDA) funding have been devoted to research on a nuclear power source. While this funding has ceased, some privately funded development continues. Should we develop a successful device that lacks only an acceptable power source, pressure may be brought to proceed with a plutonium-238-powered engine, at enormous cost and risk. Because of its dangerous qualities and its value ($1,000 per gram for a device using 50 grams of plutonium), the material must be closely guarded from manufacture, through transportation and storage, to implantation, until removal upon death. Strict safeguards must be imposed on recipients to protect them from health risks due to radiation, physical injury, or kidnapping. For this reason, a firm commitment against the use of nuclear-powered devices must be reached so that the ultimate potential for a safe and acceptable heart device may be evaluated.

Current progress in the development of the artificial heart represents not only a great responsibility but an opportunity to carefully control the introduction of circulatory device technology. At this time, a clinically effective artificial heart is still many years away; investment in artificial heart devices may be no closer to saving lives than a comprehensive, preventive program. Therefore, decisions regarding expenditures on heart disease in our society should be considered carefully. The benefits and costs of the artificial heart should be compared to other social and medical programs designed to extend and improve the quality of life. The development of the artificial heart should be weighed against the required commitment of resources. If this commitment is assumed, the following issues must be considered: (1) who should receive the device; (2) who will absorb the costs of manufacture and implantation; and (3) what opportunities will be lost through an inability to fund other social programs.

In summary, two major issues involving the development of the artificial heart must be resolved to fully comprehend the total impact of this device. First, the government must decide if it is willing and able to guarantee equitable access to the device, a step that may increase substantially the perceived cost of the program. Second, the decision must be made to accept or reject a nuclear power source.

A nuclear power source may enhance the attractiveness of the device from a clinical standpoint, but it also involves very large social costs and risks. Because these two decisions will have a marked influence on the balance of costs and benefits of the device, they should be fully debated and resolved before a final commitment to artificial heart development is reached.

THE ARTIFICIAL HEART: THE CLEVELAND CLINIC'S APPROACH

by Yukihiko Nosé

An assessment of the impact of cardiac prostheses on the health plan in the United States is extremely important. The artificial heart will have a tremendous impact in the future. When the NIH artificial heart program was initiated about fifteen years ago, the U.S. government asked the MIT group and Hittman Associates for an assessment of the future of cardiac prostheses. At that time the total artificial heart was not yet demonstrated as a feasible medical device. We could barely keep a calf alive for a few days under nonphysiological conditions. It was not certain that a pulsatile mechanical pump would be able to replace total cardiac function and be able to sustain the normal physiology. Unfortunately the study group did not obtain a realistic appraisal of the then current technical status. At that time, we were at the peak of space exploration in the United States, and it was generally felt that the enormous technology base should be utilized for the development of many technical devices, including the artificial heart. We merely had to utilize all the available technology to develop cardiac prostheses. Unfortunately, this assessment was not accurate, and many experts in the field were never consulted. There were no clear-cut objectives for the development of a program.

However, optimistic viewpoints were expressed by all investigators. We felt that a cardiac prosthesis could be made to replace total cardiac function. Theoretically it was possible, but we had no proof at that time. We should not have attempted to follow a program without additional information.

Despite this lack of knowledge, however, the study group concluded that the cardiac prosthesis was a reasonably easy task to accomplish, and on this recommendation, various contracts were given to well-recognized engineering firms to build cardiac prostheses. Unfortunately, at that time medical researchers did not know the basic requirements, such as the optimum cardiac output, the necessity for pulsatile pumping, the proper function of the control mechanism, and the kind of feedback mechanism necessary. The most expensive and modern technology was being applied to poorly defined problems. After a multimillion dollar expenditure, very little had been achieved by the end of 1970. The approach to the problem of developing a cardiac prosthesis should have been more realistic.

We knew how to support the failing heart by assisted circulation. The pump oxygenator and intra-aortic balloon pump technologies were reasonably well established. The first and most logical step should have been to help the failing left and right ventricular function for a short period of time, and then extend the duration to long-term left ventricular assist. With this procedure, we might be able to remove the natural heart and replace it by mechanical means.

In the beginning of the 1970s the artificial heart program of the National Institutes of Health (NIH) re-evaluated the previous several years' efforts and established a more conservative and systematic approach in conducting the artificial heart program. With the able guidance of Dr. Peter Frommer, Deputy Director of the National Heart, Lung and Blood Institute, a cardiology advisory committee was assembled and, based upon their recommendations, a new program was instituted. The initial aim was to develop a temporary left ventricular assist apparatus which was to function for a period of up to two weeks. This pump, which was to be used to assist a postoperative failed heart regain acceptable function, was developed after a few years of research. Then, in December of 1975, this temporary left ventricular assist became a clinical method. From 1975 to 1977, many medical teams made progress with pneumatically driven left ventricular assist devices; they were able to demonstrate performance in calves for durations of several months. Starting in 1977 the NIH established a permanent left ventricular assist program aiming for a two-year totally implantable system. Currently various groups have demonstrated that the principal components will function up to six months *in vivo*. This limitation is not due to the equipment but to the experimental model we use. Unfortunately, after six months, the body weight of the calf is about twice that at implantation, and it is not practical to have longer studies. Based on these studies and extended *in vitro* endurance tests, it appears that two years' duration is entirely feasible. Our program is now entering the system integration stage and over the next four years, we should be able to develop a complete system.

Most medical teams* working in the field of total artificial heart

---

* Teams include the University of Utah, Pennsylvania State University, and Frie University of Berlin in addition to the Cleveland Clinic.

research can keep animals alive with a total replacement pneumatically actuated pump for three to seven months in near-normal physiological condition. We know now, as we did not in 1965, that the total artificial heart is possible. We are confident from scientific and medical points of view that this pumping apparatus can maintain stable physiological circulation for a significant period of time. By using an adult animal that does not change in size, we anticipate that we would be able to maintain performance for an extended period of time, depending only on the endurance and wear characteristics of the components. When the permanent left ventricular assist program is completed, the next step should be the development of the total replacement cardiac prosthesis, because all of the components will be capable of functioning for two years.

Current Procedures

Since 1975 approximately 100 temporary left ventricular assist pump implants were attempted by various groups worldwide. About 15 percent of the patients who used the pump survived because of this procedure, and about one third of the patients were weaned from the pump. Initially we selected critical patients with extremely high mortality rates, and we are now using a patient population with a prognostic mortality rate of 100 percent. With the help of cardiac prostheses, we have achieved survival in the range of 15 percent for this population. At the Cleveland Clinic we have tried six left ventricular assists and three biventricular assist applications. During the biventricular assist procedures the patients suffered too much damage to the myocardium and did not recover. Of the six patients who received left ventricular assists, five could be weaned from the pump after up to one week of pumping. Two patients recovered completely and returned to their original jobs. One other patient recovered cardiac function but developed renal failure during a period of hypotension and subsequently died from kidney and multi-organ failure. Considering the poor prognosis for the patient population selected for assist pump application, the initial results were quite encouraging; however, we hope that a larger percentage of this patient population will be salvaged by the temporary left ventricular assist. We predict probably up to 50 percent of this patient population can be salvaged. Other technological developments, such as hemodialysis and intra-aortic balloon applications, failed initially before patients were successfully treated.

Because the permanently implantable left ventricular assist is not yet available, questions arise about the pump-dependent patient. An intermediary left ventricular assist pump is available that can maintain the circulation for up to six months, utilizing an extracorporeal pneumatically actuated system. This will allow the patient to have a heart transplantation during this period of time. Heart transplantation, therefore, would be the answer for a pump-dependent patient and would justify the risk of this complication. We hope that within four years we will have a totally implantable left ventricular assist pump which, with widespread clinical use, will present more alternatives to urgent heart transplantation. Medically and technologically, I believe that the cardiac prosthesis is a logical and practical treatment method.

Cost

A full-fledged artificial heart program will require a large financial commitment. The potential patient population in the United States can be estimated roughly at 4,000 to 40,000 patients per year. This includes postoperative patients, as well as heart failure patients requiring emergency application of the pump. Regardless of the patient number, if the costs are as high as heart transplantation, a financial problem will exist. Cost analyses are required at this point to determine the extent of the necessary expenditures. If we assume that the initial cost for a cardiac prosthesis will be substantial, the hardware will probably be in the range of $20,000 (mass production could result in cost reductions). Surgical costs may be in the range of $15,000 to $20,000. Consequently the initial cost for an artificial heart implanation might be $35,000 to $40,000 per patient.

In juxtaposition to heart transplantation, extensive medical supervision and careful immunosuppression is unnecessary for artificial heart implantation. The cardiac prosthesis at the Cleveland Clinic does not require any medication. In addition, the design and performance of the cardiac prosthesis is very reliable. Postoperative care of this patient population is therefore substantially less compared to heart transplantation care. It will be necessary, however, to have the implanted auxiliary batteries, as well as certain actuating system components, periodically checked, serviced, or replaced. Therefore, with a low cost for postoperative care, I believe the initial $40,000 investment for a heart patient will be justifiable.

If these patients were bedridden as cardiac cripples, they would be nonproductive and would require constant care by other individuals for possibly many years. With an implanted cardiac prosthesis, the recipient can be an active and productive member of society. This is different from hemodialysis. Once a patient undergoes hemodialysis, maintenance costs about $25,000 per year, with an additional cost of $15,000 per year for medical service. However, a large, recurring, annual expenditure does not exist with the cardiac prostheses. Maintenance would be much lower than heart transplantation. I believe that the patient population requiring a cardiac prosthesis ranges in age from 40 to 50 years. These patients are highly productive, and considering their potential contributions to society, this expenditure does not seem too high.

Conclusions

Research in the area of the artificial heart is becoming more restrictive in the United States, and certain current government regulations are making the development of new medical technology expensive and time consuming. At the same time, countries such as Japan and Germany are rapidly developing high technology in this area and may surpass the United States in the building of a cardiac prosthesis. We must continue to study the economic considerations involved in the development of the artificial heart, and progress should be encouraged. If this is not done, we may have to look to other countries for major changes and advances in technology.

ENGINEERING DEVELOPMENT OF THE MECHANICAL HEART

*by Gerson Rosenberg, Alan Snyder, William S. Pierce,
and David B. Geselowitz*

Artificial hearts with implantable energy converters are being developed for the permanent replacement of irreparably damaged
human hearts. Current technology has developed to the extent that
heart assist devices have been used in over 80 patients worldwide.
These assist devices are used in conjunction with the natural heart
and can take over all or any part of the pumping capabilities of the
natural heart. They have been extensively tested in experimental
animals for over eight months and have been used in humans for
periods exceeding 20 days.

The total artificial heart has been under development for the last
two decades and has progressed from a device barely capable of
supporting an animal for a few hours to a device capable of replacing a heart for over seven months. This heart replacement allows
the animal to eat, sleep, and exercise in a normal fashion although
tethered to a pneumatic driving system (Figure 1). The replacement heart may be readily available to the patient in the not too
distant future.

This work was supported in part by awards from NHLBI Research Grant
5-RO1-HL-20356, The Pennsylvania Research Corporation, McKean County
Cardiac Committee, Mr. and Mrs. Robert E. Galbraith, and the Robert J.
Kleberg, Jr. and Helen C. Kleberg Foundation.

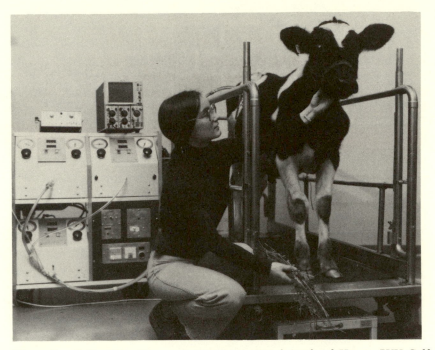

Figure 1   Calf with a Pneumatically Driven Total Artificial Heart. (WU Calf No. 250.) This animal has survived over six months.

Design Goals and Technological Problems

Although the development of a total artificial heart has been under-way for over 20 years, no clinically satisfactory device exists today. It could be argued that given enough funds during this time period, the goal could have been accomplished. This might have been the case, but we are primarily interested in the technological problems involved in developing a total artificial heart under past and current funding levels.

In designing an artificial heart the minimum requirements for such a device must be established. The requirements discussed in this paper may not be universally accepted, but they are representa-tive of the desired capabilities of a long-term (that is, greater than two years or 1×10^8 cycles) artificial heart. They are:

1. The device must fit within the body and not cause undue damage to surrounding tissue.
2. The device must last two years.
3. The device must produce an adequate and controllable cardiac output (8 to 10 1/min) while maintaining physiologic pressures.

4. The system should not produce excessive hemolysis and should be thrombus resistant to the extent that significant emboli are not created.

Other desirable traits include minimum noise, minimum vibration, and light weight, but the system requirements as stated are a minimum.

The basic technological problems encountered in developing such a device are:

1. Blood Material Interface (that is, clotting, calcification, pannus growth, hemolysis)
   a. Physiochemical phenomena
   b. Flow phenomena
2. Component Life (i.e., flex life, bearing design, lubrication)
   a. Design and material selection
   b. Fabrication
3. System Control

These technological problem areas will be discussed in terms of both the pneumatic and electrically driven systems.

Current Pneumatic Systems

A large number of artificial heart designs have been proposed and constructed. These include roller pumps, shear flow pumps, centrifugal pumps, and a variety of sac and diaphragm style pumps. Yet today several common design features can be found among the most successful developers of artificial hearts, including the use of separate pulmonary and systemic pumps to facilitate control and the retention of the biological atria to facilitate attachment of the pumps.[2-5]

The artificial hearts developed by the Pennsylvania State University are quite similar in overall appearance to those developed by other research groups (Figures 2–5). Construction of this pump includes a rigid outer polycarbonate or polysulfone case and a flexible blood sac contained within this outer case. A separate diaphragm isolates the sac from the driving fluid and prevents apposition of opposite sac walls, thus reducing damage to the formed elements in the blood. One-way tilting disc valves are employed to provide unidirectional flow (Figure 5). The driving gas alternately is forced into and evacuated from the space between the rigid case and the flexible sac/diaphragm, thus producing blood flow. The pump has a nominal stroke volume of 110 cm^3 and is capable of producing flows in excess of 14 1/min. This basic principle of operation is common to the designs of the Berlin and Salt Lake City groups.[1,6]

Figure 2 The Pneumatically Powered Artificial Heart Developed by the Penn State Group.

All three pumps employ segmented polyether polyurethane as the blood contacting material. Early blood pumps were fabricated of silica-filled polydimethyl siloxane.[7,8] This silicone rubber has a limited fatigue life and has been shown to be thrombogenic as a result of the silica filler. The segmented polyurethane was chosen because of its excellent mechanical and thromboresistance proper-

Figure 3 The Pneumatically Powered Artificial Heart Developed by the Salt Lake City Group.

ties.[9] Although other compatible materials, such as filler-free poly-dimethyl siloxane exist, their mechanical properties do not approach those of the segmented polyurethane. Thus this material has significantly reduced the blood material interface problems, provided the blood-contacting surface is kept extremely smooth and the pump has no regions of stasis where blood clotting can initiate. It is apparent that the proper choice of materials can reduce physio-chemical effects, and proper design geometry can reduce clotting due to flow phenomena. Hemolysis is not a significant problem with current designs.

Sac calcification and pannus growth are two persistent problems that have, thus far, limited experiments in calves. Calcification has been observed on certain bladders fabricated of segmented poly-urethane, segmented polyurethane-polydimethyl siloxane copolymer, and other materials.[10-13] This calcification has caused stiffening, flexion failures, and perforations of the sacs. Our observations in calves indicate that this process may be reduced with the administration of the drug warfarin sodium.[12] Studies of xenograft valves in children indicate that calcification may be more prevalent in juveniles than in adults.[14]

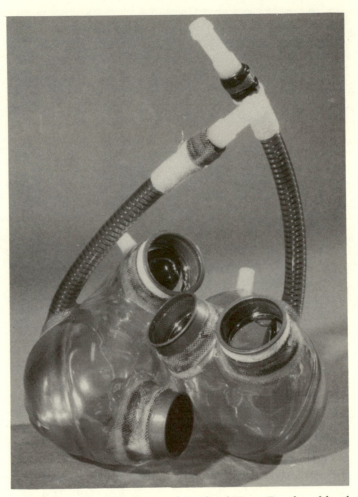

Figure 4   The Pneumatically Powered Artificial Heart Developed by the Berlin Group.

Pannus formation or tissue overgrowth at the junction of the atrial sewing ring and the inlet valve can cause a severe stenosis and reduced cardiac output. This problem may be related to infection or local flow conditions.[6,15,16] Until an adult model such as the sheep is used extensively, the causes of these two phenomena will remain speculative.

Component life in the pneumatic pumps has been a problem at various stages in the development of the heart. The use of the segmented polyurethane has increased substantially the functional life of these devices. Pump design must include careful stress analysis along with good materials selection. Stress concentration should

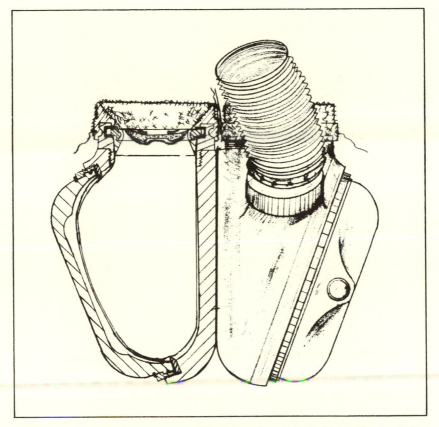

Figure 5   Engineering Drawing of The Pennsylvania State University Angle Port Pump. The flexible inner sac is held within the rigid case. One-way valves are employed to input unidirectional flow. A separate diaphragm isolates the sac from the driving gas. The gas is forced between the diaphragm and the rigid case.

be avoided and areas of flexion should follow smooth contours. The best materials and fabrication procedures cannot make up for poor design.

The pneumatic hearts are driven by a system that provides timed pulses of gas to compress the pump sac and a vacuum to aid in pump filling. The unit provides an adjustable systolic and diastolic pressure, as well as an adjustable frequency and duty cycle. Since the heart is made up of two pumps, with two drive systems, the overall cardiac output must be controlled, and the output of both the left and right pumps must be balanced. There have been various approaches to this problem (Table 1). The system developed by the Penn State group employs a negative feedback loop to control the output of each pump.[17] The rate of the left pump is varied to maintain the

Table 1   Artificial Heart Control Systems

Group	Type	Input Signals	Principle	Remarks
Salt Lake City	Intrinsic pump control	None required	Power unit setting limits complete pump filling during normal operation. Pump output increases as filling pressure increases.	Manual adjustments required to achieve maximum pump output.
Hershey	Two negative feedback servo-systems based on aortic pressure and LAP	AoP, LAP analog from left air line	Both pumps always pump full stroke. L pump rate controlled to maintain normal AoP. R pump rate maintains normal LAP.	Automatic increase in output during exercise.
Berlin	Two negative feedback systems based on right atrial pressure	RAP, LAP from implanted catheters	Pump output increased automatically to maintain normal atrial pressures.	Pump output changed by automatically varying rate and power unit pressures.

AoP—aortic pressure
LAP—left atrial pressure
RAP—right atrial pressure

aortic pressure within the desired range. The rate of the right pump is then varied to maintain the desired filling pressure of the left pump. A major feature of this system is that all control pressures are determined from the left pneumatic power unit hose pressure; thus, no implanted transducers are necessary. This eliminates a major source of infection, thrombosis, and the problem of transducer drift.

The control systems listed in Table 1 appear to work satisfactorily. Some cover greater ranges, are more responsive, and require less manual adjustment than others. At this time, control does not appear to be a limiting factor in the development of the total artificial heart.[3,10,18] Recurrent problems occur with fluid retention or high circulating blood volume in experimental animals living more than a few months. Although this could be due to control, it is much more likely due to an inadequate cardiac output since the calf grows very rapidly and can require a cardiac output in excess of the 13–14 1/min capability of current pumping systems. Also, destruction of some as yet poorly understood volume sensor mechanism in the natural atria or ventricles may be responsible.

Thus the pneumatic system appears to have reached a stage where the technological problems may be solved. Although calcification, pannus growth, and infection are recurrent problems, these problems may be overcome. While there is evidence that the immature animal has an increased propensity for calcification, calcification may occur in the adult.[11,13] Recently two goats have survived with a segmented polyurethane total biventricular by-pass for 232 and 288 days with no signs of calcification.[19] Thus it would appear that calcification of the materials employed is not a certainty in man. The use of smooth Biomer® (segmented polyether polyurethane) atrial connectors has been reported to eliminate the pannus formation found in calves with a total artificial heart.[16]

The problem of infection is due mainly to the large percutaneous tubes, the many vascular access sites, and the many venipunctures. Limiting the size and number of percutaneous tubes, as well as their proper care, can reduce infection. The use of percutaneous tubes may be totally eliminated or reduced to one small caliber tube with the motor-driven devices. In this way, a major cause of sepsis will be eliminated. Adequate system life has yet to be satisfactorily demonstrated *in vivo*; however, *in vitro* tests in excess of two years with pneumatic systems suggest that this can be accomplished. Figure 6 shows the current experience of the different groups with heart replacement in calves. These results are encouraging and show a steady increase in survival time.

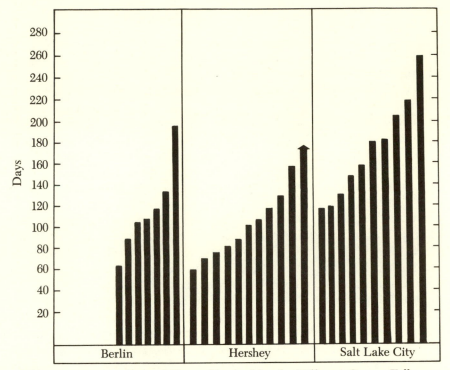

Figure 6   Longest Survival Times Obtained by the Different Groups Followng Artificial Heart Replacement in Calves.

Implantable Motor Driven Total Artificial Heart

The pneumatically driven total artificial heart requires a bulky external power unit and large diameter percutaneous air tubes that can be a source of infection. A small, portable, implantable power system for the artificial heart is under development to provide more mobility for a patient and reduce the chance of sepsis. Initially, a nuclear powered artificial heart was funded; however, due to the cost and potential danger, this system was converted to an electromechanical device which subsequently sustained a calf for 33 days.[20,21] Research using this system has since been abandoned. Only two implantable energy converters are being developed at the present time to actuate a total artificial heart, and both of these are electromechanical devices.

The unit under development by the Salt Lake City group utilizes a high-speed reversing brushless motor-driven turbine that first pumps a working fluid to compress the bladder of one pump and, after reversing, pumps the fluid to compress the other bladder. This

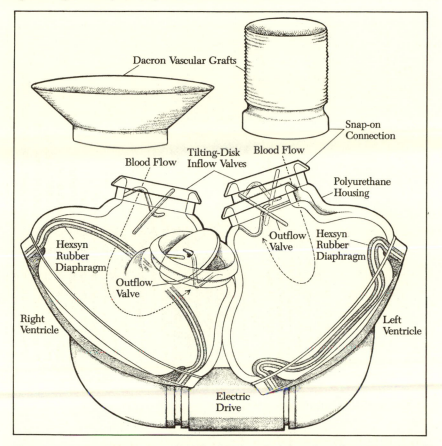

Figure 7 Motor-Driven Total Artificial Heart Developed at the University of Utah. The system utilizes a miniature axial flow pump to activate alternate pumps. *(From Jarvik, R. K., The total artificial heart, Scientific American 244:75, 1981. Copyright © 1981 by Scientific American, Inc. All rights reserved.)*

energy converter is shown in Figure 7, attached to the Jarvik 7 pump. Advantages to this system include small size and weight, but the disadvantages are that it is difficult to control and may exhibit mechanical durability problems. This system, which is still in the early developmental stages, may prove to be reliable in the future.[1]

The system under development by the Penn State group utilizes an oscillating low-speed brushless DC motor with a cam or ball-screw motion translator that is positioned between two pusher plate actuated blood pumps.[22] As the motor rotates in one direction, one pump fills and the other empties. When the motor reverses, the process is reversed. This artificial heart is an extension of the long-term, left heart assist device shown in Figures 8 and 9, which show the motor drive and one pump. This system can be expanded into a

Figure 8   Motor-Driven Left Heart Assist. This motor can easily be expanded to operate a total artificial heart by adding another pusher plate and pump.

total artificial heart by attaching another pusher plate to the other side of the motor and also a mirror image pump (see Figure 10). This system has the advantage of being able to use state of the art electronic technology and conventional engineering techniques to ensure a long functional life. The motor drive system utilizes new high-strength samarium cobalt magnets; all external and most internal components are constructed of titanium which reduces weight and eliminates corrosion. The motor drive utilizing the cam weighs

Figure 9 Motor-Driven Left Heart Assist. This motor can easily be expanded to operate a total artificial heart by adding another pusher plate and pump.

approximately 1 kg and is capable of pumping over 10 1/min at physiological pressures. The entire system, including all electronics, is approximately 20 percent efficient (efficiency = total energy supplied to the blood/total energy supplied to the pump, motor, and electronics). A new system employing a screw-type element to transfer rotary to linear motion uses a smaller but similar motor weighing only 700 gm. The projected performance is similar to that of the cam drive.

The control of this system is rather complex because the two pumps are driven by one motor. Control is accomplished by taking advantage of the nonlinear relationship between pusher plate travel and pump bladder output in early systole. The reversing electric motor drive permits variation in the beat rate and, by limiting motor oscillations, in the stroke volume of both ventricles. While pusher plate travel in both ventricles must be equal, the stroke volume of one ventricle may be varied in relation to the other by reducing the stroke and shifting the midpoint of travel. Thus, one ventricle operates on a nonlinear portion of the pusher plate travel versus pump output curve, while the other operates on the linear portion of the curve. *In vitro* tests have shown a 10 percent stroke volume difference from pump to pump. This method of control also may be augmented by a modified fill-limited method of control in which the relative duty cycle of the two pumps can be varied to produce a differential output between the two pumps. The control signals for the automatic control system are taken directly from the motor electronics. Aortic pressure is proportional to motor current; left atrial pressure is proportional to the left pump fill time determined from a Hall-effect switch mounted in the motor. Again, no implanted transducers are required.

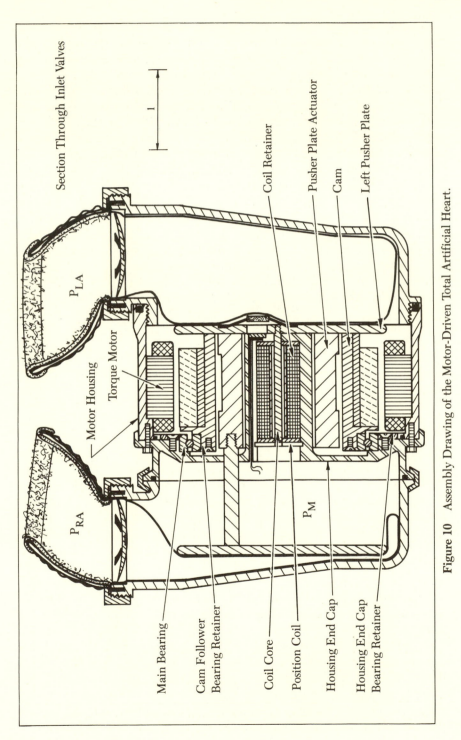

Section Through Inlet Valves

1

P_{LA}

P_{RA}

Motor Housing

Torque Motor

P_M

Coil Retainer

Pusher Plate Actuator

Cam

Left Pusher Plate

Main Bearing

Cam Follower
Bearing Retainer

Coil Core

Position Coil

Housing End Cap

Housing End Cap
Bearing Retainer

Figure 10   Assembly Drawing of the Motor-Driven Total Artificial Heart.

Depending upon the cardiac output and arterial pressure, this system requires between 10 and 20 watts of battery power. At the present time a battery pack utilizing silver zinc batteries can run the entire system for five hours, with a factor of safety of two, at a cardiac output of 5 l/min and a mean arterial pressure of 100 mm Hg. The present system weighs 15 pounds in its prototype stage and is the size of two large textbooks. Lithium-titanium disulfide batteries, although still in the developmental stages, appear to be more favorable than the silver-zinc batteries. The lithium-titanium disulfide batteries undergo no chemical deterioration after many discharge cycles and have a specific energy density as high as the silver zinc units.[23] Thus system size and weight may be reduced further and functional life increased. At the present time, only a small electrical wire must traverse the skin; in future models, energy and control signals for the system may be transmitted through the intact skin using inductive coupling techniques.

Development of a motor driven, total artificial heart (Figure 11) is still a relatively new area, but much of the technology in conven-

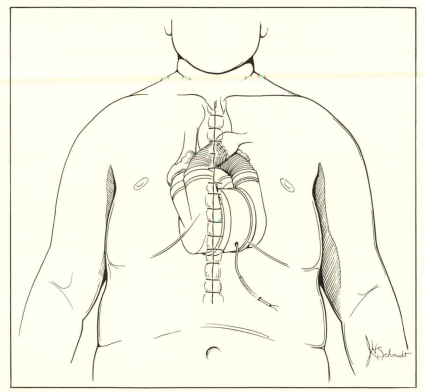

Figure 11   Diagram Showing the Motor-Driven Artificial Heart Implanted in the Chest of a Patient. An external source of electrical power must be provided.

tional mechanical design can be applied to the design of the heart. It appears that no new materials or technology are required to build an implantable motor-driven artificial heart, just creative design and good engineering development. When these systems demonstrate a satisfactory life of two years, they will be ready for human implantation.

Endnotes

1. Jarvik, R.K. (1981). The total artificial heart. *Sci. Am.* 244:74–80.
2. Bücherl, E.S., Affeld, K., Baer, P., Clevert, H.D., Frank, J., Gerlach, K., Grosse-Siestrup, C., Hennig, E., Kielbach, H., Krautzberger, W., Kuhlmann, V., Lemm, W., Mohnhaupt, A., Rennekamp, F., Unger, V., Weidemann, H., and Zartnack, F. (1979). Total artificial heart replacement. *Int. J. Artif. Organs* 2:141–52.
3. Kassai, S., Koshino, I., Washyu, T., Jacobs, G.B., Morinaga, N., Kiraly, R., and Nosé, Y. (1977). Survival for 145 days with total artificial heart. *J. Thorac. Cardiovasc. Surg.* 73:637–646.
4. Mochizuki, T., Hastings, W.L., Olsen, D.B., Lawson, J.H., Daitoh, N., Aaron, J.L., Razzeca, K.J., Jarvik, R.K., Kessler, T.R., Nielsen, S.D., and Kolff, W.J. (1980). Postoperative hemodynamic changes in calves implanted with total artificial hearts designed for human application. *Trans. Amer. Soc. Artif. Intern. Organs* 26:55–59.
5. Shaffer, L.J., Donachy, J.H., Rosenberg, G., Phillips, W.M., Landis, D.L., Prophet, G.A., Olsen, E.K., Arrowood, J.A., Pierce, W.S. (1979). Total artificial heart implantation in calves with pump of an angled port design. *Trans. Amer. Soc. Artif. Intern. Organs* 25:254–259.
6. Affeld, K. (1979). The state of the art of the Berlin total artificial heart—technical aspects. In *Assisted Circulation*, ed. F. Linger, pp. 307–334. New York: Springer-Verlag.
7. Lawson, J.H., Olsen, D.B., Hershgold, E., Kolff, J., Hadfield, K., and Kolff, W.J. (1975). A comparison of polyurethane and silastic artificial hearts in 10 long survival experiments in calves. *Trans. Amer. Soc. Artificial Intern. Organs* 21:368–372.
8. Ross, J.N., Jr., Akers, W.W., O'Bannon, W., Spargo, W.J., Serrato, M.A., Fuqua, J.M., Jr., Ruark, B.S., Weiting, D.W., Kennedy, J.H., and DeBakey, M.E. (1972). Problems encountered during the development and implantation of the Baylor-Rice orthotopic cardiac prosthesis. *Trans Amer. Soc. Artif. Intern. Organs* 18:168–175.
9. Phillips, W.M., Pierce, W.S., Rosenberg, G., and Donachy, J. (1980). The use of segmented polyurethane in ventricular assist devices and artificial hearts. *Synthetic Biomedical Polymers Concepts and Applications*, eds. M. Szychel and W.J. Robinson, pp. 39–57. Westport, Conn.: Technomic.
10. Coleman, D.L., Andrade, J.E., and Kolff, W.J. (1979). Calcification of long-term artificial heart implants. *Abst. Amer. Soc. Artif. Intern. Organs* 8:23.

11. Harasaki, H., Gerrity, R., Kiraly, R., Jacobs, G., and Nosé, Y. (1979). Calcification in blood pumps. *Trans. Amer. Soc. Artif. Intern. Organs* 25: 305–309.

12. Pierce, W.S., Donachy, J.H., Rosenberg, G., and Baier, R.E. (1980). Calcification inside artificial hearts: Inhibition by warfarin-sodium. *Science* 208:601–603.

13. Coleman, D.L., Lim, D., Kessler, T., and Andrade, J.D. (1981). Calcification of non-textured implantable blood pumps. *Abst. Amer. Soc. Artif. Intern. Organs* 10:3.

14. Kutsche, L.M., Oyer, P., Shumway, N.E., and Baum, D. (1979). An important complication of Hancock mitral valve replacement in children. *Circulation* 60 (Suppl. I): 98–103.

15. Olsen, D.B. (1979). The total artificial heart—A research tool or potential clinical reality. In *Assisted Circulation*, ed. F. Unger, pp. 283–306. New York: Springer-Verlag.

16. Jarvik, R.K., Kessler, T.R., McGill, L., Olsen, D.B., DeVries, W., Deneris, J., Blaylock, T., and Kolff, W.J. (1981). Determinants of pannus formation in long-surviving artificial heart calves, and its prevention. *Abst. Amer. Soc. Artif. Intern. Organs* 10:6.

17. Landis, D.L., Pierce, W.S., Rosenberg, G., Donachy, J.H., Brighton, J.A. (1977). Long-term *in vivo* automatic control of the artificial heart. *Trans. Amer. Soc. Artif. Intern. Organs* 23:519–525.

18. Hennig, C., Grosse-Siestrup, C., Krautzberger, W., Kles, H., and Bücherl, E.S. (1978). The relationship of cardiac output and venous pressure in long-surviving calves in total artificial hearts. *Trans. Amer. Soc. Artif. Intern. Organs* 24:616–624.

19. Imachi, K., Fujimasa, I., Miyake, H., Takido, N., Nakajima, M., Kouno, A., Ono, T., Atsumi, K. (1981). Evaluation of polyurethane sac-type blood pump after 232 and 288 days total artificial heart pumping without anticoagulant. Abst. *Amer. Soc. Artif. Intern. Organs* 10:6.

20. Cole, D.W., Holman, W.S., Mott, W.E. (1973). Status of the USAEC's nuclear-powered artificial heart. *Trans. Amer. Soc. Artif. Intern. Organs* 19:537–541.

21. Smith, L., Backman, K., Sandquist, G., Kolff, W.J., Schatten, K., Kessler, T. (1974). Development on the implantation of a total nuclear-powered artificial heart system. *Trans. Amer. Soc. Artif. Intern. Organs* 20:732–735.

22. Rosenberg, G., Landis, D.L., Donachy, J.H., Brighton, J.A., Stallsmith, J., Pierce, W.S. (1979). Design of the Pennsylvania State University artificial heart and electronic automatic control system. In *Assisted Circulation*, ed. F. Unger, pp. 344–52. New York: Springer-Verlag.

23. Murphy, D.W., Christian, P.A. (1979). Solid state electrodes for high energy batteries. *Science* 205:651–656.

COSTS AND REGULATION OF NEW MEDICAL TECHNOLOGIES: HEART TRANSPLANTS AS A CASE STUDY

by John B. Reiss, John Burckhardt, and Fred Hellinger

In the fall of 1979 the Health Care Financing Administration (HCFA), within the Department of Health and Human Services (DHHS), was asked to pay for heart transplants for Medicare beneficiaries suffering from end-stage cardiac disease or idiopathic cardiomyopathy. The HCFA made an interim decision to pay for such procedures, but only for those performed at the Stanford University Medical Center. In the spring of 1980, after considerable discussion of the potential ramifications of that interim decision, DHHS rescinded the decision as premature and denied routine reimbursement for heart transplants from Medicare trust funds. The HCFA was directed to conduct a study to develop new and better information and undertake a thorough analysis of the potential results of making routine payments for this procedure. Prior to this decision, major policy decisions concerning coverage of new medical technologies and procedures often were made on the basis of limited information and under considerable pressure from the medical community and the media.

Through regulation, the government implements policies and programs established by various statutory and budget authorities.[1] In the field of health care, federal regulations can accomplish the following: set performance standards (for example, for drugs and medical devices); affect local allocation of medical resources through health planning requirements; and establish the benefits and costs that Medicare and Medicaid will cover. If the health policy established by the government is unclear, regulatory decisions will be poor and inconsistent.

In this paper, policy and regulatory issues related to government payments for new medical procedures and technologies will be presented, using heart transplants as an example. We will review available data, identify potential budget outlays, and raise social, ethical, and legal concerns about the regulatory implications of funding heart transplantation.

The rapid growth of new and sophisticated medical technologies and procedures has increased the pressure on the federal government to recognize and pay for the latest scientific advances. At present, public monies are used to support a new technology or procedure that is judged to be medically safe and efficacious. This government reimbursement is based on an implicit interpretation of the term "ordinarily furnished" in 1861(b) of the Social Security Act, which is that if a service is provided, payment should be made. However, since new procedures and technologies never have been "ordinarily furnished," explicit judgments about making payments might be appropriate and legitimate. Alternatively, there may be a question about whether the new procedures and technologies are "reasonable and necessary" under 1862(a) of the Act.

The public must be confident that these new technologies and procedures have met threshold tests for medical effectiveness and safety, as well as economic, ethical, and legal standards. Because of the economic concern over the allocation of limited resources, each new procedure and technology must be analyzed and compared to existing health services, prior to the commitment of federal funds.

The Costs of Heart Transplantation

In this section we briefly define costs and discuss problems associated with their measurement. We present estimates of the costs of care for patients who participate in the Stanford University Medical Center program. These costs are categorized as patient evaluation costs, hospitalization costs, professional fees for the heart transplant operation, maintenance costs, other expenses, and total patient costs related to heart transplantation.

Definition of Costs

Hodgson and Mainers[2] separate the core costs associated with a new medical procedure into two components: (1) direct costs of treatment, cure, convalescence, and rehabilitation; and (2) indirect costs resulting from losses in output. In addition, other related direct and indirect costs, are mentioned such as transportation in pursuit of care, housekeeping, increases in daily living expenses, and various social and economic costs. All of these costs are borne by either society or the family unit.

Weinstein[3] proposes that the direct medical costs include: (1) medical diagnosis and treatment, including the costs of tests and treatments induced by the results of the procedure; (2) medical diagnosis and treatments for diseases or side effects produced iatrogenically; minus (3) the savings of medical tests and treatments because of morbidity avoided; plus (4) medical diagnosis and treatments incurred during years of life expectancy added. Partial measures of some of these medical costs are available for heart transplantation and are discussed; others require more data than are available currently. Nonmedical costs are discussed where data are available. Because little information exists on these costs, the total cost of illness is greater than the estimates presented here. An ideal analysis would identify the full costs of illness. However, our primary purpose is to discuss the policy and regulatory implications for government in paying for new medical procedures. Data concerning a major aspect of the personal costs of illness—the emotional and physical stress associated with continued treatment over time—are unavailable. According to Hodgson,[2] only a portion of all costs incurred can be estimated with current methodology and data; estimation of social costs and primary and secondary costs throughout the economy has only begun.

Evaluation Costs

Potential heart transplant patients are evaluated in two stages. The initial work-up is generally done at institutions near the patient's home. Selected patients are then referred to Stanford University Medical Center for further evaluation. As shown in Table 1, the number of patients who are evaluated initially is far larger than the number referred to Stanford for further evaluation. Although 50 percent of the patients undergoing further evaluation were accepted for the operation in 1977, not all of those were operated on in that year.

Various direct and indirect costs are associated with the two evaluation stages. Stanford has indicated that the direct medical costs for a patient undergoing further evaluation is about $9,100, although the range of costs is wide. For a ten-patient sample, the data show a high cost of $27,497; an average cost of $9,121; and a low cost of $3,475.[4]

Information is not available on the cost of the initial evaluation stage. However, Dr. Stinson of Stanford University writes "(f)urthermore, the bulk of costly and invasive studies have usually been performed at referring institutions (other than Stanford)."[4] The statement implies that the $9,121 average cost determined by the study

Table 1   Referrals for Cardiac Transplant—1/1/1977 to 12/31/77

Stage	No. of Patients
Initial Evaluation	234
Rejected	188
Too Old	36
Inadequate Finances	10
Premature Referral	7
Psychosocial Problems	24
Other Medical Contraindications	15
Died While under Consideration	10
Referred Elsewhere for Transplant	4
Further Evaluation	46
Rejected	23
Other Therapy Attempted	14
Died during Evaluation	6
Psychosocial Problems	2
Other Medical Contraindications	1
Accepted	23
Died While Waiting Donor	3
Deselected	0
Transplanted	14
Waiting Transplantation	6

SOURCE: Stanford University's grant proposal for a clinical study of heart transplantation submitted to the National Heart, Lung, and Blood Institute, March 1978, covering the period from May 1978 to May 1983: Grant number HL 13108–09; pp. 5–6, footnote 6.

represents less than "the bulk of the costly and invasive studies." Therefore, we posit that the direct medical costs of the initial patient evaluation are at least $10,000.

The varying costs seen in the ten-patient sample reflect differences in the medical needs of patients, particularly in regard to ability to live outside the Stanford Medical Center while awaiting a transplant. Patients who require full-time hospital care would incur higher costs than those patients who are able to live in the nearby community. However, there are costs involved with living in the community. No data are available to indicate the cost of the services received by these patients living outside the hospital; therefore, the average direct medical costs understate the total expenses incurred by, or on behalf of, all patients during the further evaluation stage.

The total direct medical costs for the evaluation stage of the heart transplant patient accepted by Stanford is the sum of costs for the two evaluation stages, which is at least $19,100. As noted earlier, this figure includes neither living costs of patients who are not inpatients at Stanford during the further evaluation stage nor transportation costs for their journey from their home to Stanford.

Hospitalization Costs at Stanford and Professional Fees

These costs are associated with the preoperative preparation of cardiac recipients, the operation, and the time in intensive care and in a ward until discharge, approximately 6–7 weeks postoperatively.[4] Also included are estimates of the costs of providing the donor heart. "Currently 60 percent of the donor hearts are obtained at sites other than Stanford . . . transported to Stanford by air and implanted immediately upon arrival . . . The care of donors (transported to Stanford for organ removal) becomes the province of the cardio-vascular team after pronouncement of brain death."[4]

The average costs of these hospital and professional services are reported by Stanford to be $77,000 for 1977. The breakdown is shown in Table 2.

Table 2   Hospitalization Costs and Professional Fees for a Heart Transplant at Stanford University in 1977, Including Donor Costs

Hospital Costs	
Recipient	$62,000
Donor	3,000
Professional Fees	
Recipient	10,000
Donor	2,000
Total	$77,000

Note: These data are the average billings for 13 patients receiving transplants in 1977 (project year 08). The numbers are from the Stanford University grant proposal and progress reports, pp. 5–28 and 5–29.

Maintenance Costs for Patients in the Stanford Program

These costs are a function of both the annual cost of maintaining a heart transplant recipient after the initial discharge from the hospital and the expected increased years of survival resulting from the operation.

The costs arise from "rehospitalization either at Stanford or at a distant hospital, . . . readmission to Stanford once a year for in-depth invasive evaluation"[4] and incidents requiring outpatient treatment. Because tracking patient data from various institutions and physicians over a wide geographic area is not done at present, we extrapolate from data provided by Stanford; the average maintenance cost per year is estimated to be about $15,000, $11,000 for

Table 3   Survival of Cardiac Transplant Patients at Stanford Undergoing
           Operation Since 1/1/74 (as of 3/18/80)

Transplant Year	No. of New Recipients Transplanted	1-year Survival	2-year Survival	3-year Survival	4-year Survival	5-year Survival
1974	13	8	7	7	7	7
1975	18	12	11	10	9	—
1976	20	14	12	12	—	—
1977	19	11	11	—	—	—
1978	25	15	—	—	—	—
1979	21	—	—	—	—	—
Total	116	60	41	29	16	7

Source: Memorandum from Lois Takaoka, Stanford University School of Medicine, on March 20, 1980, to Mary Jo Gibson, Office of Health Regulation. Table excludes transplants prior to January 1, 1974.

hospitalization, and $4,000 for professional fees. Heart transplant recipients, after the first year, receive services which cost about $8,800 per year from Stanford, $7,300 for inpatient care and $1,500 for outpatient care. We estimate that professional care and hospitalization received by heart transplant recipients in their home area may amount to $6,200. Thus, total maintenance cost after the first year are estimated to be $15,000 per year. Dr. Stinson indicates that, for a sample of 30 patients, for the first year after transplantation the average costs are $80,000 for hospitalization alone.[4]

The survival rate is calculated from the number of patients undergoing the operation at Stanford since 1974 (about one half of annual heart transplants worldwide at present), and their years of survival. These are shown in Table 3.

These data are analyzed in more detail in our discussion of the increase in patient longevity. As noted there, Stanford has derived an actuarially projected survival rate for five years of 50 percent. Based on this projection, the average maintenance costs for the patient who survives five years are estimated to be $75,000 per patient (the estimated annual maintenance cost of $15,000 multiplied by the actuarially predicted 50 percent survival period of five years).

Other Expenses

A complete measure of costs should include the societal and personal components previously identified; however, very little information is available. The relocation of a patient's family to Stanford for the duration of the treatment is a known cost.* Schroeder[5] of Stanford University estimates that families spend from "$600 to $800 a month minimum in the San Francisco Bay Area for a period ranging from six to ten months." Therefore, the minimum expense would be $3,600, the average $5,600, and the maximum $8,000.

Measuring the intangible costs is particularly difficult. For example, heart recipients experience severe emotional and physical problems. Interviewers of heart recipients from Stanford state: "Each recipient, despite enhanced quality of life after transplantation and the apparent stability of his own medical condition, was still confronted repeatedly with death of other cardiac recipients and with indications, usually in the form of infections, of his own susceptible state."[6] Accompanying emotional stress is physical pain "(s)he must take very painful steroids each day to suppress the constant stream of white blood cells . . . This, in turn, has lowered her resistance to ordinary infections . . . [she] is also taking antithymoate globulin . . . But like the steroids, it is painful."[7]

McNeil and associates[8,9] have studied the patient's role in evaluating diagnostic tests and choosing methods of treatment and have found that the majority of patients are more concerned with the quality of life than longevity. However, these studies are not concerned with measuring the cost of such choices for individuals or groups of patients. No methodology has been developed to translate the subjective responses of individuals into objective, dollar-value measures for the group, in part because no common value scale exists to compare individuals' judgments.

Total Per-Patient Costs

From the information developed in this section, for those surviving five years, the total per-patient costs related directly to the heart transplant operation are $176,700. These costs have not been dis-

---

* An attachment to the February 15, 1980, memorandum from the Assistant Secretary for Health and Surgeon General to the Administrator, HCFA, headed *"Stanford Criteria" for Selection of Heart Transplant Recipients (January 1980)* under the heading "Social Considerations" reads:

4. Financial resources must be available to the patient including:
a. Travel to and from the transplant center accompanied by a supportive family member for final evaluation of transplantation candidacy . . .
c. Living expenses for the patient and family while in the transplant center area before, during, and after transplantation, a period which may extend six to ten months.
d. Annual travel to the transplant center indefinitely for medical follow-up or when needed for severe rejection episodes or complex infections.

Table 4   Cumulative Cost for a Heart Transplant Patient ($ thousand)

	Cost of Each Stage	Cumulative Cost
Initial Evaluation (pre-Stanford)	10.0	10
Further Evaluation	9.1	19.1
Operation	77.0	96.1
Maintenance	75.0	171.1
Other	5.6	176.7

counted to provide a present value calculation since that would give only an illusion of accuracy. The bulk of the costs accrue in the first year, so are not subject to discounting, and the total costs are only a rough estimate based on available data. If the maintenance costs are discounted at the maximum rate of 10 percent proposed by the cost of illness memorandum,[2] the resulting reduction in present value (approximately $3,500) is small relative to the potential error inherent in the information available. In any case, the purpose of this section of the paper is to identify potential government outlays for the procedure. As certain cost information is unavailable, this figure is relatively conservative. The component costs are shown in Table 4.

Benefits

Benefits accrue to individuals when their productive economic and/or social life is maintained, enhanced, and extended, and thereby result in benefits to society through improvements in general economic and social well-being.

Increase in Longevity

Recent publicity has reported a 50 percent, five-year post-transplant survival rate at Stanford. This number is an actuarial projection. Table 5, which is calculated from Table 4, shows that the four-year survival rate for the 1975 cohort is close to the actuarially projected five-year, 50 percent survival rate. One problem, however, is that the actuarial projection is imprecise, due to the small sample size. An additional problem is that the survival rate varies considerably, depending on the time a group of people are selected for the Stanford program. Table 6, which is calculated from data presented in Tables 1 and 3, provides survival rates for 1977 for each stage in the evaluation process, ranging from 23.9 percent to 57.8 percent.

The group may be defined as constituting those selected for further evaluation because they initially meet the Stanford selection

Table 5   Five-Year Post-Transplant Survival Rate at Stanford

Year	New Transplants	Survival Rates (%)				
		1 Year	*2 Years*	*3 Years*	*4 Years*	*5 Years*
1974	13	61.5	53.8	53.8	53.8	53.8
1975	18	66.6	61.1	55.5	50.0	—
1976	20	70.0	60.0	60.0	—	—
1977	19	57.8	57.8	—	—	—
1978	25	60.0	—	—	—	—
1979	21	—	—	—	—	—

SOURCE: Based on data from Lois Takaoka, Stanford University School of Medicine, March 20, 1980, in a memorandum to Mary Jo Gibson, Office of Health Regulation.

Table 6   Survival Rates for 1977 Stages of Selection

Stages	Number	Percent Surviving in 1979
Initial Evaluation	234	Not in program
Further Evaluation	46	23.9
Accepted	23	47.8
Operation	19	57.8

criteria. This group of 46 people has a survival rate of 23.9 percent for two years. However, we are basing our calculations of extended survival on the Stanford actuarial projection because it reflects "best" performance at this time. This projection of survival rates is based on patients who have actually received operations because the physician community uses that base.

Economic Benefits

The economic value of additional life expectancy is dependent on the patient returning to gainful employment. Since 1968 Stinson[10] indicates that 59 of 129 patients from an original group survived one year; 90 percent achieved rehabilitation (defined as restoration of physical and psychosocial capacity to a level which gives the patient an option to return to employment or to an activity of choice), and the majority returned to active employment. Therefore, 21 percent to 25 percent of the original group may have re-entered the work force.

The income flow resulting from the return to work of individuals in the Stanford group is unknown because such information is not available. However, using the U.S. mean primary family income as an indicator, the undiscounted income flow for each family is $20,091 for one year, and $100,455 for five years.*

Social costs have been reduced somewhat. To the extent that a recipient's extended life and gainful employment precludes payment of survivor or disability benefits, the costs of administering those programs are saved. Society must pay the added administrative costs of survivors' benefits payments to families or disability payments to those who do survive but do not return to work. The payments themselves are straightforward income redistribution. (The actual payment of disability income or survivors benefits are income transfers and do not change society's gross income.) Whether the net effect of the various possible changes in the administrative costs of making Social Security Act payments is positive or negative cannot be predicted from available information.

Quality of Life

Because of the body's normal rejection of foreign tissue, a patient must be maintained permanently on immunosuppressive drugs after the transplant. Because immunosuppression leads to a greater risk of infection, careful patient management is required to balance the risks of infection against the risks of rejection. The regimen involved is not conducive to the maintenance of a normal lifestyle.

The renal transplant and dialysis program is most similar to the heart program, and has encountered similar immunosuppression and patient management problems. In the context of the renal program, Rettig[15] cautions against having high expectations with respect to the quality of life. The analyses by McNeil and colleagues[8,9] previously discussed, also suggest that caution is appropriate when equating increased longevity with improved quality of life. However, for the potential heart transplant recipient, no alternative other than death exists at the present time. It would be valuable for the federal government, in conjunction with Stanford and other medical centers, to develop measures of quality of life so that the issue can be addressed.

If the selection protocols currently used at Stanford are relaxed when the procedure is considered to be beyond the experimental stage and other institutions participate, the number of patients with significant other medical problems (previously contraindications for heart transplants), is likely to increase. As Rettig[15] points out, this

---

* The Bureau of the Census reports that primary family income (husband and wife) in 1978 was median $17,640 and mean $20,091. Similar data for unrelated individuals show $7,542 and $9,820. Because of the requirements of the Stanford protocol, we have selected the highest figures as the basis for calculating the economic benefits.

has been the case with the renal program. Thus, both the survival rates and the quality of life of individuals receiving heart transplants probably will decline.

Potential Federal Program Costs and Associated Benefits

Potential Costs

In 1978 the DHHS reported that 2,800 kidneys were transplanted from cadaveric donors. In many cases, two kidneys were taken from one donor. Consequently, we can estimate that about 2,000 cadavers were obtained through the kidney network in 1980.[11] About 50 percent of these cadavers are considered unsuitable or unavailable for heart transplantation.[12] Thus, about 1,000 hearts could be made available now. Russell and Cosimi[13] indicate that the organ network could be expanded and significantly improved "by full application of available methods throughout the country," and that the number of available cadaver organs could be increased to 12,000 annually. Federal coverage of the costs involved in this procedure would create an increased demand for organs for transplantation and an increase in the number of available organs would be likely.

Theoretically, the number of potential transplants is equal to the number of potential recipients, subject to donor organ supply. According to the 1979 report by the Ad Hoc Task Force on Cardiac Replacement,[14] in 1979 there were about 32,000 potential candidates for a heart transplant. Total costs may range therefore from those associated with 1,000 transplants to a maximum potential of more than 30,000, although the current donor system would have to be significantly changed to provide this volume. The five-year cumulative costs would range from a low figure of $146.8 million to a high of $4,401 million, as shown in Table 7.

Table 7   Cumulative Year-End Cost of Transplants by Number of Potential Patients ($ million)

	Number of Patients		
	1,000	*12,000*	*30,000*
Year 0	101.7	1220.4	3051.0
Year 1	112.2	1325.4	3366.0
Year 2	122	1422.9	3658.5
Year 3	131	1512.9	3928.5
Year 4	139.3	1595.4	4176.0
Year 5	146.8	1670.4	4401.0

SOURCE: Calculated from the cost data of Table 4 and the survival data of Table 5 using the "best case" number.

Year 0 gives the expenses incurred for each patient undergoing the procedure, from initial evaluation through discharge from the hospital after the transplant operation; namely, evaluation costs, hospitalization costs, professional fees, and other expenses discussed previously. It is assumed that 30 percent of the patients die in year 1, so maintenance costs are calculated only for the survivors. A 5 percent attrition rate for surviving patients appears to hold for the subsequent years.* These costs do not reflect any program expenses incurred for those patients who undergo initial evaluation but are not referred for further evaluation, nor for those patients who undergo further evaluation but are not accepted into the program. It is likely that when few patients are accepted for transplantation, the proportion of evaluation costs not added into the total is greater than when many patients are accepted into the program. Thus the lower program costs understate the potential costs to a greater degree than do the higher program cost figures. The extent of the underestimation cannot be calculated from available information.

Historically analysts have not taken into account the impact of the shift in incentives accompanying open-ended federal funding when deriving their cost estimates.[15] Federal funding creates incentives for any institution and physician willing to undertake a newly funded procedure, making extrapolation from past behavior difficult. In the case of heart transplants, the actual cost of a nationwide program would be influenced by this, as well as other factors which cannot be predetermined, such as growth of the donor system, change in social attitudes regarding donor gifts of organs, and breakthroughs in immunosuppression or development of other new technologies.

These costs are not attributable to the illness. Neither the personal cost of pain and family disruption nor the various social and economic costs mentioned earlier are included. The costs identified are those which would be borne by the federal government if the patients are covered under the disability provisions of the Social Security Act, or under any other medical payment plan.

---

* The cumulative cost data in this table is based on the cost information of Table 3 and the survival data of Table 4. The first year survival cost (year 1) is predicated on the best survival rate, 70 percent in 1976, on the assumption the procedure is being improved steadily. The subsequent year costs are calculated on the apparent 5 percent attrition rate for each year thereafter. The calculation is: (number of surviving patients × maintenance costs) + previous year's cost. For example, for year 2 for 1,000 patients: (650 × $15,000) + $112.2 million = $122 million.

Potential Benefits

These benefits are associated with the extension of patient life. The incremental life extension will depend, in part, on whether candidates selected for transplants have additional medical problems not experienced by the current transplant patients. As noted earlier, placing a value on the benefits of life extension is difficult,[3] and quality of life measures are not sufficiently developed to be useful in this paper.

We found earlier that estimates of the value of the return to employment for each cohort of survivors can be calculated from the potential income calculations. The potential cumulative income at the end of five years for the 25 percent of survivors in each year who are assumed to be working ranges from $11.6 million to $346.5 million, as shown in Table 8. (The income potential is based on the survival experience used for Table 7 and the mean primary family income.)

Year 0 is the year of operation, and it is assumed that no surviving patient is sufficiently fit to return to work in year 1. For year 2, the assumption (similar to the cost calculations related to Table 7) is that 65 percent of the patients are still alive. For the subsequent years, a 5 percent attrition rate is assumed.

These economic benefits to the individuals need to be supplemented by the total net effect on the administrative costs of the income redistribution programs discussed previously. Unfortunately, no means of estimating this is available at present. Even though the value of the intangible benefits, such as quality of life, cannot be measured at present, they need to be detailed and considered when judgments are being made about the total benefits of heart transplants.

Table 8 Cumulative Year-End Income Potential for Surviving Recipients Who Resume Employment ($ million)

	Number of Recipients		
	1,000	*12,000*	*30,000*
Year 0	—	—	—
Year 1	0	0	0
Year 2	3.3	32.6	97.9
Year 3	6.3	62.7	188.3
Year 4	9.1	90.3	271.2
Year 5	11.6	120.4	346.5

Policy and Regulatory Issues

Policy and regulatory issues go beyond the individual and social cost effectiveness of heart transplantation. Potential ethical problems include experimentation on human subjects, informed consent, privacy, and questions of distributive justice. Regulatory provisions can be prescriptive, controlling patient and provider activities, including medical practice, or can be patient-outcome oriented. Since there is not space to analyze such issues in depth, three examples follow.

Patient Selection and Coverage

Strict patient selection protocols usually are used during clinical research trials of new medical procedures and are designed to achieve the highest possible rate of successful treatment outcomes. This is currently true for the Stanford heart transplant program[10] and was true initially for the renal dialysis program.[16] The Stanford patient selection criteria take into consideration the nature of the cardiovascular problem; the presence of other medical problems, such as infections or diabetes requiring insulin; an age greater than 50 years; a psychological evaluation; and various economic criteria.

Clearly such selection criteria pose ethical and moral, as well as medical and financial, implications. "Because clinical trials are of necessity performed with narrowly defined populations, questions on the applicability of results to other populations are complex, involving both scientific and health care delivery considerations."[17] While medical standards might call for restricted coverage, the government must take into account the sensitive and difficult nature of exclusion processes, especially in regard to antidiscrimination statutes and other statements of congressional intent.*

The renal dialysis and transplantation program has shown that as federal funding becomes available, all constraints on participation are eliminated. Now "neither age nor pre-existing serious systemic diseases are contraindications to transplantation."[18] The only formal acceptance criteria for the program is disabling or end-stage uremia.[19] If the trend for kidney patients is a typical progression, wherein the high patient selectivity of the research sector gives way to the widespread availability of a procedure under public funding, candidates previously excluded as marginal or contraindicated prob-

---

* For example: The Age Discrimination Act of 1975, 42 USC 6102; The Civil Rights Act of 1964, 42 USC 2000d; The Health Planning and Resources Development Act of 1974, 42 USC 300k-1 (for access questions).

ably would be included in a national heart transplant program. The paradox is that as more resources are invested in this marginal group, the benefits associated with longevity improvements, as measured by survival rates, become less and less.

Regulatory Alternatives

If heart transplants are to be covered by the federal government, criteria must be developed to ensure equitable access, the quality of care provided, and proper accountability for public expenditures. Two regulatory forms could be: (1) outcome or performance-oriented regulations, or (2) command and control regulations.

Performance-oriented regulations should depend on measures of patient well-being resulting from the procedure. The problem with this approach is the lack of agreement on outcome measures, which is particularly acute for new procedures. While it is theoretically possible to establish reasonable standards for desired changes in longevity, morbidity, and rehabilitation, no professional agreement exists concerning these standards and their measurement. Measuring patient outcomes on a case-by-case basis is open to dispute because "hospitals operating on sicker patients will predictably experience poor outcomes."[20] Which variables affect hospital or physician outcomes is in dispute. Consequently, while standardization of outcomes is desirable, it is unlikely to occur at present.

Command and control regulations do appear to assure the desired results. These regulations may specify the type and skill of physicians, supporting staff (even staffing ratios), institutional conditions, and so forth. Standards are established that usually specify levels of training or skills that are thought to affect the quality of care a patient receives. Although evidence relating these precise requirements to the desired results often is not available, they are consistent with preconceptions held by many, enhance the influence and prestige of the professional groups involved, and give an appearance of certainty.

Payment could be linked to these regulatory criteria, to the extent that payment would not be made unless the standards are adopted. Payments could be restricted to a limited number of institutions and physicians certified by the government to provide the procedure, thus potentially regionalizing care. Standards could be set to determine levels of payment or create maximum limits on payments. New reports could be required if needed to provide a basis for payment and accountability. Such regulatory activities will affect the distribution of the procedure and its rate of diffusion, and could have broad ramifications for the practice of medicine.

Implications of the Regulatory Alternatives for the Industry

The rate of diffusion of a new procedure or service depends on several factors that primarily include: the nature and timing of federal coverage decisions; regulatory standards; professional acceptance and willingness to adopt the procedure; the importance to the potential recipients; the number of potential recipients; and the level at which payments will be made. Financing is particularly important since "to an important degree, it is the availability of financing, rather than medical efficacy that governs the adoption of high cost technology."[21]

Federal coverage policy affects not only the diffusion of the procedure under consideration; it may increase the rate of diffusion and innovation of associated ancillary technologies as well. For instance, the development of new scanning technologies that will aid cardiovascular surgery has been supported by the National Heart, Lung, and Blood Institute. Among these is Positron Emission Transaxial Tomography (PETT), currently one of the most expensive of the emission tomography developments.[22] The capital cost of PETTs and associated equipment, in single unit settings, is estimated at up to $2 million; annual operating charges can range to $750,000.[23] The cost analyses in this paper did not include any costs associated with potential expansion of related technologies. To fully assess potential government outlays, the costs of the ancillary technologies as well as those of new procedures must be considered. Technically, these costs can be allocated in various ways. As an example, PETT devices may be used for many different diagnostic purposes; consequently, no correct way has been established for allocating the capital and operating costs.[3]

Government policymakers, third-party payers, and health planners should be concerned with these issues. As long as the federal payment system remains cost based, no economic incentives exist for providers to make efficient use of technology, facilities, and personnel; so "utilization will tend to expand until there is little or zero benefit."[19] Increasingly planners have tried to control diffusion through use of detailed requirements for unit volume, types of procedures, types of personnel, and so forth.* Third-party payers have become sensitive to the consequences of unlimited reimbursement and have begun to establish more detailed conditions for payment. However, a potential problem with these efforts is the

---

* Examples will be found in the State of New Jersey, Department of Health, state health planning regulations covering such topics as burn centers, cardiac catheterization, cardiac surgery and neonatal intensive care.

likelihood of inconsistent standards among the various requirements imposed by planners, third-party payers, professional groups, and the government.

A question underlying many of these issues is the degree to which government-mandated criteria and standards will affect medical practice. Government is increasingly concerned with paying the least cost, consistent with maintenance of quality care, and may move toward stringent requirements governing adoption of medical techniques, physician credentials, institutional success rates, and facility equipment purchases to achieve such results.

Concluding Comments

Society must address a number of critical questions before deciding to pay for new medical procedures or technologies. For example, are the benefits worth the cost, or should the availability of benefits be limited? The judgment involved concerns the withholding or use of life-extending medical resources. Consequently, many intangibles and value-laden judgments must be weighed. The decision cannot be based solely on technical information.

If the federal government decides to pay for new procedures and technologies, the diffusion of the procedure, the quality of the care, the rate of innovation of other technologies, and the style of medical practice will be affected by the regulations designed to implement that policy. The public, the federal government, and the medical community should begin to consider the implications of a federal decision to pay for heart transplants so that the many issues raised in this paper can be examined more extensively.

Heart transplants are very costly. Within five years of the inception of a national heart transplant program, national costs could range from more than $150 million for 1,000 potential recipients to at least $4.5 billion for 30,000. Consequently, the opportunity costs of heart transplantation should be compared with costs and benefits associated with other medical procedures and other health or social programs. The Child Health Assurance Program is estimated to cost about $400 million and could give immediate benefit to several million mothers and children. A full-scale program to assist financially troubled hospitals might cost as much as $600 million but could ensure services to thousands of people who might otherwise lack access to care. Alternatively, the potential expenditures for heart transplants could be used to expand the prevention of heart disease through programs such as the Stanford Heart Disease Prevention Program and the North Karelia Project.[24]

The decision to pay for major new procedures and technologies such as heart transplantation should be made in the context of such competing priorities because "the technological imperative, when carried to its extreme, incurs fantastic expense for relatively small and, at times, counterproductive outcomes."[25]

Endnotes

1. Reiss, J.B. (1980). Cost containment: Long- and short-run strategies: Commentary. In *National Health Insurance: What Now? What Later? What Never?* ed. M.V. Pauly, pp. 374–381. Washington, D.C.: American Enterprise Institute for Public Policy Research.
2. Hodgson, T. and Mainers, N. (1978). Guidelines for PHS Cost of Illness Studies. Unpublished memorandum, National Institute of Health, Public Health Service.
3. Weinstein, M.C. (1979). Economic evaluation of medical procedures and technologies: Progress, problems, and prospects. *Medical Technology*, National Center for Health Services Research, Research Proceedings Series, DHEW Publication No. (PHS) 79-3254, September.
4. Stinson, E.B. (1980). Personal communication to Dr. Peter Frommer, Deputy Director, National Heart, Lung, and Blood Institute, October 1978 to April 1979.
5. Schroeder, J.S. (1980). *Cardiac Transplantation Guidelines.* Progress Report to NIH on HL 13108-09. Stanford: Stanford University Medical Center.
6. Christopherson, L., Guipp, R.B., and Stinson, E. (1976). Rehabilitation after cardiac transplantation. *JAMA* 236:2084.
7. Krutz, H. (1980). A new heart for Dr. Mary L. Smith. *Washington Star*, Sunday, March 23.
8. McNeil, B.J. and Pauker, S. (1979). The patient's role in assessing the value of diagnostic tests. *Radiology* 132:605–610.
9. McNeil, B.J., Weichselbaum, R., and Pauker, S. (1978). Fallacy of the five-year survival rate in lung cancer. *N. Engl. J. Med.* 299:1397–1401.
10. Stinson, E.B. (1978). Current status of cardiac transplantation. In *Heart Failure*, ed A.P. Fishman. Washington, D.C.: Hemisphere Publishing Corp.
11. Teutsch, S. (1980). Personal telephone communication to John Burckhardt, Office of Health Regulation.
12. Towray, O.B. (1979). Medicare coverage of heart transplantation. Unpublished memorandum, National Center for Health Care Technology.
13. Russell, P.S. and Cosimi, A.B. (1979). Transplantation. *N. Engl. J. Med.* 301:477.
14. Ad Hoc Task Force on Cardiac Replacement Report (1979). *Cardiac Replacement: Medical, Ethical, Technological, and Economic Implications.* Washington, D.C.: U.S. Government Printing Office.
15. Rettig, R.A. (1979). End-stage renal disease and the "cost" of medical technology. In *Medical Technology: The Culprit Behind Health Care Costs?* eds. A.H. Altman and R. Blendon, pp. 95–99. U.S. Department of

Health, Education and Welfare, Publication No. (PHS) 79-3216.

16. Fox, R.C., and Swazey, J.P. (1974). *The Courage to Fail.* Chicago: University of Chicago Press.

17. Levy, R.I., and Sondik, E.J. (1978). Decision-making in planning large-scale comparative studies. *Ann. N.Y. Acad. of Sci.* 304:455.

18. Williams, G.M. (1978). Status of renal transplantation today. *Surg. Clin. N. Amer.* 58:274.

19. Russell, L.B. (1979). *Technology in Hospitals.* Washington, D.C.: The Brookings Institution.

20. Luft, H.S., Bunker, J.P., and Enthoven A.C. (1979). Should operations be regionalized? *N. Engl. J. Med.* 301:1367.

21. Rice, D.P. and Wilson, D. (1976). The American medical economy: Problems and perspectives. *Journal of Health Politics, Policy and Law* 1:155.

22. For more information on these developments, see Grubb, P.L., Jr. (1978). Emission computed tomography. *Neurosurgery* 2:273–280. Budinger, T.F., and Rollo, D.F. (1977). Physics and instrumentation. *Progress in Cardiovascular Diseases* XX:19–51.

23. Reiss, J.B., Burckhardt, J., and Allsopp, P. (1980). Some expenditure implications of the potential diffusion of PETT scanners. A Report of the Office of Health Regulation, Washington, D.C.

24. Farquhar, J.W., Wood, P.D., Breitrose, H., Haskell, W.L., Meyer, A.J., Maccoby, N., Alexander, J.K., Brown, B.W., Jr., McAlister, A.L., Nash, J.D., and Stern, M.P. (1977). Community education for cardiovascular health. *Lancet,* June: 1192–1195.

25. Mechanic, D. (1977). The growth of medical technology and bureaucracy: Implications for medical care. *Milbank Memorial Fund Quarterly,* Winter: 64.

INDEX

419